Fibre Optics

Fibre Optics
Principles and Advanced Practices
Second Edition

Abdul Al-Azzawi

CRC Press
Taylor & Francis Group
Boca Raton London New York

CRC Press is an imprint of the
Taylor & Francis Group, an **informa** business

This material was previously published in Photonics: Principles and Practices © 2007 by Taylor & Francis Group, LLC.

CRC Press
Taylor & Francis Group
6000 Broken Sound Parkway NW, Suite 300
Boca Raton, FL 33487-2742

First issued in paperback 2019

© 2017 by Taylor & Francis Group, LLC
CRC Press is an imprint of Taylor & Francis Group, an Informa business

No claim to original U.S. Government works

ISBN-13: 978-0-4987-7432-1 (hbk)
ISBN-13: 978-0-367-87885-6 (pbk)

Visit the Taylor & Francis Web site at
http://www.taylorandfrancis.com

and the CRC Press Web site at
http://www.crcpress.com

Contents

Preface

We live in a world that is bathed in light. Light is one of the most familiar and essential things in our lives. For many thousands of years, the Sun was our only source of light. Eventually, the ability to create fire and its by-product, light, led to a profound change in the way humans managed time. Today, there are many options for creating light. Our understanding of light has spawned many applications of light. Light can be used in fibre communications; early applications included ship-to-ship communications using the Morse code. Infrared remote controls for televisions demonstrated free-space optical communications, using many of the same principles. Optical fibre has revolutionised the way we interact with the world. Fibre optics, as a waveguide technology, now provide an essential backbone for much of the world's high-speed communication networks. Photodynamic therapies use light to treat cancers. Light is used to treat those who suffer from seasonal disorders. Lasers are now used in medical applications, such as re-shaping corneas, cauterising blood vessels and removing tattoos. Lasers are also used in industrial applications, such as cutting metal, welding and sensing. New imaging technology permits the creation of flat-panel displays, night-vision devices and autonomous product inspection systems. With so many applications of light, the need for photonics technology and innovation will most certainly grow in the future as new applications emerge.

A unique approach is taken in this book to present fibre optics and their applications in photonics technology. This book covers the basic theoretical principles and industrial applications of photonics technology in a way that is suitable for students, professionals and professors. Each chapter has two parts: theoretical and practical. The theoretical part has adequate material to cover the subject. In the experimental part, students will apply the learned theoretical concepts in simple and advanced experimental works. In this way, students will learn and gain practical hands-on experience in the fibre optic subjects. This will assist the students to apply the theoretical knowledge to real-world applications. The step-by-step approach and technical illustrations in this book will guide students through each experiment. The experimental work has more than one case in most of the chapters and sometimes has sub-cases.

This book is written in a simple language and gives adequate information and instruction to enable students to achieve the maximum comprehension. An effort has been made to use the International System of Units (SI) throughout the book. The organisation of the chapters is designed to provide a solid foundation for today's photonics students and to upgrade their knowledge. Universal tools, devices and equipment, which are used throughout the experiments, are available in any photonics, physics and material labs. This book abounds in theoretical and practical aids and is an effective teaching tool, helpful to both professors and students. Simple and advanced subjects are presented by an expert author, and some new fibre optic subjects appear for the first time in this book.

Care has been taken to label parts clearly and to use diagrams wherever they will aid in understanding the subject. Some figures are drawn in three dimensions, wherever applicable, for easy understanding of the concepts.

This book is organised into 15 chapters:

- Chapters 1 through 12 cover fibre optic cables, advanced fibre optic cables, light attenuation in optical components, fibre optic cable types and installations, fibre optic connectors, passive fibre optic devices, wavelength division multiplexer, optical amplifiers, optical receivers, opto-mechanical switches, laser technology, and optical fibre communications.
- Chapter 13 presents fibre optic lighting systems, design and applications.
- Chapter 14 covers fibre optic testing procedures for optical fibre devices.
- Chapter 15 presents laboratory safety and safety ethics.

This book includes 409 figures, 62 tables, and 38 experimental cases. It was developed with generous input from members of the photonics industry, research scientists and members from academia.

Abdul Al-Azzawi
Algonquin College, Canada

Acknowledgements

This book would not have been possible without the enthusiasm and teamwork of my colleagues and the support of my family. In particular, I thank Mietek Slocinski for his support, time and energy in working long hours during the years this book took to complete, for setting up the labs, taking pictures and discussing everything thoroughly.

The author thanks Steve Finnegan, Kathy Deugo and Nicole McGahey for their support and for solving the problems.

The author thanks his daughter, Abeer, and son, Abaida, for their help in reviewing the chapters and making drawings and figures.

The author thanks Eng. Monica Havelock for her contribution in working long hours in reviewing and editing the materials and for her support.

The author is grateful to Professors Devon Galway and Rao Kollipara for their comments and feedback in reviewing some materials of this book.

The author thanks Gergely Horvath for all the hard work in reviewing and proofreading most chapters of this book. He also thanks Madeleine Camm, Andrew Lynch and Nicolas Lea for reviewing a few chapters of this book.

The author thanks Dr. Imad Hasan, Dr. Wahab Almuhtadi, Dr. Mostefa Mohamad, Eng. Nazar Rida and Eng. Mohamad Mohamad for proposing and reviewing some of the materials of this book.

The following people's participation in writing the chapters is highly appreciated:

- The optical receivers chapter has been written by Dr. Imad Hasan and Prof. Wael Ismaeil.
- The optical fibre communications chapter has been written by Dr. Wahab Almuhtadi.
- The fibre optic testing chapter has been written by Eng. Valerie Dube.

Author

Abdul Al-Azzawi, PhD, graduated from the University of Strathclyde, Glasgow, Scotland, the United Kingdom. He has worked in the photonics manufacturing industry and research at National Research Council, Ottawa, Canada, and has taught at Algonquin College, Ontario, Canada. While working at NRC, he participated in a study of energy saving in a residential building and developed the green building assessment programme. As a photonics engineer, he designed new production lines, modified products, developed manufacturing process and designed new jigs.

At Algonquin College, he has been teaching mechanical and photonics courses in the mechanical and photonics engineering programmes. He was a member of the founding team of the photonics engineering programmes. He has published five books and many papers, and he has participated in many workshops and conferences around the world.

He is the coordinator of the photonics engineering programme at Algonquin College. His special areas of interest include optic and optical fibre devices, fibre optic lighting and fibre optic sensors. He is a member of the professional photonics societies in Canada. He is the recipient of the 2005 NISOD Excellence Award from the University of Texas, Austin, Texas.

Contributors

Chapter 12 Optical-Fibre Communications
written by
Dr. Wahab Almuhtadi, P.Eng., SMIEEE,
EIC Fellow
Algonquin College
Applied Research Coordinator, Faculty of
    Technology & Trades
Professor, and Coordinator, BIT- Photonics &
    Laser Technology
Applied Science & Environmental Technology,
    School of Advanced Technology
Ottawa, Ontario, Canada

Chapter 14 Fibre-Optic Testing
written by
Eng. Valerie Dube
Patent Examiner at CIPO-Electrical Division
Ottawa, Ontario, Canada

Chapter 9 Optical Receivers and Senders
written by
Dr. Imad Hasan
Algonquin College
Business Administration Core and Service
    Courses Department
Ottawa, Ontario, Canada

Chapter 9 Optical Receivers and Senders
written by
Prof. Wael Ismaeil
Algonquin College
Robotics Department
Ottawa, Ontario, Canada

1 Fibre-Optic Cables

1.1 INTRODUCTION

Fibre-optic cables transmit data through very small cores at the speed of light. Significantly different from copper cables, fibre-optic cables offer high bandwidths and low losses, which allow high data-transmission rates over long distances. Light propagates throughout the fibre cables according to the principle of total internal reflection.

There are three common types of fibre-optic cables: single-mode, multi-mode and graded-index (GRIN). Each has its own advantages and disadvantages. There are also several different designs of fibre-optic cables, each made for different applications. In addition, new fibre-optic cables with different core and cladding designs have been emerging; these are faster and can carry more modes. Although fibre-optic cables are mostly used in communication systems, they also have established medical, military, scanning, imaging and sensing applications. They are also used in optical fibre devices and fibre-optic lighting.

This chapter will discuss the fabrication processes used in manufacturing fibre cables. The processes produce a thin flexible glass strand with a diameter even smaller than that of human hair. This chapter will also detail methods of coupling a light source with a fibre cable in the manufacturing of optical fibre devices. It will also compare fibre and copper cables and describe the applications of fibre-optic cables in many fields and sectors of modern society. Finally, this chapter will present four experimental cases, including fibre cable inspection and handling, fibre cable end preparation, numerical aperture measurements and calculations and fibre cable power output intensity measurements and calculations.

1.2 THE EVOLUTION OF FIBRE-OPTIC CABLES

The evolution of optical communication systems dates back to the early 1790s, when the French engineer Claude Chappe (1763–1805) invented the optical telegraph. His system involved a series of semaphores mounted on towers, where human operators relayed messages from one tower to the next. This was certainly an improvement over hand-delivered messages. But by the mid-nineteenth century, the optical telegraph was replaced by the electric telegraph, leaving behind a legacy of Telegraph Hills, San Francisco.

In 1841, Swiss physicist Daniel Colladen (1802–1893) and French physicist Jacques Babinet (1794–1872) showed in their popular science lectures that light could be guided along jets of water for fountain displays. Then in 1870, Irish physicist John Tyndall (1820–1893) demonstrated the light-pipe phenomenon at the Royal Society of England. Tyndall directed a beam of sunlight into a container of water and opened the spout. Water flowed out in a jet, and the pull of gravity bent the water into a parabolic shape, as shown in Figure 1.1. Light was trapped inside the water jet by the total internal reflection phenomenon. The light beam bounced off the top surface, then off the lower surface of the jet, until turbulence occurred in the flowing water and broke up the beam. This experiment marked the first research into the guided transmission of light by an interface between two optical materials.

In 1880, Alexander Graham Bell patented an optical telephone system that he called the photophone, but his earlier invention—the telephone—proved far more practical. Although Bell dreamed of sending signals through the air, the atmosphere did not transmit light as reliably as wires carried electricity. For the next several decades, though light was used for a few special applications such as signalling between ships, inventions using optical communication gathered dust on the shelf. Bell donated his photophone to the Smithsonian Institution, Washington, DC.

Ultimately, a new light-guiding technology that slowly took root solved the problem of optical transmission. The technology depended on the phenomenon of total internal reflection, which can confine light in an optical material that is surrounded by another optical material with a lower refractive index, such as glass in air. However, it was a long time before this method was adapted for communications.

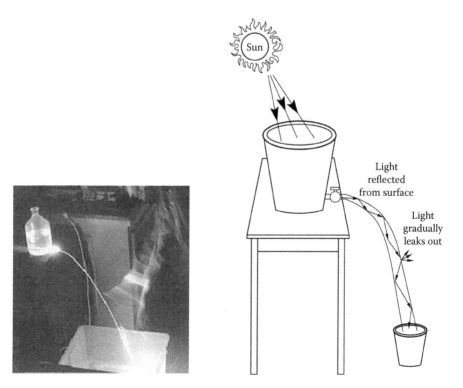

FIGURE 1.1   Total internal reflection of light in a water jet.

Optical fibres went a step further. These are essentially transparent rods of glass or plastic stretched until they are long and flexible. During the 1920s, John Logie Baird in England and Clarence W. Hansell in the United States patented the idea of using arrays of hollow pipes or transparent rods to transmit images for television or facsimile systems. British Patent Spec 20,969/27 was registered to Baird, and U.S. Patent 1,751,584 was granted to Hansell in 1930 for the scanning and transmission of a television image via fibres. Also in 1930, H. Lamm, in Germany, demonstrated light transmission through fibres. The next reported activity in this field took place in 1951, when A. C. S. van Heel in Holland and Harold H. Hopkins and Narinder S. Kapany of Imperial College in London investigated light transmission through bundles of fibres. Although van Heel coated his fibres with plastic, Kapany explored fibre alignment, and as reported in his book, *Fibre Optics*, produced the first undistorted image through an aligned bundle of uncoated glass fibres.

Neither van Heel nor Hopkins and Kapany made bundles that could carry light far, but their reports sparked the fibre optics revolution. The crucial innovation was made by van Heel, stimulated by a conversation with the American optical physicist Brian O'Brien. Although all earlier fibres were bare, with total internal reflection at a glass–air interface, Heel covered a bare fibre of glass or plastic with a transparent cladding of lower refractive index. This protected the total reflection surface from losses and greatly reduced crosstalk between fibre cables. The next key step was the development of glass-clad fibres by Lawrence Curtiss at the University of Michigan, Ann Arbor, Michigan. By 1960, glass-clad fibres had reached an attenuation of about one decibel per metre, which was fine for medical imaging, though much too high for communications.

In 1958, higher optical frequencies seemed the next logical step to Alec Reeves, the forward-looking engineer at Britain's Standard Telecommunications Laboratories, Harlow, Essex, who invented digital pulse-code modulation before the Second World War. Other people climbed on the optical communications bandwagon when the laser was invented in 1960. Then the 22 July, 1960, issue of *Electronics* magazine introduced its report on Theodore Maiman's demonstration of the first laser by saying, 'Usable communications channels in the electromagnetic spectrum may be extended by development of an experimental optical frequency amplifier.'

All was quiet in the new science until 1967, when Charles K. Kao and George A. Hockham of Britain's Standard Telecommunications Laboratories suggested the impending development of a new communications medium using a cladded fibre, with wavelengths in the millimetre range. Optical fibres had been attracting attention because they were analogous in theory to the plastic dielectric waveguides used in certain microwave applications.

Since then, the technology has grown enormously—slowly at first, but almost exponentially since the 1970s.

The Corning Company's breakthrough in fibre-optic cable manufacturing was among the most dramatic of the many developments that opened the door to fibre optic communications. In the same year, Bell Labs and a team at the Ioffe Physical Institute in St. Petersburg, Russia, made the first semiconductor diode lasers that were able to emit continuous light waves at room temperature. Over the next several years, fibre losses dropped dramatically, aided by both improved fabrication methods and the shift to longer wavelengths at which fibres have inherently lower attenuation.

Early single-mode fibres had cores several micrometres in diameter. During the early 1970s, this bothered developers, who doubted that it would be possible to achieve the micrometre-scale tolerances needed to couple light efficiently into the tiny cores from light sources, or in splices or connectors. Not satisfied with the low bandwidth of step-index multi-mode fibre, they concentrated on multi-mode fibres with a refractive-index gradient between core and cladding, and core diameters of 50 or 62.5 μm. The first generation of telephone field trials in 1977 used these fibres to transmit light at 850 nm from gallium–aluminium – arsenide laser diodes.

Government and defence agencies, telephone companies and a widening range of other private companies turned to fibre optics for telecommunications and other uses, many of which were protected as military or proprietary secrets. Each new user was attracted by the fact that optical transmission does not simply reduce but entirely avoids the risks of short circuits, wire tapping and electromagnetic crosstalk. Among the converts were General Telephone of Indiana, which had a 3-mile link between two switching centres in Fort Wayne, Indiana, carrying 5000 telephone circuits through 14 fibres only. A similar system was installed in San Angelo, Texas, before the end of 1981. Bell Canada also used fibre technology, as did Alberta Government Telephones of Canada. The list grew as companies recognised the superiority of fibre-optic cable over the old copper wire technology.

Overseas, England completed nearly 450 km of optical fibre cable under water. Two grades of fibre cables were used: one for higher bit-rate systems of 140 Mbps, which used laser injection techniques, and the other for systems of 8 to 34 Mbps. In addition, France, Germany and other European countries were entering the field, primarily in telephone work, and also in other applications.

Those first-generation systems could transmit light over several kilometres, but were limited by a loss of about 2 dB/km in the fibre cable. Thus, a second generation system soon appeared that used new lasers that emitted at 1300 nm, with fibre attenuation as low as 0.5 dB/km. The development of hardware for the first transatlantic fibre cable showed that single-mode systems were feasible, so when deregulation opened the long-distance phone market in the early 1980s, carriers built national backbone systems of single-mode fibres with 1300 nm light sources. That technology spread into other telecommunication applications. Meanwhile, a newer generation of single-mode systems found applications in submarine cables and systems serving large numbers of subscribers; these operate at 1550 nm.

More important was the development of the erbium-doped optical fibres, which serve as optical amplifiers. Submarine cables with these optical amplifiers operate at high speeds. Optical amplifiers are also attractive for fibre systems delivering the same signals to many terminals, because the fibre amplifiers can compensate for losses in dividing the signals among many terminals. The biggest challenge remaining for fibre optics now is one of economics.

By 1990, AT&T had laid a fibre cable that spanned the Atlantic Ocean. Cables between other countries were also completed, and electric utilities were installing fibre cable links for communicating with substations, replacing the venerable carrier telephone equipment that multiplexes via the power lines and through the air. Fibre cables have also been laid underground as security links.

Today, telephone and cable television companies can justify the cost of installing fibre links to remote sites that serve tens to hundreds of customers. However, terminal equipment remains too expensive to install the fibres all the way to customers' homes, at least for present services. Instead, cable and phone companies run twisted wire pairs or coaxial cables from optical network units to individual houses.

The wireless telecommunication market is growing quickly. Systems using wavelengths in the range 1625–1675 nm are also available for very fast data transmission. Recent years have seen a drastic increase in the advanced types of optic cables available. These cables offer a very low loss and operate at very high speeds for a wide range of wavelengths. Time will tell how long fibre-optic technology will last.

1.3 FIBRE-OPTIC CABLES

Fibre-optic cable is a filament of transparent material used to transmit light, as shown in Figure 1.2. Virtually all fibre-optic cables share the same fundamental structure. The centre of the cable is referred to as the core. It has a higher refractive index than the cladding, which surrounds the core. The contact surface between the core and the cladding creates an interface surface that guides the light; the difference between the refractive index of the core and cladding is what causes the mirror-like interface surface, which guides light along the core. Light bounces through the core from one end to the other according to the principle of total internal reflection, as explained by the laws of light. The cladding is then covered with a protective plastic or polyvinyl chloride jacket. The diameters of the core, cladding and jacket can vary widely; for example, a single fibre-optic cable can have core, cladding and jacket diameters of 9, 125 and 250 μm, respectively.

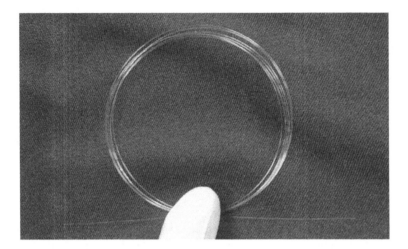

FIGURE 1.2   A fibre-optic cable.

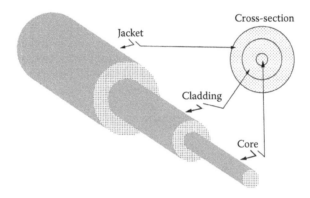

FIGURE 1.3   Schematic view and cross-section of a fibre-optic cable.

Figure 1.3 shows the structure of a typical fibre-optic cable. The cores of most fibre-optic cables are made from pure glass, whereas the claddings are made from less pure glass. Glass fibre-optic cable has the lowest attenuation over long distances but comes at the highest cost. A pure glass fibre-optic cable has a glass core and a glass cladding. Fibre-optic cable cores and claddings may be made from plastic, which is not as clear as glass but is more flexible and easier to handle. Compared with other fibre cables, plastic fibre cables are limited in power loss and bandwidth. However, they are more affordable, easy to use and attractive in applications where a high bandwidth or a low loss is not a concern. A few glass fibre cable cores are clad with plastic. Their performance, though not as good as all-glass fibre cables, is quite respectable. More details on plastic cables will be presented in Section 1.4.

The jacket is made from polymer (PVC, plastic, etc.) to protect the core and the cladding from mechanical damage. The jacket has several major attributes, including bending ability, abrasion resistance, static fatigue protection, toughness, moisture resistance and the ability to be stripped. Fibre-optic cable jackets are made in different colours for colour-coding identification. Some optical fibres are coated with a copper-based alloy that allows operation at up to 700°C and 500°C for short and long periods, respectively.

Table 1.1 shows the common diameters of the core, cladding and jacket of four commonly used fibre-optic cables in manufacturing optical fibre devices. Different types and cross-sections are also available for building advanced fibre-optic devices. These types will be presented in Sections 1.7 and 1.8.

TABLE 1.1

Common Diameters of Fibre-Optic Cables

Standard Diameter of Core (μm)	Standard Diameter of Cladding (μm)	Standard Diameter of Jacket (μm)
8	125	250
50	125	250
62.5	125	250
100	140	250

1.4   PLASTIC FIBRE CABLES

Plastic optical fibre (POF) has the highest attenuation over short distances, but it comes at the lowest cost. A plastic fibre-optic cable has a plastic core and plastic cladding. It is also quite thick, with typical core/cladding diameters of 480/500, 735/750 and 980/1000 μm. The core generally consists of polymethyl methacrylate (PMMA) coated with a fluoropolymer. POF cables are used in small optical devices, lighting applications, automobiles, music systems and other electronic systems. The cables are also used in communication systems where a high bandwidth or a low loss is not a concern. The increased interest in POF is for two reasons: (1) the higher attenuation relative to glass, which may not be a serious obstacle with the short cable runs often required in premise networks and (2) the cost advantage, which appeals to network architects faced with budget decisions. POF cables do, however, have a problem with flammability. On account of this, they may not be appropriate for certain environments, and care must be given when they are run through a plenum. Otherwise, plastic fibre is considered extremely rugged, with a tight bend radius and the ability to withstand mechanical stress.

Plastic-clad silica (PCS) fibre-optic cable has an attenuation—and cost—that lie between those of glass and plastic. PCS fibre-optic cable has a glass core that is often made of vitreous silica; the cladding is often plastic, usually a silicone elastomer with a lower refractive index. In 1984, the International Electrotechnical Commission (IEC) standardised PCS fibre-optic cable to have the following dimensions: a core of 200 μm, a silicone elastomer cladding of 380 μm, and a jacket of 600 μm.

Plastic fibre cables are fabricated using the same principles as glass fibre cables. A core with a higher index of refraction is surrounded by a cladding with a lower index of refraction. The cladding is then coated with a coloured jacket for coding purposes; glass and plastic cables are similarly colour coded. POF cables are available in single- and multi-step index, as well as GRIN.

Recent developments in the polymer industry have led to improvements in plastic fibre-optic cables. Plastic fibre cables will eventually replace glass fibre cables because of their many advantages, including their ease in connection using epoxy as well as their lower price, durability, lower weight and smaller bending radii.

1.5   LIGHT PROPAGATION IN FIBRE-OPTIC CABLES

Figure 1.4 shows the principle of light propagation through a fibre-optic cable. To understand the principle, consider when light is injected into an optical material that is surrounded by another optical material with a different index of refraction. The first and second optical materials will have a higher and lower index of refraction, respectively. Figure 1.4a shows light incident at an angle θ to the normal. The light refracts by an angle β from the normal. If the angle of incidence is increased, the angle of refraction will also increase, as shown in Figure 1.4b. In Figure 1.4c, light incident at an angle θ equal to the critical angle θ_c gives an angle of refraction of 90 degrees. The refracted light lies on the interface of the first-second optical materials. Snell's law is applicable here. In Figure 1.4d, when light lands incident at an angle θ greater than the critical angle, the light

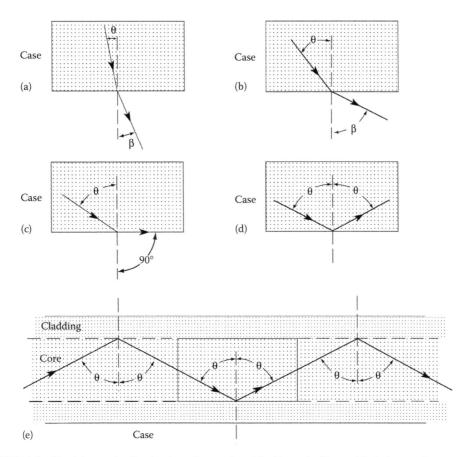

FIGURE 1.4 Total internal reflection in a fibre-optic cable (a) ray incident with θ close to the normal line (b) ray incident with θ further away from the normal line (c) ray incident with θ equals the critical angle (d) ray incident with θ close to equal to reflection angle and (e) ray propagates through a fiber optic cable.

will reflect by the total internal reflection phenomenon. In this case, the angle of incidence equals the angle of reflection, which is defined by the first law of light.

Assume the first optical material is very long. It has a circular cross-section with a very small diameter like the core, and at the same time is surrounded by another optical material resembling the cladding in a fibre-optic cable, as shown in Figure 1.4e. Now, light injected into a fibre-optic cable and striking the core-to-cladding interface at a greater angle than the critical angle reflects back into the core. As the angles of incidence and reflection are equal, the reflected light will again be reflected. The light will continue to bounce through the length of the fibre cable. Light in Figure 1.4, Cases (a), (b) and (c) show that the light passes into the cladding. The cladding is usually inefficient as a light carrier compared to the core of the fibre cable. Light in the cladding becomes part of the losses, some of which will be presented in Section 1.16, that usually occur in any fibre-optic cable. Therefore, light propagation in a fibre-optic cable is governed by the following:

- The wavelength of light
- The angle of incidence of the light at the input of the fibre cable
- The indices of refraction of the core and cladding
- The composition of the core and cladding
- The length of the cable
- The bending radius of the cable
- The sizes of the core and cladding

- The design of the core and cladding
- The transmission modes
- The temperature and environmental conditions of the fibre cable
- The strength and flexibility of the fibre cable

1.6   REFRACTIVE-INDEX PROFILE

As explained above, the fibre-optic cable's core and cladding are made from different optical materials, each of which has a different index of refraction. The refractive-index profile describes the relationship between the indices of refraction of the core and cladding. When the core is made from uniform optical material, the refractive-index profile is flat. This is called a step-index profile, as shown in Figure 1.5a and b. The step-index fibre cable has a core with a uniform index of refraction throughout the cable. The profile shows a sharp step at the junction of the core and cladding. When the core is made from multi-layered optical materials and each layer has a differing index of refraction, the refractive-index profile has a curved shape. This type of profile is called a graded-index, as shown in Figure 1.5c. In contrast to the step-index profile, the GRIN profile has a non-uniform core. The index is highest at the centre and gradually decreases until it matches that of the cladding. There is no sharp step between the core and the cladding in GRIN fibre cables.

1.7   TYPES OF FIBRE-OPTIC CABLES

There are, thus, three common types of fibre-optic cables, as listed below. The suitability of each type for a particular application depends on the fibre-optic cable's characteristics.

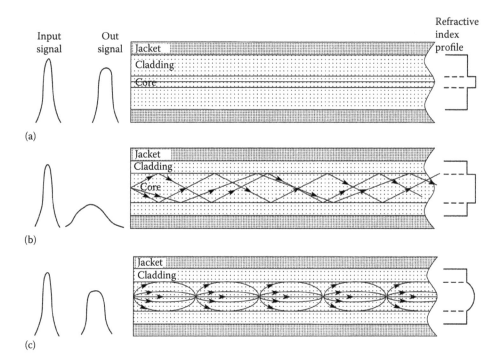

FIGURE 1.5   Types of fibre-optic cables: (a) single-mode step index fibre cable, (b) multi-mode step index fibre cable and (c) multi-mode graded-index fibre cable.

1.7.1 SINGLE-MODE STEP-INDEX FIBRE CABLE

The single-mode step-index fibre cable, sometimes called a single-mode fibre cable, is shown in Figure 1.5a. The single and multi-mode step-index fibre cables are the simplest types. Single-mode fibre cables have extremely small core diameters, ranging from 5 to 9.5 μm. The core is surrounded by a standard cladding diameter of 125 μm. The jacket is applied on the cladding to provide mechanical protection, as shown in Figure 1.3. Jackets are made of one type of polymer in different colours for colour-coding purposes. Single-mode fibres have the potential to carry signals for long distances with a low loss, and are mainly used in communication systems. The number of modes that propagate in a single-mode fibre depends on the wavelength of light carried. The number of modes will be given in Equation 1.9. A wavelength of 980 nm results in multi-mode operation. As the wavelength is increased, the fibre carries fewer and fewer modes until only one mode remains. Single-mode operation begins when the wavelength approaches the core diameter. At 1310 nm, for example, the fibre cable permits only one mode. It then operates as a single-mode fibre cable.

1.7.2 MULTI-MODE STEP-INDEX FIBRE CABLE (MULTI-MODE FIBRE CABLE)

The multi-mode step-index fibre cable, sometimes called a multi-mode fibre cable, is shown in Figure 1.5b. Multi-mode fibre cables have bigger diameters than their single-mode counterparts, with core diameters ranging from 100 to 970 μm. They are available as glass fibres (a glass core and glass cladding), PCS (a glass core and plastic cladding) and plastic fibres (a plastic core and cladding). They are also the widest ranging, although not the most efficient in long distances, and they experience higher losses than the single-mode fibre cables. Multi-mode fibre cables have the potential to carry signals for moderate and long distances with a low loss (when optical amplifiers are used to boost the signals to the required power).

As light rays bounded through a fibre cable reflect at different angles for different ray paths, the path lengths of different modes will also be different. Thus, different rays take a shorter or longer time to travel the length of the fibre cable. The ray that goes straight down the centre of the core without reflecting arrives at the other end faster. Other rays take slightly longer and thus arrive later. Accordingly, light rays entering a fibre at the same time will exit at the other end at different times. In time, the light will spread out because of the different modes. This is called modal dispersion. Dispersion describes the spreading of light rays by various mechanisms. Modal dispersion is that type of dispersion that results from the varying modal path lengths in the fibre cable.

1.7.3 MULTI-MODE GRADED-INDEX FIBRE (GRADED-INDEX FIBRE CABLE)

Multi-mode GRIN fibres are sometimes called GRIN fibre cables, as shown in Figure 1.5c. GRIN and multi-mode fibre cables have similar diameters. Common GRIN fibres have core diameters of 50, 62.5 or 85 μm, with a cladding diameter of 125 μm. The core consists of numerous concentric layers of glass, somewhat like the annular rings of a tree or a piece of onion. Each successive layer expanding outward from the central axis of the core until the inner diameter of the cladding has a lower index of refraction. Light travels faster in an optical material that has a lower index of refraction. Thus, the further the light is from the centre axis, the greater its speed. Each layer of the core refracts the light according to Snell's law. Instead of being sharply reflected as it is in a step-index fibre, the light is now bent or continually refracted in an almost sinusoidal pattern. Those light rays that follow the longest path by travelling near the outside of the core have a faster average velocity. The light ray travelling near the centre of the core has the slowest average velocity. As a result, all rays tend to reach the end of the fibre at the same time. Thus, one way to reduce modal dispersion is to use GRIN fibres. This type of fibre-optic cable is popular in applications that require a wide range of wavelengths, in particular telecommunication, scanning, imaging and data-processing systems.

Fibre cables are designed for a specific wavelength, called the cut-off wavelength, above which the fibre carries only one mode. A fibre designed for single-mode operation at 1310 nm has a cut-off wavelength of around 1200 nm. Although optical power is confined to the core in a multi-mode fibre, it is not so confined in a single-mode fibre. This diameter of optical power is called the *mode field diameter*. It is usually more important to know the mode field diameter than the core diameter.

1.8   POLARISATION MAINTAINING FIBRE CABLES

Polarisation maintaining (PM) fibres are constructed by placing specially designed asymmetries into the core. PM fibres guide only one possible mode of propagation. They also maintain the electromagnetic field vector direction. This type of single-mode fibre is used in building optical fibre devices that work with polarised light, such as polarisation beam splitters, couplers, modulators and interferometric sensors. PM fibres are desirable, as most lasers emit highly polarised light, and the polarised properties are highly desirable in many measurement applications.

There are four common designs for PM fibres, as shown in Figure 1.6. Figure 1.6a shows the PANDA fibre. The PANDA fibre employs a stress technique that stresses the core of the fibre to create two propagation paths within the fibre core, with two stress rods placed within the cladding in the same plane on opposite sides of the core. Linearly polarised light aligned to either the slow or fast axis of the fibre will remain linearly polarised as it propagates through the fibre.

Figure 1.6b shows a cross-section of the bow-tie fibre. A pair of wedges on opposite sides of the core generates the optimum stress distribution within the fibre. This patented design offers

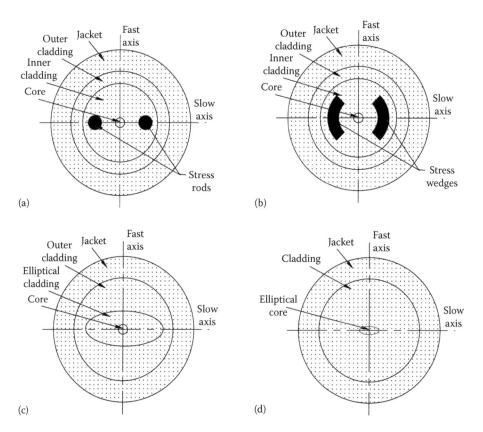

FIGURE 1.6   Polarisation maintaining fibre cables: (a) PANDA fibre, (b) bow-tie fibre, (c) elliptical core fibre and (d) elliptical core.

the best in terms of both performance and handling, with minimum stress breakout when cleaved, connectorised or polished.

Figure 1.6c shows a circular core surrounded by an elliptical boron-doped cladding. Figure 1.6d shows a PM fibre with a very high level of doping in an elliptical core, which causes polarised waves along the major and minor axes of the ellipse to have different effective indices of refraction. Thus, the fibres contain non-symmetrical stress-production parts.

1.9 SPECIALTY FIBRE CABLES

Section 1.7 presented common fibre cables that are used to guide light over relatively long distances in communication systems, imaging, scanning and medical applications. Fibre optic cable technology and its applications have experienced a diversity of technological advancements that make optical fibres able to fulfil every possible application. Other types of fibres are optimised for a variety of specific applications and for research and development. Chapter 2 will discuss the advanced fibre-optic cables used in many applications requiring a low loss, low chromatic dispersion and high transmission rate over long distances.

1.10 FIBRE CABLE FABRICATION TECHNIQUES

A variety of fabrication techniques are used to produce fibre-optic cables. This section will discuss two common methods used to manufacture optical glass fibre cables. The first method involves directly drawing the fibre from two molten glass rods or *preforms*, which are placed in two concentric crucibles; this is called the double crucible method. The second method involves forming a preform.

There are several processes for producing preforms to be used in the fibre-drawing method. The manufacture of fibre cable requires sophisticated and highly accurate techniques. The biggest challenge is the purification of the materials used in the construction of the core and cladding. In particular, the value of the index of refraction must be very precise, especially when manufacturing GRIN fibres. During the glass fibre optic cable fabrication process, impurities are intentionally added to the pure glass to obtain the desired indices of refraction needed to guide light. Germanium or phosphorous are added to increase the index of refraction, whereas boron or fluorine are added to decrease the index of refraction. Unfortunately, adding these residual impurities may increase the attenuation by either scattering or absorbing the light. The fabrication methods also need to be extremely precise regarding fibre dimensions and tolerances. The diameter of the core must be controlled to high precision, and the core must be located at the centre of the cladding. These same issues are also present when applying the cladding to the core. In the final step, the cladding is coated with a polymer jacket layer for mechanical and environmental protection.

1.10.1 DOUBLE CRUCIBLE METHOD

The double crucible method is illustrated in Figure 1.7. A pair of platinum crucibles sits one inside the other. Molten core glass is then placed in the inner crucible, and molten cladding glass is fed into the outer crucible. The crucibles are kept at a high temperature, typically between 1850°C and 2000°C. Using a precision-feed mechanism, the two glasses come together at the base of the outer crucible, forming a core-cladding fibre. The fibre is drawn out from the crucibles. Then the fibre passes through high-precision diameter measurement and control equipment and is monitored by imaging or x-rays. The control equipment detects any non-homogeneity or bubbles in the drawn fibre. The fibre is then covered by a coloured layer of jacket. Again, the jacket layer goes through diameter control and monitoring equipment. The end of the fibre cable is attached to a rotating spool, which turns steadily. The fibre is then tested for the attenuation (dB per kilometre), dispersion and any other requirements specified by the customers or industry. Industry typically requires fibre

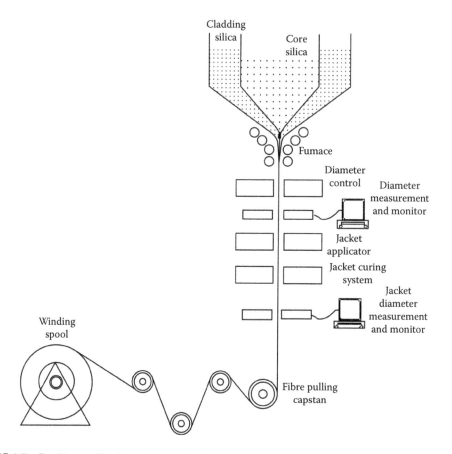

FIGURE 1.7   Double crucible fabrication method.

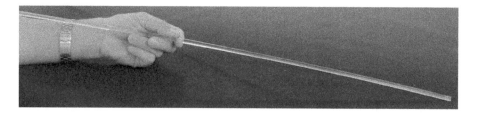

FIGURE 1.8   A preform.

cables to be a few hundred kilometres in length. Fibre cables are also available in different types and working wavelengths.

The rod tube is one of the simplest methods of fibre fabrication. In the rod-tube procedure, a rod of core glass is placed inside a tube of cladding glass. This arrangement forms the preform that is required for the drawing process of the fibre. A preform is shown in Figure 1.8.

1.10.2 CHEMICAL VAPOUR DEPOSITION PROCESSES

The preforms used in the fibre-drawing method are fabricated using chemical vapour deposition (CVD) processes. The CVD method is similar to the fabrication steps explained in the double crucible method. A preform is again needed for the fibre-drawing process. The method for preparing the preform is called the rod-tube procedure. A rod of core glass is placed inside a tube of cladding glass.

All these processes are based on thermal chemical vapour reactions that form oxides. These oxides are deposited as layers of glass particles called soot, which is deposited on the outer rotating rod or inside glass tube to produce the preforms. Starting materials are solutions of O_2 mixed with $SiCl_4$, $GeCl_4$, $4POCl_3$ or gaseous BCl_3. These liquids are evaporated within an oxygen stream at a high temperature to produce silicon dioxide (SiO_2) and other oxides, known as dopants. Chemical reactions proceed as follows:

$$SiCl_4 + O_2 \rightarrow SiO_2 + 2Cl_2$$

$$GeCl_4 + O_2 \rightarrow GeO_2 + 2Cl_2$$

$$4POCl_3 + 3O_2 \rightarrow 2P_2O_5 + 6Cl_2$$

$$4BCl_3 + 3O_2 \rightarrow 2B_2O_3 + 6Cl_2$$

SiO_2, or pure silica, is usually obtained in the form of small (submicron) particles called soot. This soot is deposited on the target rod or tube. The deposition of the silica soot, layer upon layer, forms a homogeneous transparent material. The manufacturer can control the exact amount of dopant added to each layer, thus controlling the refractive-index profile. For example, germanium dioxide (GeO_2) and phosphorus pentoxide (P_2O_5) increase the refractive index of glass, whereas boron oxide (B_2O_3) decreases it. Changing the composition of the mixture during the process influences the refractive-index profile of the preform. To change the value of a cladding's refractive index, some dopants are used. For example, fluorine (F) is used to decrease the cladding's refractive index in a depressed-cladding material.

The vapour process produces an extremely pure material whose characteristics are under the absolute control of the manufacturer. The preforms prepared in the vapour deposition processes will be explained next.

1.10.3 OUTSIDE VAPOUR DEPOSITION

This was the first successful mass-fabrication process that produced preforms used by the fibre-drawing method. The outside vapour deposition (OVD) process, also called the soot process, was developed by Corning Company in 1972. This process consists of four phases: lay down, consolidation, drawing and measurement. During the lay-down phase, the materials that make up the core and cladding are vapour deposited around the rotating target rod. The result of this process is a soot preform. The refractive-index profile and fibre geometry are formed during this phase, as shown in Figure 1.9.

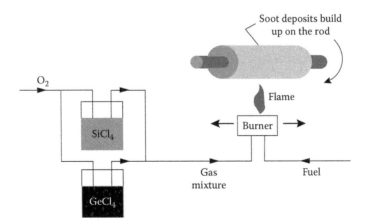

FIGURE 1.9 Outside vapour deposition (OVD).

In the consolidation phase, the target rod is removed, and the soot preform is placed inside a consolidation furnace. Here, the soot preform is consolidated into a solid, clear glass preform, and the centre hole is closed. During consolidation, a drying gas flows through the preform to remove residual moisture.

Then, in the drawing method, the preform is attached to a precision-feed mechanism that feeds it into a furnace at a controlled speed, producing a fibre with the required diameter. Later, a colour jacket layer is applied. Diameter measurement and control equipment are also constantly checking the core, cladding and jacket for diameter sizes and quality.

Finally, in the measurement phase, each fibre reel is tested for compliance with the fibre characteristics given in the data sheets. Specifically, the fibre is tested for strength, attenuation and dimensional characteristics. Fibres are also tested for bandwidth, numerical aperture (NA), dispersion and cut-off wavelength.

1.10.4   VAPOUR AXIAL DEPOSITION

The vapour axial deposition (VAD) is another form of outside deposition. This method was developed in 1977 by Japanese scientists. Figure 1.10 illustrates the VAD process. Silica particles obtained from a reaction among gases in a heated zone are deposited at the bottom end of a target, or seed rod, that rotates and moves upward. This deposition forms a porous preform, the upper end of which is heated in a ring furnace to produce a silica preform. The drawing and measurement steps are similar to those of the other deposition processes; however, the VAD process does not involve a central hole. The profile of the refractive index is formed by using many burners,

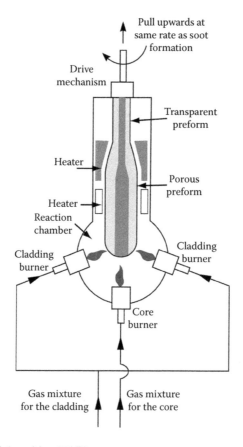

FIGURE 1.10   Vapour axial deposition (VAD).

a technique that allows the manufacturer to change the direction of the flow of a specific gas mixture. The VAD process produces both step-index and GRIN fibre profiles, as the deposited particle density varies due to temperature gradients produced in the plane perpendicular to the core axis.

1.10.5 MODIFIED CHEMICAL VAPOUR DEPOSITION

The modified chemical vapour-deposition process was developed by Bell Laboratories in 1974 and has since been widely used in the production of GRIN fibre. Early in the development, the process was called inside vapour deposition (IVD). Later, this was significantly improved and renamed as modified chemical vapour deposition (MCVD). This is the major process used in fibre production throughout the world. It is illustrated in Figure 1.11.

In MCVD, a mixture of $SiCl_4$, $GeCl_4$, $POCl_3$, O_2 and H_2 gases flow though the inside of the tube, which is surrounded on the outside by a heat source. The heat source converts the gases into snow-like, high-surface-area silica soot inside the tube, as shown in Figure 1.11. The soot deposits on the tube downstream of the flame. The burner moves along the outside of the quartz tube, creating the fine soot particles and sintering the soot into a thin layer of doped glass on the inside. After the layer is deposited, the mixture of reactive gases is changed and the burner is brought back to the starting position. The above step is repeated and a subsequent layer is deposited. This process is continued, layer by layer, to construct the complex core structure of the optical fibre. Varying the concentration of dopants with each layer changes the refractive index, creating a GRIN profile. Once the glass is deposited, the tube is collapsed into a solid rod called a preform. The preform manufactured on the MCVD lathe is heated and drawn down to the standard diameter of 125 μm. Each preform generates many kilometres of fibre.

This operation is performed on a draw tower. The tower has a furnace at the top to melt the silica preform. Gauges are used to measure and control the diameter of the glass fibre to within sub-microns, as the fibre is pulled from the preform. Very fine control of the profile can be obtained using this technique. These MCVD layers are designed to be much thinner than the wavelength of light travelling down the fibre. An acrylate coating is applied during the draw process, which protects the pristine silica fibre from the environment.

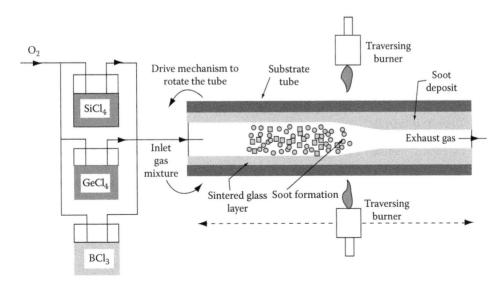

FIGURE 1.11   Modified chemical vapour deposition (MCVD).

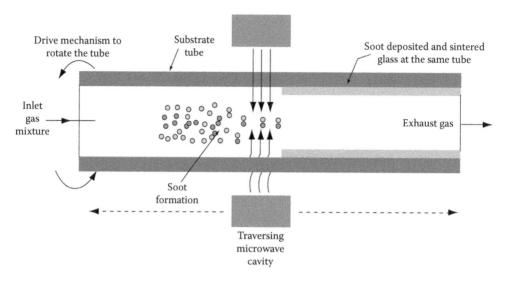

FIGURE 1.12   Plasma activated chemical vapour deposition (PCVD).

After the fibre is manufactured, each spool of fibre is tested to ensure that it meets strict industry and internal specifications. These tests include the measurement of mechanical strength, geometric properties and optical properties. One key test for laser-optimised multi-mode fibres is the high-resolution differential mode delay (HRDMD).

1.10.6   PLASMA CHEMICAL VAPOUR DEPOSITION

The plasma activated chemical deposition (PCVD) process was developed in 1975 by Phillips Research Laboratories, a Dutch consumer electronics and telecommunications company. The PCVD process is very similar to the MCVD process. However, instead of heating the outside of the silica tube, the energy source here is provided by a high-power microwave, resulting in ionised gas–plasma inside the tube. Figure 1.12 illustrates the PCVD process. Non-isothermal plasma in the microwave frequency range is used instead of a torch or flame. The plasma makes the reaction proceed at about 1000°C to 1200°C. This results in very thin layers deposited inside the tube. Although this method allows layers to be grown at relatively low temperatures, the deposition rate is rather slow compared to other methods. However, this process can produce large preforms capable of producing a few hundred kilometres of fibre.

1.11   FIBRE DRAWING

Optical fibres are obtained by drawing from the preform at a high temperature. The drawing process must be integrated with the jacket-coating process to avoid contamination of the fibre surface. These two processes are shown in Figure 1.13. The preform is heated in a furnace to a molten state. Then it is attached to a precision-feed mechanism that pushes the preform at the proper speed into the furnace at a high temperature. The drawing process is very precise and continuously controlled to check the diameter of the fibre; diameter drift cannot exceed 0.1%. The filament fibre then passes through a series of coating and jacket applicators and curing processes, depending on the customer's requirements. The outside diameter of the jacket is measured and monitored for defects. The fibre, in its final shape, is pulled down and wound around a winding drum or spool.

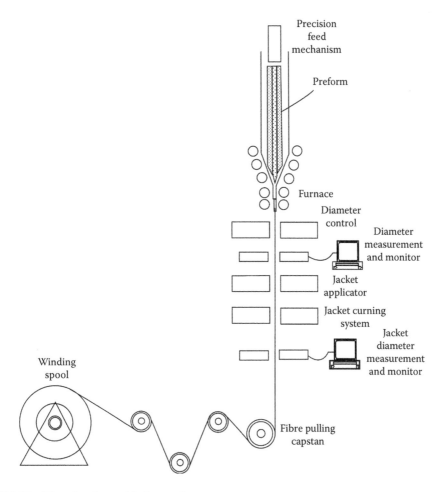

FIGURE 1.13 Fibre drawing and jacket coating.

1.12 NUMERICAL APERTURE

Figure 1.14 illustrates a light ray incident on the fibre cable input passing through the fibre core. The cone is known as the acceptance cone for the fibre cable. Any ray outside this cone will not propagate through the fibre core, as the incident angle is too large. The angle (θ) is known as the half cone angle for this fibre cable. The full cone angle is also known as the acceptance angle (β) of the cable. Thus, the full-acceptance angle (β) can be calculated by

$$\beta = 2\theta \tag{1.1}$$

The NA is a measurement of the amount of light that can be collected by, or that emerges from, the core of a fibre optic cable. The NA is the product of the index of refraction of the incident medium (n_i) and the sine of the maximum ray angle (θ) of the light incident on the core, as shown in Figure 1.14.

$$NA = n_i \sin(\theta) \tag{1.2}$$

In air, the sine of this half cone angle is the value of the NA of the cable.

Therefore, the half-acceptance angle (θ) is

$$\theta = \sin^{-1}(NA) \tag{1.3}$$

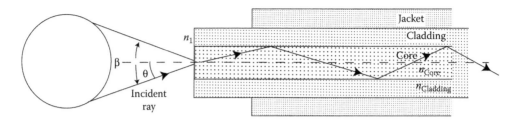

FIGURE 1.14 Light ray in a fibre-optic cable.

Recall that the refractive index of a material (n) is defined as the ratio between the speed of light in a vacuum (c) and the speed of light in the medium (v):

$$n = \frac{c}{v} \tag{1.4}$$

For example, the refractive index for air, water and glass crown is 1.0003, 1.33 and 1.52, respectively.
It is convenient to define the fractional refractive index change as

$$\Delta = \frac{n_{core} - n_{cladding}}{n_{core}} \tag{1.5}$$

We can also calculate the NA from the following formula:

$$NA = n_{core}\sqrt{2\Delta} \tag{1.6}$$

To understand ray tracing in a fibre cable, recall the principles of Snell's law, as explained by the law of refraction. The standard measure of acceptance angle is the NA, which is the sine of the half-acceptance angle (θ). Therefore, the NA can be calculated by

$$NA = \sqrt{n_{core}^2 - n_{cladding}^2} = \sin\theta \tag{1.7}$$

where n_{core} and $n_{cladding}$ are the refractive indices for the core and cladding, respectively.

1.13 MODES IN A FIBRE-OPTIC CABLE

A mode is a mathematical and physical concept that describes the propagation of electromagnetic waves through an optical medium. In its mathematical form, mode theory derives from Maxwell's equations. James Maxwell, a Scottish physicist (1831–1879), first gave mathematical expression to the relationship between electric and magnetic energy, showing that they were both a single form of electromagnetic energy, not two different forms, as was then believed. His equations also showed that the propagation of this energy followed strict rules. Maxwell's equations form the basis of the electromagnetic theory of light.

A mode is an allowed solution to Maxwell's equations. A mode is simply a path that a light ray can follow when travelling through a fibre cable. Currently, the number of modes supported by a fibre cable ranges from one to hundreds of thousands. Thus, a fibre cable provides a path for one or thousands of light rays, depending on its size, design and properties.

Each mode also carries a characteristic amount of optical power. Most fibre cables today support many modes. When light is first injected into the fibre cable, a mode may carry too much or too little power, depending on the injected power of light. Over the length of the fibre cable, power transfers between modes until all modes carry their characteristic power. When a fibre cable reaches this point, it is said to have reached the steady state, or equilibrium mode distribution (EMD). Achieving

EMD in plastic fibres requires only a few hundred metres. For high-performance glass fibre cables, EMD often requires hundreds of kilometres.

The number of modes supported by a fibre cable can be calculated from the V number or normalised frequency. The V number is a fibre cable parameter that takes into account the core diameter (d), wavelength of the propagated light (λ) and NA of the fibre cable. The V number is calculated by

$$V = \frac{2\pi d}{\lambda}(\mathrm{NA}) \tag{1.8}$$

From the V number, the number of modes (N) in a fibre-optic cable can be calculated for the following cases:

1. For a single-mode step-index fibre cable, the number of modes (N) can be approximated by

$$N = \frac{V^2}{2} \tag{1.9}$$

2. For a multi-mode step-index and GRIN fibre cable, the number of modes (N) can be approximated by

$$N = \frac{V^2}{4} \tag{1.10}$$

1.14 LIGHT SOURCE COUPLING TO A FIBRE CABLE

There are several techniques used for coupling a light source into a fibre-optic cable. The coupling process can be done during the manufacture of a laser light source, which is connected to a fibre cable at one end and a pig tail connector at the other, as shown in Figure 1.15. Light-emitting diodes and laser diodes have edge or top-surface light emission. Therefore, coupling of a cable to the emitting side of a light source depends on many factors, such as the design and the packaging size of the device. There are also similar processes used in coupling a fibre cable to an optical component, such as a prism or a GRIN lens.

For an efficient coupling process, the emitting area of the light source should be equal to or slightly larger than the core of the fibre cable to ensure that all the light passes through the cable with minimum loss. Figure 1.16a shows one of the most common coupling processes: direct coupling.

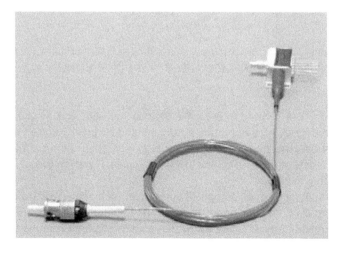

FIGURE 1.15   A light source coupled to a fibre cable.

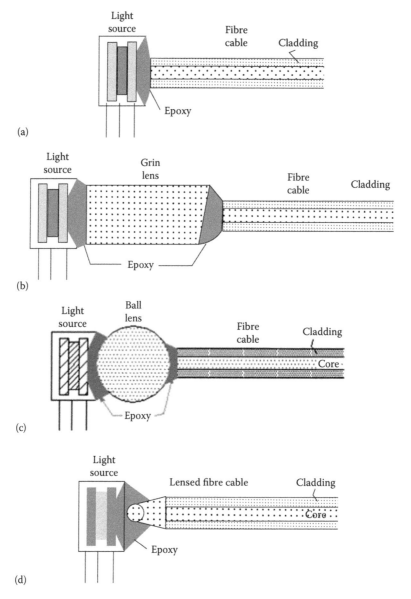

FIGURE 1.16 Common methods of coupling a light source to a fibre cable: (a) direct fibre coupling, (b) GRIN lens coupling, (c) ball lens coupling and (d) lensed fibre cable coupling.

Here, an optical epoxy can be used to bond the fibre cable to the light source. Application of the epoxy and curing process are very important during manufacturing to reduce losses. This method is the cheapest form of coupling because of the short alignment time and the simple assembly design that enables the alignment of the light source to the fibre cable. However, this method is less efficient than the other coupling methods.

To improve the coupling process, an optical component such as a GRIN lens or ball lens, a grating or a prism can be used. A problem arises when the light-emitting area and the fibre cable are not the same size, creating losses. Using a lens that reduces the light beam size to match the fibre cable can solve this problem. Figure 1.16b illustrates the use of a GRIN lens placed between the light source and the cable. GRIN lenses are commonly used in building optical fibre devices, such as switches and polarisation beam splitters, because of easy alignment by a fairly skilled employee. A ball lens can also be used in the coupling process, as shown in Figure 1.16c. Optical devices using ball lenses take more time for alignment to the light source. They are also more complicated to design and bigger in size than the devices using GRIN lenses. In addition, light coupling suffers high loss because more epoxy is needed to bond the ball lens into position. Figure 1.16d illustrates the coupling of a lensed fibre cable to a light source. This arrangement has an easy coupling alignment with a low loss and a small packaging size.

1.15   LAUNCHING LIGHT CONDITIONS INTO FIBRE CABLES

The amount of light carried by each mode in a fibre cable is determined by the light-input or light-launch conditions. If the angular spread of the rays from the light source is greater than the angular spread that can be accepted by the fibre cable (i.e., the NA of the light input to the fibre cable is greater than the NA of the fibre), then the radius of the light input is greater than the radius of the fibre cable. This case is called overfilled, as shown in Figure 1.17a. Here, a portion of the light source will be launched into the cladding and will be considered light loss. Conversely, when the input light NA is less than the NA of the fibre cable, that is, the radius of the light input is smaller than the radius of the fibre cable, this is called underfilled, as shown in Figure 1.17b. These two coupling conditions yield different attenuations, with the overfilled case having a higher loss than the underfilled case.

Other factors must also be taken into consideration in coupling: the condition the fibre cable end, the index of refraction of the epoxy, the epoxy curing process, the alignment method, the type of optical components and the packaging. Similarly, it is necessary to consider some of these factors in any type of connection between two fibre cables. Common factors include the angular alignment, the air gap, the different NAs or core diameters and the cable end conditions. Fibre cable connections will be discussed in more detail in Chapter 5.

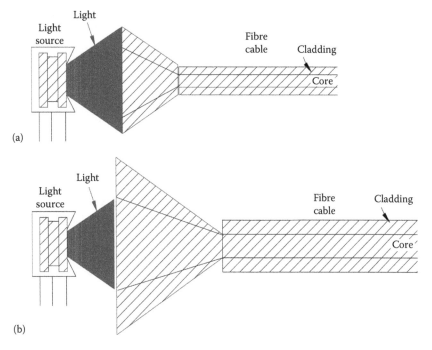

FIGURE 1.17   Launching light into a fibre cable: (a) overfilled case and (b) underfilled case.

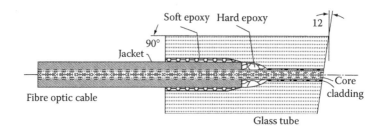

FIGURE 1.18 A fibre tube assembly/single fibre cable.

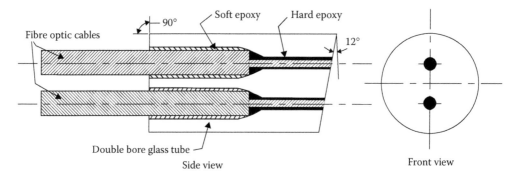

FIGURE 1.19 A fibre tube assembly/double fibre tube assembly (PANDA type).

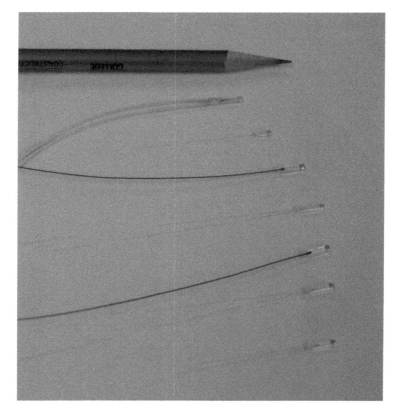

FIGURE 1.20 Fibre tube assemblies.

1.16 FIBRE TUBE ASSEMBLY

The fibre tube assembly is typically the first step when manufacturing an optical device. The fibre tube assembly provides a means to handle, position and glue the fibre optic cable to an optical component, such as a GRIN lens or a beam splitter. The fibre tube assembly is manufactured with a single fibre optic cable and a glass tube, as shown in Figure 1.18 or double fibre optic cable and a glass tube, as shown in Figure 1.19. The fibre optic cable is attached to the tube using two types of epoxy. A hard type of epoxy fills the front part of the glass tube and a soft type of epoxy fills the side of the fibre cables. Air bubbles in the epoxy or cracks in the tube will introduce insertion loss, especially under temperature variations. Fibre optic cable must also not be cracked or stressed, as this would increase losses. The fibre tube assembly end is polished at an angle of 2, 6, 8 or 12 degrees. These angles depend on the design of an optical fibre device. The assembly end must be polished precisely and carefully to maintain a surface free of cracks and scratches. Figure 1.20 shows the fibre tube assemblies ready to be joined to a GRIN lens or an optical component. The fibre tube assembly has a polished-end angle matching the polished-end angle of the GRIN lens. An accurate optical alignment is achieved when the fibre tube assembly and the GRIN lens are facing each other in perfect alignment. The end of a fibre tube assembly is polished to the required angle using an industrial polishing machine for mass production.

1.16.1 MANUFACTURING OF A FIBRE TUBE ASSEMBLY

Normally, there are many manufacturing, test, quality control and packaging processes available for each manufacturing step of a production line. These processes are written by expert engineers to control products as per customer specifications. Every employee must follow the process in their workstation, precisely. By these processes, a standard product can be controlled and achieved having the required specifications.

The following steps are used to manufacture a double fibre tube assembly:

1. Prepare and inspect a glass tube, as shown in Figure 1.21.
2. Prepare and cut a fibre cable to the required length. A few centimetres must be added to the required length by the customer, to give allowance for the following manufacturing steps.
3. Clean and inspect the fibre cable for dirt, cracks, punches, deformations and missing jacket. If the fibre cable passes the test, put it to the next manufacturing station in the line. If the fibre cable fails the test, leave it in a recycle bin.
4. Winding the fibre cable to the required diameter. Normally, winding diameter between 9–10 and 11–12 centimetres for a cable having a diameter of 125 and 900 µm, respectively.
5. Prepare and clean one end of the fibre cable, if one fibre cable in the fibre tube assembly. Prepare two fibre cable ends if double fibre tube assembly is required. The prepared end of the fibre cable should be two to three times the length of a glass tube.
6. Insert the fibre cables prepared end into a glass tube.
7. Mount the fibre tube assembly onto a heater jig, as shown in Figure 1.22.
8. Apply a soft and hard epoxy on the cable side and end of the glass tube, respectively. The best way to apply the epoxy is by using a micro-spatula for small manufacturing quantities or an automatic epoxy dispenser for large quantities.
9. Put on the power of the heater jig for a certain time depending on the type of epoxy.
10. Use a manual cleaver to cleave the protruded fibre cable with the glass tube end.

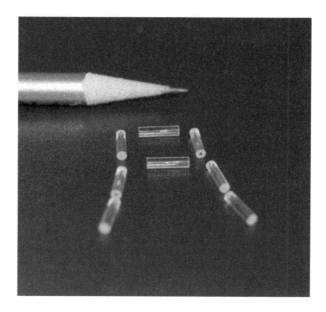

FIGURE 1.21   Glass tubes.

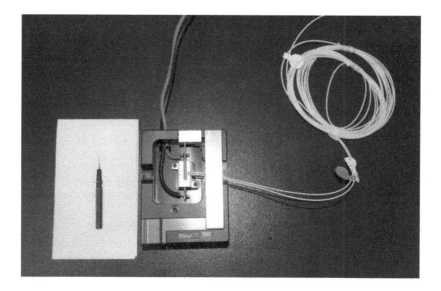

FIGURE 1.22   Fibre tube assembly mounted on a heater jig.

11. Use an adaptor/holder for the required polished angle of the end of the glass tube. Mount the fibre tube assembly on the polishing machine, as shown in Figure 1.23. Choose the appropriate polishing pad grade. There are four polishing pad grades. The grey colour size is for 63.0 µm, blue colour 9.0 µm, violet colour 1.0 µm and white colour 0.3 µm. Use polishing liquid during the polishing process.
12. Rinse the glass tube end with water.
13. Follow the test procedure for the fibre tube assembly end using an inspection microscope. Look at the quality of the polished end of the glass tube, according to required customer specifications. If the fibre cable assembly passes the test, put it to the next manufacturing station in the line. If the fibre tube assembly fails the test, leave it in a recycle bin.

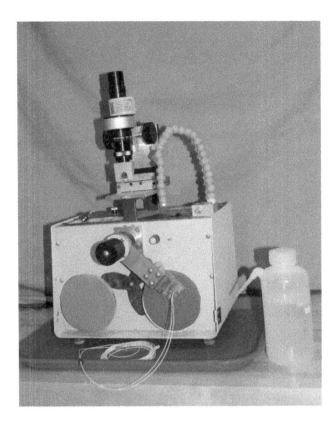

FIGURE 1.23   Fibre tube assembly mounted on a polishing machine.

14. The final product can be checked by quality control process stage. Quality control process can be applied at any stage in the production line.
15. Put the fibre tube assembly in a plastic bag inside a cardboard box with production papers and ship it to the next manufacturing stage or to a customer.

1.16.2   EXPERIMENTAL WORK FOR TESTING FIBRE TUBE ASSEMBLY

The purpose is to determine light losses of a single fibre tube assembly or a double fibre tube assembly power output. A He–Ne laser light passes through the fibre tube assembly. The light losses through a fibre tube assembly are tested by measuring the power input and power output at one end of the fibre cable or power at both ends of the double fibre cables.

Any incident light power passing through an optical component, such as a fibre optic cable, optical device and epoxy layer, is subjected to losses. Attenuation measures the reduction in light signal strength by comparing output power with input power. Measurements are presented in decibels (dB). The decibel is an important unit of measurement in fibre optic components, devices and systems loss calculations. The loss is calculated by

$$\text{Loss}(dB) = -10 \log_{10}\left(\frac{P_{\text{out}}}{P_{\text{in}}}\right) \tag{1.11}$$

1.16.2.1   Technique and Apparatus

Appendix A presents the details of the devices, components, tools and parts.

1. A 2 ft × 2 ft optical breadboard
2. He–Ne laser light source and power supply

3. Laser mount assembly
4. Laser light power metre
5. Hardware assembly (clamps, posts, screw kits, screwdriver kits, sundry positioners, etc.)
6. Double fibre tube assembly
7. Two fibre cable end holder/positioner assembly
8. Heating jig
9. Denaturated ethanol
10. Tissue
11. Swabs
12. Micro-spatula
13. Timer
14. Inspection microscope
15. Target board

1.16.2.2 Procedure

Follow the laboratory procedures and instructions given by the professor and/or instructor.

1.16.2.3 Safety Procedure

Follow all safety procedures and regulations regarding the use of fibre optic cables, GRIN lens, laser light source device, instruments and cleaning chemicals.

1.16.2.4 Apparatus Set-Up

This experiment measures the power of the laser beam entering into the fibre tube assembly and leaving through the two fibre cables, using a laser light sensor located at a fixed distance from the laser source and the two ends of the fibre cables. The output power will be measured for the input power and each fibre optic cable and the data recorded for loss calculations.

This experiment can be repeated by adding a quarter-pitch GRIN lens at the glass tube assembly input and at the fibre cable's ends. The laser light passes through GRIN lens that is aligned with the fibre tube glass. The alignment means the end face of the lens is parallel face-to-face with the fibre glass tube within a gap of 2–3 microns. The focal area of the GRIN lens is incident on the fibre tube glass face covering the two fibre cable ends. The advantage of using the GRIN lens is to have concentrated laser light at the input and outputs. This additional experiment would present the GRIN lens theory, alignment process and hands-on experience.

The following steps illustrate the experimental lab set-up for this case:

1. Figure 1.24 shows the apparatus set-up.
2. Bolt the laser short rod to the breadboard.
3. Bolt the laser mount to the clamp using bolts from the screw kit.
4. Put the clamp on the short rod.
5. Place the He–Ne laser into the laser mount and tighten the screw. Follow the operation and safety procedures of the laser device in use.
6. Check the laser alignment with the line of bolt holes on the breadboard and adjust when necessary.
7. Place the laser sensor in front of the laser source at a fixed distance for measuring the power output from fibre optic cable's ends.
8. Turn off the lights of the lab.
9. Turn on the laser power. Measure the laser input power (P_{in}). Fill out Table 1.2.
10. Mount a fibre tube assembly holder/positioner to the breadboard facing the laser source.
11. Mount a fibre tube assembly on the holder/positioner. The glass tube faces the laser beam, so that the laser beam passes through the centre of the glass tube, as shown in Figure 1.24.

FIGURE 1.24 Testing a fibre tube assembly.

TABLE 1.2
Laser Light Power Loss through a Double Fibre Optic Cable Assembly

Situation	P_{in} (unit)	P_{out} (unit)	Loss (dB)
Laser power supply input (P_{in})	–	–	–
Power output from the double fibre optic cable assembly end (P_{out1})		–	–
Power output from the double fibre optic cable assembly end (P_{out2})	–	–	–

12. Mount two fibre cable holder/positioners. Mount the output end of the fibre cables into the brass fibre cable holders and place it in the hole of the fibre cable holder/positioners. Mount the holder/positioners in front of the laser sensor at a fixed distance.
13. Verify the alignment of the fibre cable assembly end with the laser beam.
14. Check to ensure that you have a red circular output from the output end of the fibre optic cables on the target board. Point the fibre cable output ends towards the centre of the laser sensor.
15. Measure the laser output power (P_{out}) from the two ends of the fibre optic cables. Fill out Table 1.2.
16. Turn off the laser power.
17. Turn on the lights of the lab.

1.16.2.5   Data Collection

Table 1.2 presents laser light power loss through a double fibre optic cable assembly.

1. Measure the laser output power (P_{in}).
2. Measure the laser output power $\left(P_{out_1}\right)$ and $\left(P_{out_2}\right)$ from the double fibre tube assembly ends.
3. Fill out Table 1.2 for power measurements.

1.16.2.6   Calculations and Analysis

1. Calculate the double fibre optic cables power using Equation 1.11.
2. Fill out Table 1.2 for loss calculations.

1.16.2.7   Results and Discussions

1. Report the calculated result for the double fibre optic cable power loss (dB).
2. Compare and discuss the calculated losses for each fibre cable output.

1.16.2.8   Conclusion

Summarise the important observations and findings obtained in this lab experiment.

1.16.2.9   Suggestions for Future Lab Work

List any suggestions for improvements using different experimental equipment, procedures and techniques for any future lab work. These suggestions should be theoretically justified and technically feasible.

1.16.2.10   References

List any references that were used in the report. Use one format in writing the references. Never mix reference formats in a report.

1.16.2.11   Appendix

List all of the materials and information that are too detailed to be included in the body of the report.

1.17   FIBRE-OPTIC CABLES VERSUS COPPER CABLES

There is still a place for fibre optic and copper cables in communication systems, but the shrinking price gap, coupled with increasing bandwidth demands, makes fibre cables worth using in more situations than ever before. The customers of small and large communication providers are dispersed all over the world, and they need to send and receive lot of data. Customers require large amounts of bandwidth, and fibre cables are the only medium that can support this. Some of those customers' bandwidth requirements have been growing exponentially since the beginning of the twenty-first century. In reality, all fibre networks have a lot of room for future growth. The ultimate choice is whether to use fibre-optic cables. As mentioned above, the fibre-optic technology is moving forward to create high-capacity fibres with low production and installation costs and increasing bandwidths. The overall cost difference between optical fibre and twisted-pair copper cabling has been reduced. Now the choice between optical fibre and twisted-pair copper cabling has shifted in favour of the fibre cable.

Desktop computers require very high bandwidths. One way to meet this need is to wire them with fibre optic cables. However, copper cable has continued to prove more capable than expected; every time that new, higher-speed network standards appear to be forcing a move to fibre cable, someone has found a way to pump more data through the old copper cables.

Still, fibre cables can be made even more economically attractive by rethinking the way the network is physically laid out. As fibre cable can be run for longer distances than copper, the

TABLE 1.3
Fibre-Optic Cables versus Copper Cables

No.	Fibre-Optic Cables	Copper Cables
1	Fibre-based systems are more expensive to buy and install	Copper-based systems are less expensive to buy and install
2	Fibre is clearly the superior technologically. Installing fibre ensures performance, as even higher speed networks will emerge in the future	Installing copper cable ensures performance for low-speed networks
3	Carry high data capacity over long distances	Carry low data capacity over short distances
4	Wide bandwidth	Limited bandwidth
5	Low loss per cable length	Conventional loss per cable length
6	Immune to electrical interference and tapping	Not immune to electrical interference and tapping
7	Small size and lightweight	Large size and heavyweight
8	New technology reduces installation time	Conventional technology keeps the same installation time
9	High safety	Low safety
10	Fast-developing technology	Steady-state developing technology

networks could be laid out without the wiring closets full of additional gear that are common in copper-based networks. Instead, fibre might run directly from the desktop back to the server or to the backbone connecting floors of a building, and the savings on the intermediate gear might more than cancel out the higher cost of installing the fibre cable. Running fibre cable to small enclosures close to users, and then covering the last short distances with copper, provides an economical alternative that minimises the length of the twisted-pair cable used.

The increasing use of wireless networking also opens up a variation on the fibre cable network layout. Fibre can be run to a wireless access point that can then be used to serve a group. Thus, the copper cable can be eliminated altogether without actually taking fibre cable to every machine.

Many companies are removing existing copper cables and replacing them with fibre cables. When facilities are built or refurbished, fibre cables are installed. The choice between copper and fibre cables depends on several factors, including the applications being run on the network, the company's future plans and the demands of customers.

In particular, there are some specific situations in which fibre has advantages over copper cable. First, fibre cable is immune to electrical interference and tapping. It also carries high data capacity over long distances and is small and lightweight. When it comes to testing, fibre may still require some fairly sophisticated equipment—but the newer standards for copper cabling present the same issues.

In the end, the choice between fibre and copper cables comes down to the company's networking requirements, the needs of individual users and the budget. Table 1.3 presents a side-by-side comparison of the important differences between fibre and copper cables.

1.18   APPLICATIONS OF FIBRE-OPTIC CABLES

Since the discovery of the laser, fibre-optic cables and optical fibre devices have seen increased applications in every sector of industry. Light is a very important element in our lives, controlling and operating many types of devices, instruments and systems. Fibre optics has emerged as a practical technology that is easy to work with. Fibre-optic cables with other optic components are used in building optical fibre devices and systems. One of the large-scale applications of fibre optics is its use in communication systems. The small sizes and wide bandwidths, as well as their capacity to carry large amounts of information, make optical fibres very attractive for use in these systems. Later chapters in this book will explain in more detail communication systems that use optical fibre technology. Video, including broadcast television, is one of the main telecommunication applications. Other applications include cable television, high-speed Internet, wireless transition, remote monitoring and surveillance. Fibre-optic video transmission is successfully used around the world in surveillance and remote monitoring systems with many applications. Fibre-optics applications in the military include communications, command-and-control links on ships and aircrafts, data links for satellite earth stations and transmission lines for tactical command-post communications.

Fibre cables can be used throughout the communication network, including in the final link into the subscriber's home to wall outlets. This field has continued to develop since the discovery of optical amplifiers, dense wavelength division multiplexers (DWDM), fibre Bragg grating (FBG) and photonics crystal fibres.

One particular advantage of fibre-optic cables is that they are immune to electromagnetic interference (EMI) from electricity. Therefore, optic cables can be placed near high-voltage power cables without any effect on data transmission. Similarly, the cables can also be laid along railway lines without suffering from EMI.

Optical fibres are applied in building night-vision viewing devices, scanning and sensing instruments and vibration sensors, which are extensively used in military, medical and other applications. Imaging techniques have been rapidly developed for a variety of medical applications, such as viewing inside human tissues and scanning microscopic particles.

Fibres are also used in monitoring and sensing technology. They are used as sensors to monitor the vibration in the structures of bridges and high buildings. They are also used as gas and DNA sensors.

The use of fibre-optic cables in lighting systems can reduce energy consumption. Fibre-optic lighting systems are developing quickly with wide applications. Fibre-optic lighting systems can be applied to the interior and exterior of commercial, retail and residential buildings. New applications are being explored in landscaping, waterscaping, medical lighting instruments and theme environments. More details will be discussed in the fibre-optic lighting chapter.

Fibre-optic cables can also be coated for special handling requirements and resistance to temperature, chemicals or radiation. Radiation-resistant fibre is suitable for use in environments where electronics-based optical solutions are not viable, such as monitoring nuclear waste disposal in storage facilities. To make the fibres heat resistant, a chemical-resistant polyimide coating that can withstand temperatures of up to 300°C is applied. This is especially useful for manufacturers designing medical equipment for applications in which autoclave sterilisation is necessary.

1.19   EXPERIMENTAL WORK

In this experiment, the student will perform the following cases:

1.19.1   Case (a): Fibre Cable Inspection and Handling

After completing this lab experiment, students will be able to inspect, clean and handle fibre optic cables. A full, detailed procedure is presented in the apparatus set-up section.

1.19.2   Case (b): Fibre Cable Ends Preparation

The students will also be able to prepare fibre-optic cable ends. A full, detailed procedure is presented in the apparatus set-up section.

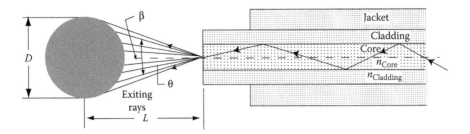

FIGURE 1.25 Numerical aperture (NA).

1.19.3 Case (c): Numerical Aperture and Acceptance Angles Calculation

The students will calculate the NA of a fibre-optic cable. Figure 1.25 shows the laser beam emerging from a fibre-optic cable core when the fibre output is projected onto a black/white card. The following formulas are used to calculate the half-acceptance (θ) and full-acceptance (β) angles, as well as the NA of the fibre-optic cable used in this lab experiment. From the geometry of Figure 1.25, it is easy to calculate the half-acceptanceangle (θ) and the full-acceptance angle (β) by

$$\tan\theta = \frac{D}{2L} \tag{1.12}$$

Thus,
$$\theta = \tan^{-1}\left(\frac{D}{2L}\right) \tag{1.13}$$

$$\beta = 2\theta \tag{1.14}$$

Assuming the refractive index (n_i) for air = 1, use Equation 1.2 to calculate the NA of a fibre-optic cable:

$$NA = \sin\theta \tag{1.15}$$

1.19.4 Case (d): Fibre Cable Power Output Intensity

Students will practice measuring the output intensity of a laser beam at a distance (L) from the fibre cable end, using a laser light sensor. The students will also understand how the intensity on the card varies with distance. They will determine the laser beam intensity on the card as a function of distance.

Light wave motion involves the propagation of energy. The rate of energy transfer is expressed in terms of intensity, which is the energy transported per unit time across a unit area. Energy/time is power, and we usually express intensity as power per unit area.

$$\text{Intensity}(I) = \frac{\text{Energy/Time}}{\text{Area}} = \frac{\text{Power}}{\text{Area}} \tag{1.16}$$

The standard units of intensity (power/area) are watts per square metre (W/m^2).

The geometric shape of the laser beam, when emerging from the fibre cable end, is defined by the NA, as explained in Case (c). The intensity of the laser beam emerging from the fibre cable end and incident on a card is, therefore, inversely proportional to its distance from the fibre cable end. The geometric diagram of the laser beam emerging from the fibre cable and incident on the black/white card will be defined by the Equations 1.12 and 1.13.

Again, as explained in Case (c), the laser beam spot has a diameter (D) incident on the black/white card at a distance (L) from the fibre cable end. Equations 1.12 and 1.13 give the relation between (D) and (L). Therefore, (D) is directly proportional to $2L \tan \theta$, as given by Equation 1.12. The area (A) of the laser beam spot is calculated by

$$A = \frac{\pi D^2}{4} \tag{1.17}$$

1.19.5 TECHNIQUE AND APPARATUS

Appendix A presents details of the devices, components, tools and parts. Figure 1.26 shows the fibre-optic cable ends preparation and cleaning kit.

1. Fibre-optic cable, 250 µm outside diameter, 500 m in length.
2. Safety goggles.
3. Clean room wipers.
4. Finger cots.
5. Rubber gloves.
6. Cleaver.
7. Fibre-optic strippers.
8. Scissors.
9. Cleaner (denatured ethanol).
10. Microscope with a minimum of 10X magnification.
11. Thermal cleaver, as shown in Figure 1.31.
12. Mechanical cleaver, as shown in Figure 1.33.
13. A 2 ft × 2 ft optical breadboard.

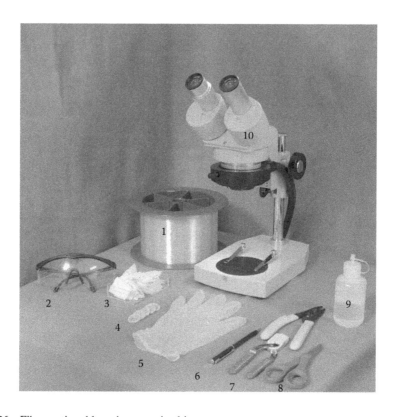

FIGURE 1.26   Fibre-optic cable end preparation kit.

14. He–Ne laser light source and power supply.
15. Laser mount assembly.
16. Laser light sensor.
17. Laser light power meter.
18. 20X microscope objective lens. Use the 20X lens to focus the collimated laser beam onto the fibre-optic cable core. The lens will increase the input power into the fibre cable core and consequently increase the output power.
19. Hardware assembly (clamps, posts, screw kits, screwdriver kits, lens/fibre cable holder/positioner, sundry positioners, brass fibre cable holders, fibre cable holder/positioner and so on).
20. Lens/fibre cable holder/positioner assembly.
21. Fibre cable holder/positioner assembly.
22. Black/white card and cardholder.
23. Vernier.
24. Ruler.

1.19.6 PROCEDURE

Follow the laboratory procedures and instructions given by the professor and/or instructor.

1.19.7 SAFETY PROCEDURE

Follow all safety procedures and regulations regarding the use of fibre-optic cables. You must wear safety glasses and finger cots or gloves when working with and handling fibre optic cables, optical components and optical cleaning chemicals.

1.19.8 APPARATUS SET-UP

1.19.8.1 Case (a): Fibre Cable Inspection and Handling

1.19.8.1.1 Fibre-Optic Cable Defect Types

The following are the types of fibre-optic cable defects:

1. Missing jacket: The fibre-optic cable jacket is removed, damaged, cracked or shows signs of abrasion.
2. Pinch: The fibre-optic cable has an indentation, but there are no breaks in the jacket.
3. Delamination: There is a separation between the fibre-optic cable jacket and cladding or a change in the cladding colour caused by heat or chemicals, but there are no breaks in the jacket.
4. Contamination: There is a contaminate or any foreign material present on the outer surface of the fibre-optic cable jacket.

1.19.8.1.2 Fibre-Optic Cable Inspection

To find fibre-optic cable defects, the following are the most common inspection procedures to use:

1. Visual inspection: The jacket of the fibre-optic cable should be inspected using an illuminated magnifying aid.
2. Touch inspection: The jacket of the fibre-optic cable should be inspected using your fingers. To perform touch inspection of the cable jacket, start from one end of the fibre-optic cable. Gently squeeze the cable between the tips of the thumb and index finger, then slide the fibre-optic cable, pulling by hand in one direction to the other end.
3. Combined inspection: Fibre-optic cable should be inspected using an illuminated magnifying aid and your fingers to perform the combined (visual and touch) inspection. Visual inspection procedure can be performed under a magnifying aid.

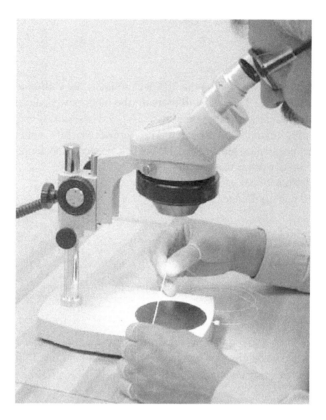

FIGURE 1.27 Fibre-optic cable inspection.

4. Inspection under a microscope. Fibre-optic cables should be inspected under a powerful microscope (20X), as shown in Figure 1.27, to find any small crack or damage on the cable jacket that cannot be diagnosed using the above inspection procedures.
5. Test inspection: This procedure is an expensive and precise inspection. The fibre-optic cable should be tested by sending a signal from a light source and measuring the signal output power. This procedure determines the loss as well as the location of any light leakage from the fibre-optic cable jacket.

1.19.8.1.3 Fibre-Optic Cable Cleaning

The following steps are necessary in the fibre-optic cable cleaning process:

1. The fibre-optic cable should be inspected and kept clean during the manufacturing process.
2. If any type of contamination is found on the fibre-optic cable, follow the recommended cleaning procedure. Each type of fibre-optic cable has a cleaning procedure that uses a specific cleaning solvent. Check the fibre-optic cable cleaning process before applying any cleaning solvent.
3. Use a swab or fibre-optic cable cleaning pad dampened with solvent to remove the foreign material. Finger cots should be used when handling solvents. Gently rub the swab/pad along the fibre-optic cable, in one direction only, to remove the contamination, as shown in Figure 1.28.
4. When cleaning do not use aggressive rubbing to remove the contamination. Do not use fingernails or any other hard-surfaced objects to scrape off the contamination.
5. If you are unable to remove the foreign material using the above methods, reject the fibre-optic cable.

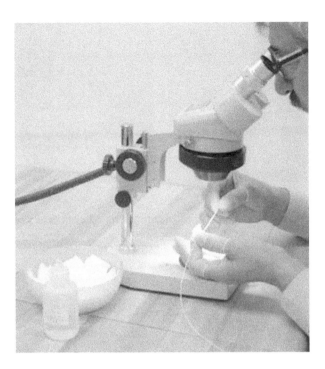

FIGURE 1.28 Fibre-optic cable cleaning.

1.19.8.1.4 Fibre-Optic Cable Handling
When handling fibre-optic cable, observe the following rules:

1. Do not allow the fibre-optic cable to rest on the floor.
2. Do not allow the fibre-optic cable to be crushed or pinched.
3. Do not allow the fibre-optic cable to drop directly over sharp edges, such as a table edge, tools or handling trays.
4. Do not allow the fibre-optic cable to come in contact with hot surfaces, such as a soldering iron, hot air pencil or hot handling trays from an oven.
5. Do not allow anything to be placed on top of the fibre-optic cable during the handling.
6. Do not permit any kind of macro bends to be present in a fibre-optic cable. Bends increase attenuation and decrease the tensile strength of the fibre-optic cable.
7. The fibre-optic cable should be coiled, as shown in Figure 1.29a. The coil diameter depends on the fibre-optic cable diameter and type. The fibre cable can also be coiled around a rubber ring, as shown in Figure 1.29b. The size of the rubber ring depends on the fibre-optic cable diameter and type.
8. Always separate paper from the fibre-optic cable. Put paper in a plastic bag when handling the fibre-optic cable.

1.19.8.2 Case (b): Fibre Cable Ends Preparation
1.19.8.2.1 Manual-Stripping Procedure
1. Hold and open the hand stripper with one hand *only.*
2. Hold the fibre-optic cable very tightly between the thumb and forefinger. Place the stripper on the fibre-optic cable, making sure to insert the fibre-optic cable through the *V's* in the heads, as shown in Figure 1.30. The stripper should be perpendicular to the fibre-optic cable. Approximately one inch of the fibre-optic cable should pass through to the other side of the stripper.

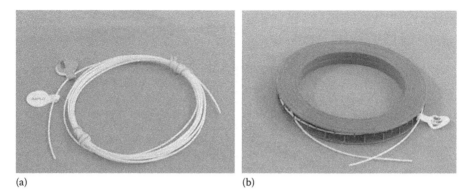

(a)   (b)

FIGURE 1.29 Fibre-optic cable handling: (a) coiled fibre and (b) coiled fibre on a rubber ring.

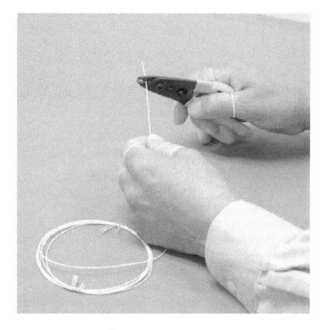

FIGURE 1.30 Manual stripping.

3. Gently squeeze the handles until the stripper closes completely. Keep the handles in this position.
4. While holding the fibre-optic cable tightly, pull the stripper as straight as possible along the length of the fibre-optic cable, towards the fibre-optic cable end.
5. Always keep the hand stripper clean.

1.19.8.2.2 Thermal-Stripping Procedure

In contrast to mechanical strippers, which create a mechanical stress and leave scratches on the surface of the cladding, thermal strippers are easy, quick and safe tools used for the stripping of jackets. The following steps summarise the procedure:

1. Turn on the thermal stripper.
2. Hold and open the thermal stripper with one hand *only*.

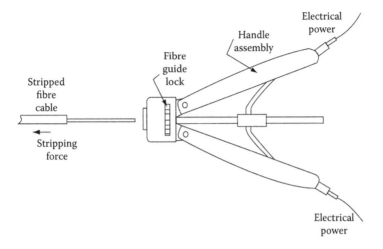

FIGURE 1.31 Thermal stripping.

3. Hold the fibre-optic cable very tightly between the thumb and forefinger. Insert the fibre cable into the stripper for the stripping length that is required, as shown in Figure 1.31. The stripper should be lined up with the fibre-optic cable. Approximately one inch of the fibre-optic cable should pass into the stripper.
4. Gently squeeze the handles until the stripper closes completely. Keep the handles in this position.
5. While holding the fibre-optic cable tightly, pull the stripper as straight as possible along the length of the fibre-optic cable, away from the fibre.
6. Always keep the thermal stripper clean. Follow the cleaning procedure of the device.

1.19.8.2.3   Cleaning Procedure after Stripping

1. Take a piece of cotton tissue dampened with denatured ethanol or any cleaning chemical suitable for the fibre.
2. Wrap it around the stripped end of the fibre-optic cable.
3. Pull in one direction towards the end.
4. Clean more than once, if necessary.

1.19.8.2.4   Manual-Cleaving Procedure

1. Place the stripped length of fibre-optic cable firmly on the inside part of your forefinger.
2. Hold the cleaver with your other hand perpendicular to the fibre-optic cable. Make a gentle scratch across the cladding surface of the fibre-optic cable at a distance of about a half of an inch from the end of the fibre-optic cable, as shown in Figure 1.32.
3. Use the same tool to break off the fibre-optic cable at the scribed mark.

1.19.8.2.5   Mechanical-Cleaving Procedure

A mechanical cleaver is used in most production lines. The mechanical cleaver gives a precise cut at a cleave angle of 90 degrees to the fibre-optic cable end. Figure 1.33 shows a mechanical cleaver that cleaves the fibre-optic cable end to a required length; this is typical of most manufacturing and test processes. To use the mechanical cleaver perform the following steps:

1. Get a mechanical cleaver ready.
2. Press the red lever (a), as shown in Figure 1.33.

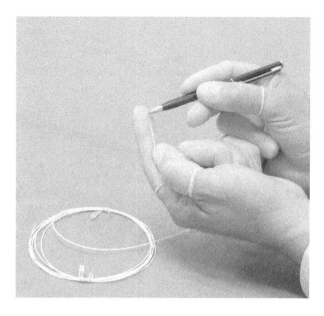

FIGURE 1.32 Cleaving.

FIGURE 1.33 Mechanical cleaving.

3. Open the top mechanism (b).
4. Open the small cover (c).
5. Place the stripped fibre-optic cable at the required cleave length on the disc blade on the block (d). This is where the cleaving will take place.
6. Close the small cover (c) and top mechanism (b).
7. Push block (d) in the direction of the arrow to cleave the fibre.
8. Press knob (e) to break the fibre.
9. Press the red lever (a), open the top mechanism (b) and open the small cover (c).
10. Remove the fibre-optic cable.
11. The fibre optic cable is cleaved automatically to the required cleave length.

1.19.8.3 Case (c): NA and Acceptance Angles Calculation

1. Figure 1.34 shows the experimental apparatus set-up.
2. Bolt the laser short rod to the breadboard.
3. Bolt the laser mount to the clamp using bolts from the screw kit.
4. Put the clamp on the short rod.
5. Place the He–Ne laser into the laser mount and tighten the screw. Turn on the laser device. Follow the operation and safety procedures of the laser device being used.
6. Check that the laser is aligned with the line of bolt holes and adjust, if necessary.
7. Mount a lens/fibre cable holder/positioner to the breadboard so that the laser beam passes over the centre hole.
8. Add the 20X microscope objective lens to the lens/fibre cable holder/positioner.
9. Prepare a fibre cable with a good clean cleave at each end, as described in Cases (a) and (b).
10. Insert one end (input) of the fibre cable into the brass fibre cable holder and place it in the hole of the lens/fibre cable holder/positioner.
11. Extend the fibre cable so that the end is at the centre of the lens/fibre cable holder/positioner. This is a very important step for obtaining an accurate value of fibre cable NA.
12. Recheck the alignment of the light-launching arrangement by making sure that the input end of the fibre cable remains at the centre of the laser beam.
13. Mount the output end of the fibre cable into the brass fibre cable holder and place it in the hole of the fibre cable holder/positioner.
14. Place a black/white card on a cardholder in front of the output end of the fibre cable at a distance (L).
15. Turn off the lights of the lab and observe the circular red laser beam spot with the diameter (D) on the black/white card, as shown in Figure 1.35.
16. Measure the diameter (D) of the spot on the black/white card, and the distance (L) from the fibre cable end to the black/white card. Fill out Table 1.4.
17. Repeat this procedure five times for five different distances.
18. Turn on the lights of the lab.

FIGURE 1.34 Numerical aperture apparatus set-up.

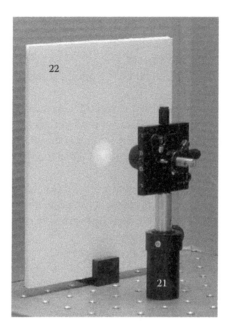

FIGURE 1.35   Red laser beam spot on the black/white card.

TABLE 1.4
Numerical Aperture Data Collection and Calculations

Measured Values		Calculated Values		
Distance (L) (unit)	Diameter (D) (unit)	Half-Acceptance Angle (9) (degrees)	Full-Acceptance Angle (3) (degrees)	NA
$L_1 =$ ( )	$D_1 =$ ( )			$NA_1 =$
$L_2 =$ ( )	$D_2 =$ ( )			$NA_2 =$
$L_3 =$ ( )	$D_3 =$ ( )			$NA_3 =$
$L_4 =$ ( )	$D_4 =$ ( )			$NA_4 =$
$L_5 =$ ( )	$D_5 =$ ( )			$NA_5 =$

1.19.8.4   Case (d): Fibre Cable Power Output Intensity

Repeat the set-up procedure explained in Case (c) of this experiment. Figure 1.36 shows the experimental apparatus set-up for fibre cable power output intensity.

For Case (d), add the following steps:

1. Place the black/white card in front of the output end of the fibre cable at distance (L).
2. Measure the distance between the fibre cable end and the black/white card. This is the distance (L). Fill out Table 1.5.
3. Measure the diameter (D) of the spot on the black/white card from the fibre cable output end at distance (L), as shown in Figure 1.35. Fill out Table 1.5.

FIGURE 1.36 Fibre cable power output intensity apparatus set-up.

TABLE 1.5
Data Collection and Calculations

Measured Values		Calculated Values		
Distance (L) (unit)	Diameter (D) (unit)	Output Power (P_{out}) (unit)	Area (A) (unit²)	Intensity (I) = (P_{out}/A) (unit) (unit²)
L_1 = ( )	D_1 = ( )	P_{out1} = ( )	A_1 = ( )	I_1 = ( )
L_2 = ( )	D_2 = ( )	P_{out2} = ( )	A_2 = ( )	I_2 = ( )
L_3 = ( )	D_3 = ( )	P_{out3} = ( )	A_3 = ( )	I_3 = ( )
L_4 = ( )	D_4 = ( )	P_{out4} = ( )	A_4 = ( )	I_4 = ( )
L_5 = ( )	D_5 = ( )	P_{out5} = ( )	A_5 = ( )	I_5 = ( )

4. Place the laser sensor in front of the output end of the fibre cable at the same distance (L).
5. Measure the laser output power (P_{out}) at that distance (L). Fill out Table 1.5.
6. Repeat the measurement of the laser output power (P_{out}) and spot diameter for each of the five distances (L).

1.19.9 DATA COLLECTION

1.19.9.1 Case (a): Fibre Cable Inspection and Handling

No data collection is required for this case.

1.19.9.2 Case (b): Fibre Cable Ends Preparation
No data collection is required for this case.

1.19.9.3 Case (c): NA and Acceptance Angles Calculation
1. Measure the diameter (D) of the laser red spot at the distance (L) from the black/white card. Report these measurements for the five distances to get five diameters.
2. Fill out Table 1.4 with the measured data.

1.19.9.4 Case (d): Fibre Cable Power Output Intensity
1. Measure the laser output power (P_{out}) at that distance (L).
2. Repeat the measurement of the laser output power (P_{out}) and spot diameter (D) for each of the five distances (L).
3. Fill out Table 1.5 with the measured data.

1.19.10 CALCULATIONS AND ANALYSIS

1.19.10.1 Case (a): Fibre Cable Inspection and Handling
No calculations and analysis are required for this case.

1.19.10.2 Case (b): Fibre Cable Ends Preparation
No calculations and analysis are required for this case.

1.19.10.3 Case (c): NA and Acceptance Angles Calculation
1. Plot the graph of (L) versus (D).
2. Calculate the fibre cable half-acceptance angle (θ), full-acceptance angle (β) and NA using Equations 1.13 through 1.15, respectively.
3. Fill out Table 1.4 with the calculated values.

1.19.10.4 Case (d): Fibre Cable Power Output Intensity
1. Calculate the area (A) of the laser beam spot and the output intensity of the laser beam using Equations 1.17 and 1.16, respectively.
2. Plot the graph of the laser power output (P_{out}) as a function of the distance (L).
3. Plot the graph of the laser power intensity (I) as a function of the distance (L).
4. Fill out Table 1.5 with the calculated values.
5. Determine the relationship between the laser power intensity (I) and the distance (L).

1.19.11 RESULTS AND DISCUSSIONS

1.19.11.1 Case (a): Fibre Cable Inspection and Handling
Discuss fibre cable inspection and handling procedure and your observations.

1.19.11.2 Case (b): Fibre Cable Ends Preparation
Discuss the fibre cable inspection and handling procedures and your observations.

1.19.11.3 Case (c): NA and Acceptance Angles Calculation
1. Determine the NA value from the graph.
2. Report the NA value.
3. Compare the graphical and calculated values of the NA.
4. Compare the graphical and calculated values with the actual value of the NA for the fibre cable, which is provided by the manufacturer specifications.

1.19.11.4 Case (d): Fibre Cable Power Output Intensity

1. Present the plotted graph of the laser power output (P_{out}) as a function of the distance (L).
2. Present the plotted graph of the laser power intensity (I) as a function of the distance (L).
3. Report the relationship between the laser power intensity (I) as a function of distance (L).

1.19.12 CONCLUSION

Summarise the important observations and findings obtained in this lab experiment.

1.19.13 SUGGESTIONS FOR FUTURE LAB WORK

List any suggestions for improvements using different experimental equipment, procedures and techniques for any future lab work. These suggestions should be theoretically justified and technically feasible.

1.20 LIST OF REFERENCES

List any references that were used in the report. Use one format in writing the references. Never mix reference formats in a report.

1.21 APPENDIX

List all of the materials and information that are too detailed to be included in the body of the report.

FURTHER READING

Agrawal, G. P., *Fiber-Optic Communication Systems*, 2nd ed., Wiley, New York, 1997.
Beiser, Arthur, *Physics*, 5th ed., Addison-Wesley Publishing Company, Reading, MA, 1991.
Boyd, Waldo T., *Fiber Optics Communications, Experiments and Projects*, 1st ed., Howard W. Sams & Co., Washington, DC, 1987.
Buckler, G., Fiber versus copper, *CNS Magazine*, May/June, 10–14, 2004.
Chen, Kevin P., In-fiber light powers active fiber optical components, *Photon. Spectra*, April, 78–90, 2005.
Cole, M., *Telecommunications*, Prentice Hall, Englewood Cliffs, NJ, 1999.
Cornsweet, T. N., *Visual Perception*, Academic Press, New York, 1970.
Derickson, D., *Fiber Optic Test and Measurement*, Prentice Hall PTR, Englewood Cliffs, NJ, 1998.
Dutton, H. J. R., *Understanding Optical Communications*, IBM Prentice Hall, New Jersey, 1998.
Eggleton, B. J., Kerbage, C., Westbrook, P. S., Windeler, R. S., and Hale, A., Microstructured optical fiber devices, *Opt. Express Dec.*, 17, 698–713, 2001.
Goff, D. R., and Hansen, K. S., *Fiber Optic Reference Guide: A Practical Guide to the Technology*, 2nd ed., Butterworth-Heinemann, London, 1999.
Green, P. E., *Fiber Optic Networks*, Prentice Hall, Englewood Cliffs, NJ, 1993.
Groth, D., Mcbee, J., and Barnett, D., *Cabling—The Complete Guide to Network Wiring*, 2nd ed., Sams & Company, Washington, DC, 2001.
Hecht, J., *City of Light: The Story of Fiber Optics*, Oxford University Press, New York, 1999.
Hecht, J., *Understanding Fiber Optics*, 3rd ed., Prentice Hall, Englewood Cliffs, NJ, 1999.
Hecht, E., *Optics*, 4th ed., Addison-Wesley Longman, New York, 2002.
Horng, H. E., Chieh, J. J., Chao, Y. H., Yang, S. Y., Chin-Yih, Hong, and Yang, H. C., Designing optical-fiber modulators by using magnetic fluids, *Opt. Lett.*, 30(5), 543–545, 2005.
Hoss, R. J., *Fiber Optic Communications—Design Handbook*, Prentice Hall, Englewood Cliffs, NJ, 1990.
Kao, C. K., *Optical Fiber Systems: Technology, Design, and Applications*, McGraw-Hill, New York, 1982.
Kasap, S. O., *Optoelectronics and Photonics Principles and Practices*, Prentice Hall PTR, Englewood Cliffs, NJ, 2001.
Keise, G., *Optical Communications Essentials*, 1st ed., McGraw-Hill, New York, 2003.
Keise, G., *Optical Fiber Communications*, 3rd ed., McGraw-Hill, New York, 2000.

Kolimbiris, H., *Fiber Optics Communications*, Prentice Hall, Englewood Cliffs, NJ, 2004.

Lerner, R. G., and Trigg, G. L., *Encyclopedia of Physics*, 2nd ed., VCH Publishers, New York, 1991.

Malacara, D., *Geometrical and Instrumental Optics*, Academic Press Company, Boston, MA, 1988.

Mazzarese, D., Meeting the OM-3 challenge—Fabricating this new generation of multimode fiber requires innovative state of the art designs and testing processes, *Cabling Syst.,* December, 18–20, 2002.

Mynbaev, D. K., and Scheiner, L. L., *Fiber-Optic Communication Technology—Fiber Fabrication*, Prentice Hall PTR, Englewood Cliffs, NJ, 2005.

National Research Council, *Harnessing Light Optical Science and Engineering for the 21st Century*, National Material Advisory Board, National Academy Press, Washington, DC, 1998.

Newport Corporation, *Projects in Fiber Optics Applications Handbook*, Newport Corporation, Irvine, CA, 1986.

Newport Corporation, *Photonics 2000 Catalog*, Newport Corporation, Irvine, CA, 2000.

Palais, J. C., *Fiber Optic Communications*, 4th ed., Prentice Hall, Englewood Cliffs, NJ, 1998.

Romine, G. S., *Applied Physics Concepts into Practice*, Prentice Hall, Inc., Englewood Cliffs, NJ, 2001.

Salah, B. E. A., and Teich, M. C., *Fundamentals of Photonics*, Wiley, New York, 1991.

Senior, J. M., *Optical Fiber Communications*, *Principle and Practices*, 3rd ed., Personal Education Limited, Essex, England, 2009.

Shamir, J., *Optical Systems and Processes*, SPIE Optical Engineering Press, Washington, DC, 1999.

Shashidhar, N., Lensing technology, *Corning Incorporated Fiber Product News*, April, 14–15, 2004.

Shotwell, R. A., *An Introduction to Fiber Optics*, Prentice Hall, Englewood Cliffs, NJ, 1997.

Sterling, D. J. Jr., *Technician's Guide to Fiber Optics*, 2nd ed., Delmar Publishers, New York, 1993.

Sterling, D. J. Jr., *Premises Cabling*, Delmar Publishers, New York, 1996.

Torrey, S., Gel-free outdoor loose tube and ribbon fiber optic cable, *Fiberoptic Product News*, August, 14–16, 2004.

Ungar, S., *Fiber Optics: Theory and Applications*, Wiley, New York, 1990.

Vacca, J., *The Cabling Handbook*, Prentice Hall, Englewood Cliffs, NJ, 1998.

Walker, R., and Bessant, N., Image fiber delivers vision of the future, *Photon. Spectra*, March 96–97, 2005.

Weisskopf, V. F., How light interacts with matter, *Sci. Am.,* September, 60–71, 1968.

Yeh, C., *Handbook of Fiber Optics: Theory and Applications*, Academic Press, San Diego, CA, 1990.

Yeh, C., *Applied Photonics*, Academic Press, New York, 1994.

2 Advanced Fibre-Optic Cables

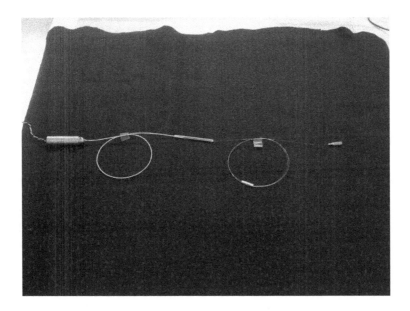

2.1 INTRODUCTION

The scaling down of fibre light sources brings the benefits of this fibre technology to a wide range of applications. These developments have led to a surge of interest in advanced fibre cables for both industrial and military applications. Advanced fibre cables are also used for transmitting high volumes of data in communication systems over long distances for getting very clear images, and in building many sophisticated instruments for a variety of applications. By creating new core designs, adding dopants to the fibre core and cladding, and developing manufacturing processes, engineers achieve advanced fibre-optic cable technology.

For example, the core of the holey fibre consists of many air holes; each hole acts as a single fibre. This fibre enables a high data transmission rate and capacity, and, consequently, reduces the cost of the network.

As explained in the preceding chapter, light propagates throughout the fibres by the principle of total internal reflection. The advanced fibres also have the same characteristics and suffer the same kind of losses as the common types of fibres, but in general terms, they have very low loss.

This chapter proposes a few experiments, such as fibre cable end preparation, and the measurements of numerical aperture (NA) aperture, power output intensity, and calculations of loss. The experimental results from this chapter can be compared with the preceding chapter. Some facts surrounding the advanced fibres can be concluded.

2.2 ADVANCED TYPES OF FIBRE-OPTIC CABLES

The advanced core and cladding designs of newer fibre cables attempt to achieve lower loss and low dispersion over a wide range of wavelengths, so the cable can operate at very high speeds over very long distances. These types are also called *holey* fibres because they have tubes or spaces in the core along the fibre's length. Some types of these advanced fibre-optic cables are presented below:

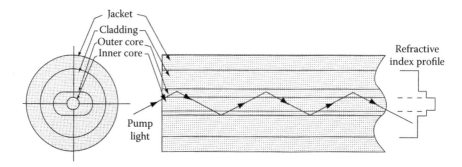

FIGURE 2.1   Dual-core fibre for high-power laser.

2.2.1   DUAL-CORE FIBRE FOR HIGH-POWER LASER

Fibre lasers—also called fibre sources—are light sources made from fibre. They are built around fibres cores, which are doped with materials that can be stimulated to emit light. Light stimulation and amplification are generally explained by the laser principles.

Single-mode, rare-earth-doped fibre lasers and amplifiers are widely used in telecommunications and other applications requiring compact, rugged optical sources with high beam quality. Fibre sources provide high electrical-to-optical conversion efficiency. The glass host broadens the optical transitions in the rare-earth ion dopants, gives continuous tunability. The variety of possible rare-earth dopants and co-dopants, such as erbium (Er^{3+}), neodymium (Nd), ytterbium (Yb), aluminium (Al) and germanium (Ge), yields broad wavelength coverage in the near-IR spectral region.

The developments of power scaling of continuous wave (CW) and pulsed fibre sources bring the benefits of this technology to a wide range of applications. These developments have led to a surge of interest in fibre-based laser systems for both industrial and military applications.

High-power fibre sources incorporate double-clad fibre, as shown in Figure 2.1, in which the rare-earth-doped core is surrounded by a much larger and higher NA outer core. The inner core has a higher index of refraction than the outer core, and the outer core, in turn, has a higher index of refraction than the cladding. The cladding may be glass or polymer, always with an index lower than the outer core. The inner core is normally doped with a high light-emitting rare-earth element. Light from high-power multi-mode pump diodes can be launched efficiently into the outer core. The pump light is absorbed only in the inner core, where it operates in a single mode. When the pump light goes through the inner core many times, it can excite light-emitting atoms efficiently. This type of fibre is used in building a femtosecond switch in a dual-core-fibre non-linear coupler and some devices that need higher laser power.

2.2.2   FIBRE BRAGG GRATINGS

Fibre Bragg grating (FBG) technology has gained favour in a wide range of applications because of its all-fibre configuration, great flexibility and highly-efficient filtering functions. FBGs are most commonly used for stabilising pump lasers in optical amplifiers, wavelength division multiplexing, filtering and chromatic dispersion compensation, because their efficiency reduces the cost of optical networking. FBGs are made from a simple ordinary single-mode fibre. The grating is constructed by varying the index of refraction of the core along the fibre's length. The index of refraction rises and falls along the fibre's length periodically in a pattern that would look like a uniform wave, as shown in Figure 2.2. All wavelengths are scattered by the regularly-spaced index variation, which causes the waves to interfere with each other in the fibre. Light at the wavelength that hits each index peak at the same phase experiences constructive interference, and is selectively reflected back towards the source. These are called resonant waves. The spacing of the grating matches the resonant wavelength. Light scattered at other wavelengths by the grating is out of phase and cancels out by destructive

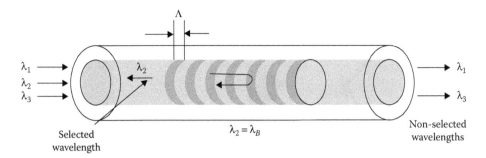

FIGURE 2.2 A fibre Bragg grating (FBG).

interference. The waves transmitted though the fibres are called non-resonant waves. The FBG performance is influenced by the centre wavelength, bandwidth and reflectance peak. The FBG has basic parameters that control the performance, such as the grating period, the grating length, the modulation depth, the index of refraction profile and the wavelengths of the incoming light.

The grating selective reflected light at the Bragg wavelength (λ_B) is given by

$$\lambda_B = 2n_{\text{eff}}\Lambda \tag{2.1}$$

where:

    n_{eff} is the effective index of refraction of the core

    Λ is the pitch of the grating in the core

FBGs are sensitive to temperature variation. Therefore, they must be kept at an operating temperature between $-5°C$ and $+70°C$ to operate accurately. When used in building wavelength-division multiplexing (WDM) systems, temperature compensation techniques are necessary. Emerging manufacturing improvements—such as complex holographic phase masks—combined with automated processes will drive down costs and further improve the performance of FBG-based components.

2.2.2.1 Manufacturing Method

The simplest manufacturing method of FBGs is exposing a photosensitive fibre to an intensity pattern of ultraviolet radiation to write a FBG. The complete manufacturing process of FBG has the following four steps:

1. *Preparing the photosensitive optical fibre*: The preparation consists of stripping the coating (jacket) from the region to be exposed. Using chemicals for coating removal instead of mechanical stripping helps to avoid mechanical damage to the fibre.
2. *Recording of the grating*: Recording FBG inside a fibre usually involves interference between two coherent UV laser beams that precisely define the spacing between the grating's planes. For symmetric incidence and an angle (θ) between the UV beams, the pitch of the recorded grating is given as $\Lambda = \lambda/2 \sin(\theta/2)$, where λ is the illumination wavelength. The basic interferometric set-up splits the laser beam to create an interference pattern on the fibre, as shown in Figure 2.3a. Another interferometric set-up uses a mirror positioned perpendicularly to the fibre to fold half of the laser onto its other half, as shown in Figure 2.3b. This process is adequate for short FBGs. A third method uses a phase mask to generate multiple beams. Two of the beams carry approximately 40% of the total energy and closely overlap the phase mask surface to create an interference pattern at the desired Bragg wavelength, as shown in Figure 2.3c. The fibre-phase mask assembly can be illuminated either by a large beam to cover the full

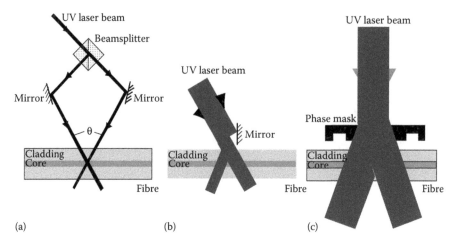

FIGURE 2.3 Several configurations can be used to make FBGs.

FBG length or by a small scanning beam. As the fibre is usually close to the phase mask, the assembly is a very stable mechanical system, better than the process shown in Figure 2.3a.

3. *Thermal annealing*: Subjecting the FBG to thermal annealing at high temperatures (>250°C) helps ensure that their optical properties remain stable for at least 25 years. Annealing techniques must heat the grating locally without damaging the coating of the fibre.

4. *Packaging*: The final step of the manufacturing process consists of packaging the grating. Depending on the application, there are three packaging processes: (1) recoating the fibre with coating similar to the fibre jacket, (2) protecting the grating with a mechanical sleeve and (3) packaging the grating in a thermal module.

2.2.3 Chirped Fibre Bragg Gratings

Chirped fibre Bragg gratings (chirped FBGs) are suitable for compensating the chromatic dispersion that occurs along a fibre link. The chirped FBG is composed of an optical core that varies the period of the grating along the length of the grating, as shown in Figure 2.4. Chirped FBGs can be made by two methods: either by varying the period of the grating or by varying the average index of refraction of the grating. Both methods have the effect of gradually changing the grating period.

The grating reflects light propagating through the fibre when its wavelength corresponds to the grating period. As the grating period varies along the axis, the different wavelengths are reflected by different portions of the grating and, accordingly, are delayed by different amounts of time. The net effect is a compression (or a broadening) of the input pulse that can be tailored to compensate for the chromatic dispersion accumulated along the fibre link. Although entering the grating at different times, all the wavelength components of broadened pulses return to the entrance at the same time.

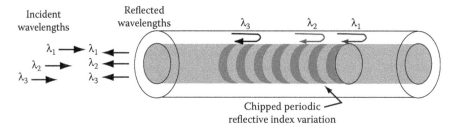

FIGURE 2.4 A chirped fibre Bragg grating (chirped FBG).

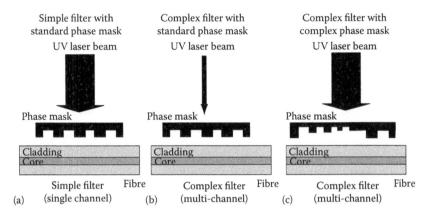

FIGURE 2.5 Several configurations can be used to make chirped FBGs: (a) Simple writing scheme, (b) Complex writing scheme and (c) Simple writing scheme.

2.2.3.1 Manufacturing Method

When fabricating a simple FBG, a standard phase mask is usually used, along with a fixed UV laser beam, as shown in Figure 2.5a. This technique is similar to the one used for the fabrication of simple FBG devices, such as pump laser stabilisers. This recording technique is very efficient, but can be applied only to simple FBG filters. One of the most common techniques for fabricating an FBG filter that requires a complex profile, and that uses a standard phase mask, involves a complex recording scheme wherein the UV laser beam is scanned along the phase mask, as shown in Figure 2.5b. A complex phase mask can be used for the fabrication of a complex FBG filter by transferring the pattern of the recording scheme directly into the phase mask, as shown in Figure 2.5c.

FBG technology is also well-suited to tunable dispersion compensation. Submitting the FBG to a temperature gradient allows the grating chirp to be changed, and, accordingly, the dispersion level to be tuned. Single gratings can be used for producing negative dispersion or for producing a similar positive dispersion range.

FBGs are a mature technology that is very well-adapted to dispersion compensation in optical communication systems. With all of the aforementioned developments, FBG-based technology can be considered a cost-effective alternative to competing technologies, while providing major advantages, such as small size, no non-linear optical effects and low insertion loss.

2.2.4 BLAZED FIBRE BRAGG GRATINGS

A blazed FBG is constructed when the grating is written at an oblique angle to the centre of the core, as shown in Figure 2.6. The selected wavelengths are reflected out of the fibre, and the non-selected wavelengths are passed through the fibre. In this case, a filter is constructed. FBGs are used in erbium-doped fibre amplifiers (EDFAs). Manufacturing configurations are similar to the processes used in manufacturing the FBGs.

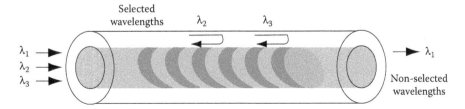

FIGURE 2.6 A blazed FBG.

2.2.5 Non-Zero-Dispersion Fibre-Optic Cables

Non-zero-dispersion fibre was introduced to the market in 1993. This fibre type provides optimum performance for optically-amplified systems, over longer distances and with higher capacity. This cable can be used with optimum dispersion slope over the S-band (1480–1520 nm), C-band (1520–1570 nm) and L-band (1570–1620 nm).

The advanced core designs of newer single-mode fibres attempt to make lower loss and low dispersion coincide at the same wavelength, so the system can operate at very high speeds over very long distances. Dispersion-shifted fibres have a structure that changes the zero-dispersion wavelength, typically from 1300 to 1550 nm. Dispersion-flattened fibres have a structure that lowers dispersion over a wide range of wavelengths.

2.2.6 Photonic Crystal Fibre Cables

Recently, there has been a dramatic increase of interest in photonic crystal fibre (PCF), also called microstructure fibre (MOF) or holey fibre, which features an array of air holes running along the fibre length, as shown in Figure 2.7. Photonic crystal fibres use a micro-structured cladding region with air holes to guide light in a pure silica core, giving rise to novel functionalities. The PCFs are divided into two classes: photonic band gap fibres associated with the photonic band gap structures, and index-guiding holey fibres based on the modified total internal reflection. The latter form their cores by filling one or

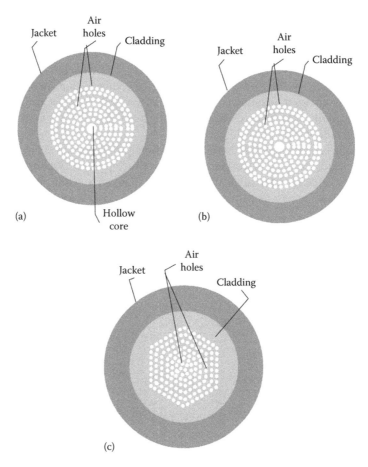

FIGURE 2.7   Cross-sections of the PCF cables: (a) Hollow core photonic band-gap fiber, (b) Endlessly photonic crystal fiber and (c) Photonic crystal fiber.

several air holes in the array with fibre material. This fibre exhibits many unique properties of light guidance, such as single-mode operation over a wide range of wavelengths, highly adjustable effective mode area and non-linearity, anomalous dispersion at visible and near-infrared wavelengths and high birefringence. They are capable of high-capacity transmission over transoceanic distances, and extremely high-efficiency transmission over a wide range of wavelengths.

Photonic crystal fibre keeps light confined to a hollow core that is about five times thinner than a human hair. The fibre cross-section is an air core of about 15 μm in diameter surrounded by a web like structure of glass and air. Photonic crystals work on the principle of refraction, or the bending of light as it passes from one material to another. Light travels through the core of the material just as it travels through the middle of hollow fibre-optic lines. However, photonic crystal is better than a reflective coating at keeping light from scattering or being absorbed, which keeps signals stronger over longer distances. Photonic crystal fibres are widely used in telecommunications—to filter light signals, to represent data in quantum communications, implementing fibre-based polarisation splitters, transmitting exotic wavelengths in the IR or UV range, for applications in medical equipment and sensors, diode-pumped Nd^{3+} laser sources and measurements in scientific experiments.

Figure 2.8 illustrates the stack-and-draw process, the most widely used method for fabricating PCF. Silica capillary tubes are stacked by hand to make a preform. The preform has a diameter

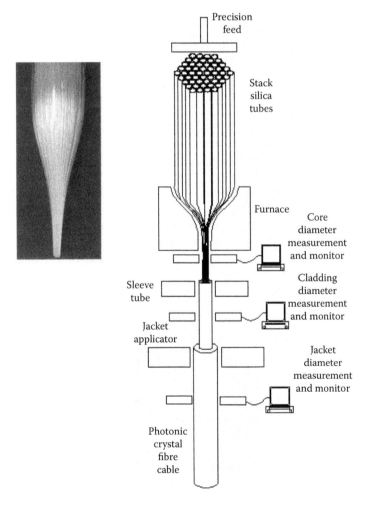

FIGURE 2.8   In the stack-and-draw process for fabricating PCFs.

of 20–40 mm. The core is embedded by omitting several capillaries from the centre of the stack. Typically, several hundred capillaries are stacked in a close-packed array and inserted into a jacketing tube to create a fibre preform. This preform is drawn into fibre in two stages, adding a second jacketing tube before the final draw. The extra jacket enables the creation of fibres with a standard outer diameter while providing independent control over the photonic band gap pitch, and hence, the operating wavelength. The preform is drawn into fibre, as shown in Figure 2.8. The preform is attached to a precision feed that moves it into the furnace at a proper speed to often and shrinks the silica to a diameter of 2–4 mm. The drawing fibre passes through high-precision diameter control and is monitored by imaging or x-ray to detect any non-homogeneity or bubbles in the drawing fibre. It is then inserted into a silica sleeve tube and drawn down again to a fibre of typically 125 μm in diameter. After this stage, the diameter is continuously monitored by an accurate measurement and monitoring device. The fibre is then covered by a coloured jacket layer. Again, the jacket layer undergoes diameter control and monitoring. The end of the fibre cable is attached to a rotating spool, which turns steadily. The fibre cable is then tested for attenuation in dB per kilometre, for dispersion, and any other requirements specified by the customer. Drawn PCF lengths of a few kilometres are typical.

Photonic crystal fibres are available in different types, such as highly non-linear, double-cladding high NA, large mode-field area and polarisation-maintaining. These types have different mode-field areas, numbers of holes and working wavelengths. Some types of PCFs have inner and outer cladding layers that are used in highly-polarisation-maintaining cables.

2.2.7 Microstructure Fibre Cables

Microstructure fibres can be used to create a supercontinuum (SC). This is a broadening of the narrow laser spectrum into a broad continuous spectrum with many of the properties of coherent light. SCs have applications in frequency metrology and optical coherence tomography. The addition of Bragg gratings to the SC light enhances the SC spectrum and alters the dispersion profile, thus making gratings an effective tool in tailoring SC sources.

A grapefruit fibre is one type of MOF. In the grapefruit MOF, the lower-order modes are confined by the air holes and are sensitive to surrounding media. Figure 2.9 illustrates a cross-section of the grapefruit MOF. After the addition of long period gratings (LPG) fibre, the higher-order modes interact with the holes and with the material with which they may be infused. The manufacturing process of this fibre is similar to the manufacturing process of the crystal fibres. This fibre has the same manufacturing process as that of holey fibres. These processes are explained in the above sections.

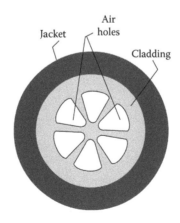

FIGURE 2.9   A cross-section of the grapefruit microstructure fibre (MOF).

2.2.8 POLYMER HOLEY-FIBRE CABLES

The polymer holey-fibre has same cross-section as the photonics crystal fibres. A polymer fibre is fabricated by first drilling holes into a polymethyl methacrylate (PMMA) preform in which the holes drill down to an intermediate preform. The holes of this intermediate preform are filled with a solution of the dye Rhodamine 6 G, which permeates the PMMA. The intermediate preform is heated to drive out the solvent molecules and to lock the dye molecules into the structure. Finally, the preform is drawn into a fibre with a 600-μm outer diameter and an 18-μm core diameter. Polymer fibres can be used in replacing the silica fibres in limited applications.

2.2.9 IMAGE FIBRE CABLES

Image fibre cables are used for remote or semi-flexible viewing systems, and ultra-thin image fibre delivers vision to previously inconceivable locations. Until recently, applications for this type of optical fibre have been in the medical industry, primarily in endoscopes. On account of the versatility of the image fibre, however, it is meeting a variety of specialised demands.

Figure 2.10 shows an image fibre made of silica glass that transmits a coherent picture from one end to the other. The technology behind ultra-thin image fibres is not new, but manufacturers and designers are increasingly seeking to expand into new imaging applications. Figure 2.11 shows the cross-section of an image fibre cable. The multiple cores are pure silica or silica doped with germanium, and they share a common cladding that is doped with fluorine. The cores, or pixels, are approximately three times thinner than the core of a single-mode telecom fibre; a 1600-pixel fibre has an outer diameter of approximately 210 μm, and a 10,000-pixel fibre has an outer diameter of 1700 μm. The fibre is drawn in a manner similar to that used in the production of communication fibre cables, then cut and bundled into a silica tube to make a preform. It can bend from 20 to 200 mm in radius. The versatility of image fibre is partially the result of its ability to function at wavelengths from the UV to the near-IR range.

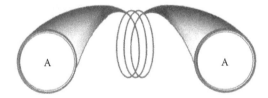

FIGURE 2.10   An image fibre transmits a picture from one end to the other.

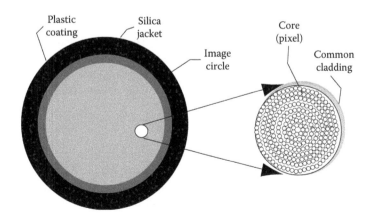

FIGURE 2.11   A cross-section of an image fibre cable.

With the customisable features of the image fibre cable, the fibre can be used to meet a variety of specific application requirements. One important area is in the field of biomolecular research. Scientists are now using the fibre in confocal microscopes for fluorescence imaging of cells and proteins at the molecular level. In the automotive sector, the image fibre is used to perform quality checks inside an engine before initial testing. Previously, the manufacturers employed rigid rodlens-based or video-imaging systems. Similarly, in the testing and maintenance of car and jet engines and steam and gas turbines, industrial endoscopes can be used where access is limited or when there is no time to dismantle components. The advantage is that image fibrescopes are flexible, whereas rigid rodlens-based systems are not. The flexibility helps them reach into previously-inaccessible areas. There are other possible applications for image fibres; for instance, using infrared wavelengths, it could be employed to take snapshots of the blades inside gas or steam turbines, which would minimize, if not eliminate, the need to shut down power stations or refineries for such maintenance checks.

2.2.10 LIQUID CRYSTAL PHOTONIC BAND GAP FIBRE CABLES

As explained above, PCFs—usually made from silica—have a micro-structured cross-section of air holes running along their length. Recently, much attention has been focused on the development and design of these fibres for various applications, including SC generation, nonlinear signal processing in high-speed optical time domain multiplexing communications systems and high-power double-clad fibre lasers. Such applications use fibres with static properties.

In PCFs, the air holes provide access to the guided light in the core and allow liquids to be infiltrated into the capillaries. The optical properties of the liquids are usually easier to modify than those of the silica. This approach, therefore, paved the way to components based on tunable PCFs.

These fibres guide light using two different principles. The first principle is modified total internal reflection, which relies on guiding light in a high-index core surrounded by a low-index cladding material. The PCF guides light by the principle of modified total internal reflection when the holes are filled with air or with a material with a lower refractive index than that of silica. The second principle involves guiding light in a low-index core surrounded by a cladding arrangement of high-index rods. Infiltrating the air holes with a high-index oil or liquid crystal transforms the fibre into a photonic band gap guiding type, which guides light in the low-index silica core.

Light coupled into the core can be guided through anti-resonant reflection from the rods, which forms a sort of two-dimensional Fabry–Perot resonant cavity surrounding the core.

The optical modes of a single liquid-crystal-filled hole determine the spectral properties of a PCF infiltrated with a pneumatic liquid crystal. As liquid crystals are highly anisotropic in molecular orientation, the optical modes are determined not only by the refractive indices of the liquid crystal, but also by the alignment within the hole. Therefore, changing the alignment of the liquid crystal and/or the refractive indices of the liquid crystal can modify the spectral properties of the fibre.

Pneumatic liquid crystals are known for exhibiting very large thermo-optic effects as a result of the existence of crystalline and isotropic liquid phases, which depend on temperature. When the liquid crystal is doped with a small amount of an absorbing dye, the high thermo-optic effect modifies its refractive index, allowing the tuning of spectral properties of the liquid crystal photonic band gap fibre.

2.2.11 LENSED AND TAPERED FIBRE CABLES

Lensed fibres are used to couple light between optical fibres and lasers, semiconductor optical amplifiers, waveguides and other fibre optical devices. Lensing technology enables highly-effective coupling within a small space. In recent years, substantial changes in lensed fibre cable technology have increased the flexibility and extended the optical performance. Figure 2.12 shows lensed fibre variations. Two significant developments have occurred: the advancement of fibre-polishing technologies and the development of novel lens designs.

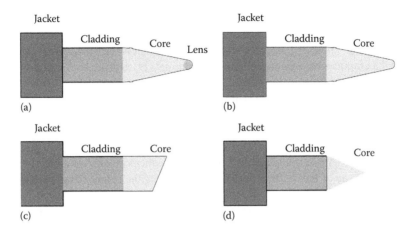

FIGURE 2.12 Lensed/tapered fibre cable variations: (a) A lended fibre, (b) A tapered fibre, (c) An angle cleaved/polished fibre and (d) A wedged fibre.

Advances in polishing technologies include process and equipment improvements and the use of laser polishing instead of mechanical polishing. The development of novel lens designs has significantly improved lens performance. A clear understanding of the optics of tapered lenses, their performance, and the various methods of manufacture and specifications will ensure that the most appropriate lensing technology is used to meet specific design requirements.

2.2.11.1   Advantages of Lensing Technology

A number of industry requirements are driving advancements in lensed fibre technology. These include the following:

- **Lower pricing**
  The price of lensed fibres has decreased because new advanced manufacturing methods are being developed to address the demand for lower pricing.
- **Higher consistency**
  Tapered lenses are made by polishing fibres into a conical shape with a spherical surface at the tip. These lenses, called conical lensed fibres, were inconsistent in quality, resulting in significant performance limitations. It was not uncommon, for example, for customers to have to test the lenses and select the ones that met optical criteria. Select rates for the lenses were as low as 60%.
- **Longer working distance**
  The working distance for lenses with a small spot size is very limited. A polished lensed fibre having a 3-μm mode-field diameter at the waist (MFDw) has a distance-to-beam waist (DBW) between 5 and 8 μm. Such a small DBW makes packaging difficult, increasing the risk of the lens hitting the facet of the device. This translates into longer process times and lower yields in component assembly.
- **Higher return loss**
  A polished lens with anti-reflective (AR) coating typically has a return loss of 30 dB. In some applications, where light is coupled into the waveguide, this could be a limitation, because the returned light acts as a feedback, increasing noise in the communication systems containing amplifiers or lasers.
- **Smaller beam diameter**
  Waveguides on materials, such as indium phosphide (InP) and gallium arsenide (GaAs), have very small MFDw. Using lenses with an MFDw that does not match the device decreases coupling efficiency, so efforts have been made to decrease the MFDw of the lens.

2.2.11.2 Manufacturing Technologies

There are a number of alternatives for manufacturing lensed fibres. Each alternative has advantages and disadvantages, and some are progressive improvements on earlier technologies. Like all technologies, not all techniques have matured or even sometimes progressed to commercial availability. These methods include mechanical polishing, chemical etching, thermally drawing the fibre to a taper and lensing the fibre. These methods are explained briefly below.

2.2.11.2.1 Mechanical Polishing

Lensed fibres are made by mechanically polishing a fibre into a conical shape with the tip shaped into a spherical lens, as shown in Figure 2.13. The lens is spherical with a radius of curvature (Rc). A challenge for this type of lens is the misalignment between the centre of curvature of the spherical lens and the optical axis. The core-to-cladding concentricity for single-mode fibre (<0.5 μm) and the geometric repeatability of the polishing method defines the degree of misalignment. This misalignment can cause deviations from Gaussian beam profiles, resulting in decreased coupling efficiency. Additionally, significant aberrations can occur while trying to polish lensed fibres to the small centre of curvature required for lensed fibres with MFDw less than 3.5 μm.

Advances in fibre-polishing equipment have increased the quality of the lensed fibres. Manufacturers can use laser ablation instead of mechanical polishing. This technique has the potential to better control the radius of curvature with mechanical polishing. It is also claimed that there is greater product consistency when a far-field monitoring feedback loop is added to the system.

2.2.11.2.2 Chemical Etching

Another manufacturing technique is chemical etching, wherein immersion in an acid solution chemically polishes the fibre. The cladding is dissolved away, leaving behind a bit of the core. Melting this stub forms a spherical lens with the desired radius of curvature. An advantage of this method is that several fibres can be shaped at a time, as opposed to other techniques where only one fibre can be shaped at a time. Although this technology has been reported on, commercially-available lensed fibres based on this technology are rare.

2.2.11.2.3 Drawn Tapered Fibres

Tapered fibres are manufactured by heating the fibre cladding and core beyond the softening point of the glass, and then applying a force to make a taper. The tip is then melted back to create a spherical lens. This technique enables greater flexibility in MFDw, as well as improving the DBW compared to mechanical polishing. In this kind of tapered lens, the amount of heat and heating time determine the Rc of the spherical lens. During manufacturing, the core of the fibre is also tapered, causing deviations from the expression relating MFDw to Rc.

In thermal methods, such as the drawn tapered fibre and lensed methods, the tapered fibre tip is made by melting the glass in such a way that the surface tension of molten glass forms a smooth surface. As the tip is frozen liquid, the surface finish can be far smoother than polishing methods.

2.2.11.2.4 Lensed Tapers

Perhaps the most promising technology is tapered lensed fibres, also called lensed fibres. Figure 2.14 shows a lensed tapered fibre-optic cable. These fibres are distinguished from tapered fibres because

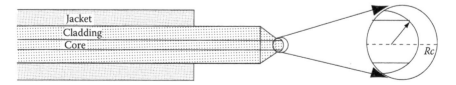

FIGURE 2.13 Lensed fibre-optic cable.

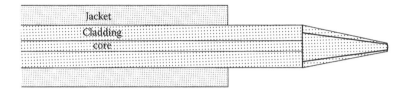

FIGURE 2.14   Lensed tapered fibre-optic cable.

they are a two-element lens with a gradient index and a geometric (spherical/hyperbolic) lens at the tip. The combination of the two elements has enabled greater flexibility for meeting a variety of optical specifications.

In lensed tapers, a fibre with a gradient index profile is fused to a fibre pigtail. This spliced fibre is drawn into a taper with a thermally-formed spherical lens at the end. The advantage of this technology is that the gradient index (GRIN)-tapered region directs the light to the spherical lens, giving the lensed fibre more consistency in the MFD. There are several factors that can be controlled in this kind of lens, including the GRIN profile, the radius of curvature, the length of the gradient index fibre and the taper length. As the GRIN fibre is fused to the pigtail, the lenses can be made on PM fibres.

Another advancement is the development of lensed tapers with a hyperbolic refractive lens. This technology has enabled a four-fold increase in the DBW, and has also increased the return loss. In this type of lensed fibre, a fibre with a GRIN is first spliced to the fibre. The GRIN fibre is then thermally shaped to make a hyperbolic shaped lens. The lensing action is a combination of the GRIN lens and the hyperbolic lens. This combination allows for a greater DBW.

Lensed fibre manufacturing has evolved significantly. Better methods for polishing and the development of more-sophisticated lens designs have increased flexibility, enabling lenses to meet a wide specification range for MFD and DBW. Each lensed fibre must be loaded individually into a fixture for anti-reflection coating. In addition to the MFDw specification, a DBW, and return loss (RL) specification may be added if important to an application.

2.2.12   BEND-INSENSITIVE FIBRE CABLES

Bend-insensitive fibre cables are designed for improved bend performance in reduced-radius applications. The fibre cables employ a moderately higher NA than standard single-mode telecommunication fibre cables, and offer improved bend performance for applications in the 1310- and 1550-nm range.

2.2.13   NANORIBBON FIBRE-OPTIC CABLES

Crystalline oxide nanoribbon fibre cables are used as waveguides. They can be used with other nanowire optical components to create elementary systems that will suit the feasibility of building a nanowire-based photonic integrated circuitry. Photonic integrated circuits are suitable for applications as active and passive components in telecommunications, optical computing and a wide variety of other sectors in photonics. Doping materials, such as SnO_2 and ZnO, are added in the core and cladding to enable a large amount of light guiding from end to end, and allow a small amount of light to escape from the fibre by scattering, as shown in Figure 2.15.

Nanoribbons with sizes on the order of 100–400 nm are suitable for wave guiding visible and ultraviolet radiation, although researchers found that synthesis process predominantly yielded ribbons for blue and green wavelengths. Measured losses depend on the area of a given ribbon in cross-section and the presence of scattering centres.

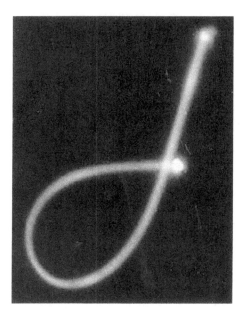

FIGURE 2.15 Nanoribbon fibre-optic cables.

To demonstrate the potential applications of waveguides in photonic circuitry, it is necessary to examine the approaches to coupling the ribbons and linking them to nanowires serving as input and output components. It has been found that staggering the ribbons for several microns with an air gap between them led to efficient evanescent tunnelling and outperformed butt-coupling. As observed using a near-field scanning optical microscope, light could be injected into the ribbons from an optically pumped ZnO nanowire source and detected by a similar wire.

2.3 APPLICATIONS OF ADVANCED FIBRE CABLES

Fibre-optic cables are used in many applications, some of which are listed in the previous Fibre-Optic Cable Chapter. One of many applications for different fibre cable types is fibre-optic sensing. Fibre-optic sensing can be classed as intrinsic and extrinsic. In intrinsic sensors, the fibre simply conducts light from a sensing head to a detector; the interaction between light and the environment takes place outside the fibre cable. In intrinsic sensors, the interaction between the environment, the fibre and the light itself generates information about a specific measurement. The key advantage of intrinsic fibre-optic sensors is the fundamental ability of a fibre to guide light around bends and over long distances. This enables the fibre to be confined within small physical volumes and magnifies the effects of very fine environment changes to a level that can be measured accurately and quantified.

FBGs are commonly employed as passive temperature and pressure sensors. The ability to adjust these optical devices introduces a new dimension for the design of very fast, precise and multi-functional fibre sensors without compromising their intrinsic advantages. As explained in the previous sections, an FBG consists of a short length of fibre-optic having a periodic modulation of refractive index along its length. This periodic structure causes the FBG to act as a narrow line filter with a peak reflectivity at a specific wavelength, called Bragg wavelength (λ_B), which is determined by the period length of the FBG and the refractive index of the fibre. The Bragg wavelength can be calculated by Equation 2.1. If the refractive index of the FBG is changed by a temperature shift, or the period is altered in some way (e.g., by compression or expansion), a shift in the peak reflected wavelength results. Detecting this shift is the basis of fibre sensing.

FBGs are also used in many other applications, such as tunable FBGs. Tuning can be accomplished by several mechanisms, such as electric heating, the piezoelectric effect, mechanical

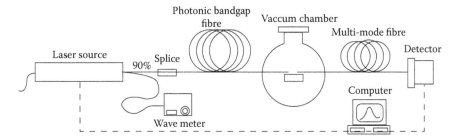

FIGURE 2.16   Band gap fibre used in gas sensor.

stretching and bending and acoustic modulation. These sensors are available in the market in different types, sizes and wavelength ranges.

Photonics band gap and photonics crystal fibres are both members of the family of MOFs, called holey fibres. Both types can be used in gas sensors. The sensors are made of a holey fibre for the 1550-nm spectral range. One end of the fibre is spliced to conventional single-mode fibre from an optical source, such as a laser or an LED, and placed on the outer end in a V-groove in a vacuum chamber, as shown in Figure 2.16. A multi-mode fibre leading to a detector is installed 50 µm away from the end in the V-groove. The separation allows gas to flow into or out of the hollow core of the band gap fibre, while still allowing for efficient optical coupling between the fibres. A one-metre length of the fibre is filled with acetylene (C_2H_2) to a pressure of 10 m bar and illuminated with a tunable laser source. The expected spectral changes of acetylene are observed and displayed on a monitor.

Beyond pressure and temperature sensor technology, advanced fibre sensor types include chemical, strain, biomedical, electrical and magnetic, rotation, vibration and displacement as major applications. One more application of the advanced fibre sensor technology is in the area of DNA analysis. Fibre-optic biosensors have the ability to detect the presence of short DNA sequences called oligonucleotides. This ability gives the diagnostic some excellent tools for early and accurate diagnosis. These fibres are efficiently constructed as waveguides with novel properties for communications and sensing applications.

2.4   EXPERIMENTAL WORK

This chapter presents delicate and expensive fibre cables. Testing these types of fibre cables in the lab is very difficult using low- or moderate-accuracy tools and measuring instruments. Therefore, students should test only the least-delicate fibre cables, such as non-zero-dispersion fibre-optic cables and the dual-core fibre for high-power laser. Students can perform the regular loss tests as explained in the Fibre-Optic Cables Chapter. A comparison between single- or multi-mode fibres can be conducted, for example, on one type of the advanced fibre cables. Finally, conclusions can be summarised for the performance of the two types that are used in the experiments and compared with the data provided by the manufacturers.

2.4.1   CONCLUSION

Summarise the important observations and findings obtained in this lab experiment.

2.4.2   SUGGESTIONS FOR FUTURE LAB WORK

List any suggestions for improvements using different experimental equipment, procedures and techniques for any future lab work. These suggestions should be theoretically justified and technically feasible.

2.5   LIST OF REFERENCES

List any references that were used in the report. Use one format in writing the references. Never mix reference formats in a report.

2.6   APPENDIX

List all of the materials and information that are too detailed to be included in the body of the report.

FURTHER READING

Agrawal, G. P., *Nonlinear Fiber Optics. Optics and Photonics*, 2nd ed., Academic Press, New York, 1995.

Alkeskjold, T. T., Laegsgaard, J., Bjarklev, A., Hermann, D., Anawati, A., Broeng, J., Li, J., and Wu, S.-T., All optical modulation in dye-doped nematic liquid crystal photonic bandgap fibers, *Opt. Express*, 12 (24), 5857–5871, 2004.

Astle, T. B. et al., *Optical Components—The Planar Revolution*, Merrill Lynch & Co., In-depth Report, New York, May 17, 2000.

Bise, R. T., Windeler, R. S., Kranz, K. S., Kerbage, C., Eggleton, B. J., and Trevor, D. J., Tunable photonic bandgap fiber, *Proc. Opt. Fiber Commun. Conf. Exhibit*, 17, 466–468, 2002.

Blaze Photonics Limited, *Photonic Crystal Fibers, Product Summary*, Blazephotonics, Bath, UK, 2003.

Burgess, D. S., Nanoribbon waveguides suited for photonics circuits. Photonics technology world, *Photon. Spectra*, 38 (11), 28, 2004.

Chen, K. P., In-fiber light powers active fiber optical components, *Photon. Spectra*, 39 (4), 78–90, 2005.

Crystal Fiber, *Photonic Crystal Fibers—Revolutionizing Optical Fiber Technology*, Crystal Fiber, Copenhagen, Denmark, 2004.

Derickson, D., *Fiber Optic Test and Measurement*, Prentice Hall PTR, Englewood Cliffs, NJ, 1998.

Deveau, R. L., *Fiber Optic Lighting: A Guide for Specifiers*, 2nd ed., Prentice Hall PTR, Englewood Cliffs, NJ, 2000.

Eggleton, B. J., Kerbage, C., Westbrook, P., Windeler, R., and Hale, A., Photonics crystal fiber, honey-comb fiber, *Opt. Lett.*, 24, 1460, 1999.

Eggleton, B. J., Kerbage, C., Westbrook, P., Windeler, R., and Hale, A., Grapefruit fiber, *J. Lightwave Devices*, 18, 1084–1100, 2000.

Eggleton, B. J., Kerbage, C., Westbrook, P., Windeler, R., and Hale, A., Microstructured optical fiber devices, *Opt. Express*, 9 (13), 698–713, 2001.

Fedder, G. K., Santhanam, S., Reed, M. L., Eagle, S. C., Guillou, D. F., Lu, M. S.-C., and Carley, R., Laminated high-aspect-ratio microstructures in a conventional CMOS process, *Sens. Actuators* A, 57 (2), 103–110, 1996.

Goff, D. R., and Kimberly, S. H., *Fiber Optic Reference Guide: A Practical Guide to the Technology*, 2nd ed., Butterworth-Heinemann, London, 1999.

Guy, M., and François, T., Fiber Bragg gratings: Better manufacturing—Better performance, *Photon. Spectra*, 9(3), 106–110, 2002.

Guy, M., and Yves, P., Fiber Bragg gratings: A versatile approach to dispersion compensation, *Photon. Spectra*, 38(8), 96–101, 2004.

Hitz, B., Holey fibers connect to conventional fibers with low loss, *Photon. Spectra*, 84–85, 2005.

Hood, D. C., and Finkelstein, M. A., Sensitivity to light, in *Handbook of Perception and Human Performance, Sensory Processes and Perception*, Boff, K. R., Kaufman, L., and Thomas, J. P., eds., Vol. 1, Chapter 5, Wiley, New York, 1986.

Kao, C. K., *Optical Fiber Systems: Technology, Design, and Applications*, McGraw-Hill, New York, 1982.

Kashyap, R., *Fiber Bragg Gratings*, Academic Press, New York, 1999.

Larsen, T., Bjarklev, A., Hermann, D., and Broeng, J., Optical devices based on liquid crystal photonic band-gap fibers, *Opt. Express*, 11 (20), 2589–2596, 2003.

Mynbaev, D. K., and Lowell, L. S., *Fiber-Optic Communication Technology—Fiber Fabrication*, Prentice Hall PTR, Englewood Cliffs, NJ, 2005.

OFS Leading Optical Innovations, Nonzero dispersion optical fiber, True Wave Reach Fiber, Product Catalog 2003, USA, 2003.

Opto-Canada, SPIE Regional Meeting on Optoelectronics, and Imaging, in *Technical Digest of SPIE—The International Society for Optical Engineering*, Vol. TD01, Ottawa, Canada, pp. 9–10, May 2002.

Razavi, B., *Design of Integrated Circuits for Optical Communications*, McGraw-Hill, New York, 2003.

Sabert, H., and Jonathan, K., Hollow-core fibers seek the "holey" grail. BlazePhotonics Ltd., *Photon. Spectra*, 37 (8), 92–94, 2003.

Sabert, H., Hollow-core fibers allow light to travel by air, *Laser Focus World*, 40 (5), 161–164, 2004.

Savage, N., Into thin air: Photonic bandgap fibers hold promise as new types of waveguides, *SPIE's oemagazine*, 4 (1), 44, 2004.

Savage, N., Holy-y waveguides. Product trends, *SPIE's oemagazine*, 5 (5), 36, 2005.

Shashidhar, N., Lensing technology, *Corning Incorporated, Fiber Product News*, 14–15, April 2004.

Shen, L. P., Huang, W. P., Chen, G. X., and Jian, S. S., Design and optimization of photonic crystal fibers for broad-band dispersion compensation, *IEEE Photon. Technol. Lett.*, 15 (4), 540–542, 2003.

The 2004 Photonics Circle of Excellence Award Winners, Crystal fiber A/S, *Photon. Spectra*, 39 (1), 25, 2005.

VonWeller, E. L., Photonics crystal fibers advances in fiber optics, *Appl. Opt*, March 1, 2005.

Walker, R. and Neil, B., Image fiber delivers vision of the future. Fujikura Europe Ltd., *Photon. Spectra*, 96–97, 2005.

West, J. A., Venkataramam, N., Smith, C. M. Gallangher, M. T., Photonic crystal fibers. in *Proceedings of the 27th European Conference on Optical Communication (ECOC'01)*, Vol. 4, pp. 582–585, Corning, New York, 2001.

Yeh, C., *Handbook of Fiber Optics: Theory and Applications*, Academic Press, San Francisco, CA, 1990.

Yeh, C., *Applied Photonics*, Academic Press, New York, 1994.

Zhang, L., and Changxi Y., Polarization selective coupling in the holey fiber with asymmetric dual cores, LEOS2003, Tucson, AZ, paper ThP1, October 26–30, 2003.

3 Light Attenuation in Optical Components

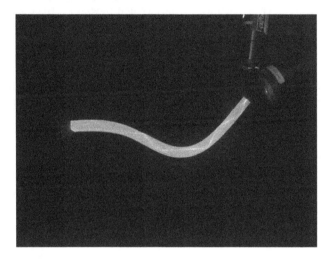

3.1 INTRODUCTION

Attenuation is the loss of power in a fibre-optic cable or any optical material, and can result from many causes. During transit, light pulses lose some of their energy. Light losses occur when the fibre-optic cables are subjected to any type of stress, temperature change or other environmental effects. The most important source of loss is the bending that occurs in the fibre-optic cable during installation or in the manufacturing process.

Losses that occur in a fibre-optic cable, optical fibre devices and systems of many devices can be calculated. The decibel (dB) is the standard unit to express the losses in optical fibre cables, devices and systems. Attenuation for commercially-available fibre cables is specified in decibels per kilometre (dB/km). Attenuation also depends on the types and specifications of the fibre.

Attenuation in optical fibres varies with the wavelength of light. There are three low-loss windows of interest: 850, 1300 and 1550 nm. The 850-nm window is perhaps the most widely used, because 850-nm devices are inexpensive. The 1300-nm window offers lower loss, but at a modest increase in cost for light-emitting diodes, the main light sources. The 1550-nm window today is mainly of interest to long-distance communication applications.

The experimental work in this chapter will enable students to practice calculating the light loss in optical components, such as a glass slide, an epoxy layer and a fibre cable. Light loss due to microscopic bending in a fibre cable can also be calculated.

3.2 LIGHT LOSSES IN AN OPTICAL MATERIAL

When light passes through an optical component, power is lost. The light loss in any optical component is dependent on the accumulative losses due to internal and external losses. Internal losses are caused by light reflection, refraction, absorption, dispersion and scattering. External losses are caused by bending, stresses, temperature changes and overall system losses. Losses due to refraction and reflection (such as Fresnel reflection, microscopic reflection, surface reflection and back reflection) are generally explained by the laws of light. Common losses due to absorption, dispersion and scattering mechanisms as well as light losses in parallel optical surfaces, and in epoxy that occur in any optical material are explained below.

3.2.1 ABSORPTION

Every optical material absorbs some of the light energy. The amount of absorption depends on the wavelength of the light and on the optical material. Absorption loss depends on the physical characteristics of the optical material, such as transitivity and index of refraction. The wavelength of the light passing through an optical material is a function of the index of refraction of the material.

3.2.2 DISPERSION

Dispersion is caused by the expansion of light pulses as they travel through optical components. This occurs because the speed of light through the optical medium is dependent on the wavelength, the propagation mode and the optical properties of the materials along the light path.

3.2.3 SCATTERING

Scattering losses occur in all optical materials. Atoms and other particles inevitably scatter some of the light that hits them. Rayleigh scattering is named after the British physicist Lord Rayleigh (1842–1919), who stated that such scattered light is not absorbed by the particles, but simply redirected. Light scattering in the core of the fibre-optic cable is a common example, as illustrated in Figure 3.1. The further the light travels through a material, the more likely scattering is to occur. Rayleigh scattering depends on the type of material and the size of the particles relative to the wavelength of the light. The amount of scattering increases quite rapidly as the wavelength decreases.

Scattering loss also occurs in optical material inhomogeneities introduced during glass preparation and the addition of dopants in the manufacturing process. Imperfect mixing and processing of chemicals and additives can cause inhomogeneities within the preparation of a preform. When the preform is used in the fibre-drawing method, rough areas will form in the core and thus increase the scattering of light in the fibre.

3.2.4 LIGHT LOSS IN PARALLEL OPTICAL SURFACES

Loss of light due to reflection at a boundary between two parallel optical surfaces comprises a large portion of the total optical losses in a system. The simplest case of reflection loss occurs when an incident ray travels normal to the boundary, as shown in Figure 3.2. The reflection coefficient (ρ) is the ratio of the reflected electric field to the incident electric field. For a ray incident at the normal:

$$\rho = \frac{n_1 - n_2}{n_1 + n_2} \tag{3.1}$$

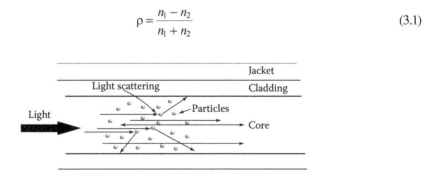

FIGURE 3.1 Light scattering in the fibre core.

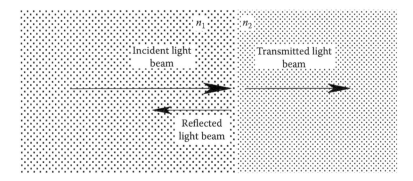

FIGURE 3.2   Light beam passing through two media.

where:

   n_1 is the refractive index of the incident medium
   n_2 is the index of the transmitted medium.

If $n_2 > n_1$, then the reflection coefficient becomes negative. This indicates a 180 degree phase shift between the incident and reflected electric fields. The reflectance (R) is the ratio of the reflected ray intensity to the incident ray intensity. As the intensity in an optical beam is proportional to the square of the beam's electric field, the reflectance is equal to the square of the reflection coefficient (ρ). The reflectance is calculated as

$$R = \left[\frac{n_1 - n_2}{n_1 + n_2} \right]^2 \tag{3.2}$$

3.2.5   Light Loss in an Epoxy Layer

Adhesives are used in manufacturing optical devices and are a key technology in the fibre-optic communications market. In order to produce low-cost and highly reliable optical components and devices, an easy-to-use, durable adhesive is necessary. Requirements for optical adhesives are extremely dependent on the specific application. A specialised group of companies develops and markets adhesives and adhesive resins for fibre-optic applications. These adhesives and resins are designed for a specific refractive index. They have high transmittance, precise curing time, heat-resistance, high elasticity and permeability.

Epoxy adhesives come in several forms. The most commonly used types are one-part, two-part and UV-curable systems. One-part systems typically require heat to cure the adhesive. Refrigeration of the liquid adhesive typically prolongs its shelf life. Two-part systems are based on a chemical reaction and thus must be used immediately after mixing. Setting times range from several minutes to several hours. UV-curable epoxy systems are one-part mixtures, which are activated by a UV-light source. As such, UV-systems do not require refrigeration. These can also be heat treated to stabilise the cure.

Epoxy adhesives are very important components in manufacturing fibre-optic devices. Each epoxy has a unique index of refraction and unique optical properties, like any other optical material. All epoxy adhesives affect the light passing through the epoxy. Light can be absorbed, dispersed and scattered by the epoxy. The optical performance of an epoxy is dependent on the epoxy preparation procedure. Cleaning the optical contact surfaces where the epoxy is to be applied, epoxy curing time, epoxy curing temperature and epoxy cooling rates all impact the optical epoxy performance. It is very important to understand an epoxy's physical properties and performance in relation to the signal wavelengths and the maximum signal power.

3.2.6 BENDING AND MICRO-BENDING

Any bending in a fibre-optic cable generates loss. Fibre-optic cable bending losses are caused by a variety of outside influences. These influences can change the physical characteristics of the cable and affect how the cable guides the light. Certain modes are affected and losses are accumulated over long distances. However, significant losses can arise from any kind of bending in a fibre cable. The cause of bending loss is easier to envisage using the ray model of light in a multimode fibre cable. When the fibre cable is straight, the ray falls within the confinement angle (θc) of the fibre cable. However, as shown in Figure 3.3, a bend will change the angle at which the ray hits the core-cladding interface. If the bend is sharp enough, the ray strikes the interface at an angle outside of the confinement angle, and the ray is refracted into the cladding and then to the outside as loss. These are referred to as leaky modes, whereby the ray leaks out and the attenuation is increased.

In another class of modes, called radiation mode, power from these modes radiates into the cladding and increases the attenuation. In radiation mode, the electromagnetic energy is distributed in the core and the cladding; however, the cladding carries no light.

When light is launched into a fibre cable, the power distribution varies as light propagates down the fibre cable. The power distribution decreases over long distances and eventually stabilises. This characteristic of optical fibre is referred to as stable-mode distribution. Stable-mode distribution can be observed in short fibre cables by introducing mode-filtering devices. Mode filtering may be accomplished through the use of mode scrambling, which can be achieved by bending the fibre cable to form a corrugated path, as shown in Figure 3.4. The corrugated path introduces a coupling, which leads to the existence of both radiation and leaky modes. In high-power applications, stable-mode distribution can be achieved because the effective portion of the signal that *leaks* is small in

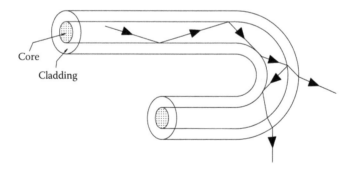

FIGURE 3.3   Light propagation around a bend in a fibre cable.

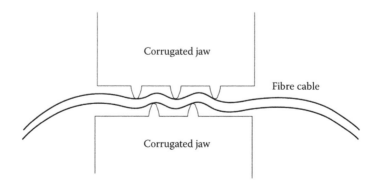

FIGURE 3.4   Mode scrambler for fibre cable micro-bends.

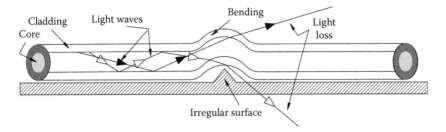

FIGURE 3.5 Fibre cable micro-bend.

comparison to the full signal strength. Mode scrambling allows repeatable laboratory measurements of signal attenuation in fibre cables.

Figure 3.5 shows a microscopic bend in a fibre-optic cable. Micro-bends can be a significant source of loss. When the fibre cable is installed and pressed onto an irregular surface, tiny bends can be created in the fibre cable. Light is lost due to these irregularities.

3.3 ATTENUATION CALCULATIONS

Any incident light power passing through an optical component—such as a glass microscope slide, fibre-optic cable and epoxy layer—is subjected to losses. Attenuation measures the reduction in light signal strength by comparing output power with input power. Measurements are made in decibels (dB). The decibel is an important unit of measurement in fibre-optic components, devices and systems loss calculations.

$$\text{Loss (dB)} = -10\log_{10}\left(\frac{P_{\text{out}}}{P_{\text{in}}}\right) \tag{3.3}$$

Equation 3.3 is also used to calculate the loss between the input and output of the cable. In industry, fibre-optic cable loss is calculated as the loss per unit length. Therefore, the fibre-optic cable loss is calculated by dividing the cable loss (dB) by cable length (km), as given in the following equation:

$$\text{Loss}_{\text{cable}}(\text{dB/km}) = \frac{-10\log_{10}\left(\frac{P_{\text{out}}}{P_{\text{in}}}\right)}{\text{Cable length}} \tag{3.4}$$

There is another type of loss, called excess loss. It is defined as the ratio between the sum of all power outputs (P_1 to P_n) and input signal power. Excess loss specifies the power lost within the system. It includes dispersion, scattering, absorption and coupling loss. Excess loss is calculated using the following formula:

$$\text{Loss (dB)} = -10\log_{10}\left(\frac{(P_1 + P_2 + P_3 + \ldots + P_n)}{P_{\text{in}}}\right) \tag{3.5}$$

Figure 3.6 shows fibre-optic attenuation as a function of wavelength. In the 1970s, communication systems operated in the wavelength range of 800–900 nm. At that time, fibre-optic cables exhibited a local minimum in the attenuation curve, and optical sources and photo detectors operating at this range were available. In addition, the fibre-optic cables were faster than their counterpart, the copper cabling systems. This region is referred as the first window.

By reducing the concentration of hydroxyl ions and metallic impurities in the fibre core material, in the 1980s manufacturers were able to fabricate fibre-optic cables with very low loss in the 1100–1600-nm region. At the same time, the demand increased for high-speed data rate transmission over long distances. Thus, the second window was defined, concentrated around 1310 nm. This spectral

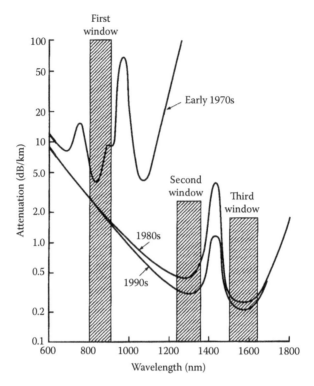

FIGURE 3.6   Fibre-optic attenuation as a function of wavelength.

band is referred to as the long-wavelength region. Due to further successful development in the fabrication of fibre-optic cables and optical amplifiers, the third window was defined around 1550 nm.

The most important aspect of the fibre-optic communication link is that many different wavelengths can be sent along a fibre simultaneously, without interference, in the 1310–1625-nm spectrum. The technology of combining a number of wavelengths onto the same fibre cable is known as wavelength division multiplexing (WDM). Furthermore, this technology experienced advanced development by using the dense wavelength division multiplexing (DWDM). WDM technology is presented in detail in the wavelength division multiplexing/demultiplexing chapter.

3.4   EXPERIMENTAL WORK

This experiment is designed to determine the power loss of light through optical components, such as glass microscope slides, epoxy layer and fibre-optic cables, in the following cases:

1. To measure and analyse the power loss of laser light through a single and up to five glass microscope slides.
2. To measure and analyse the power loss of laser light through a single microscope slide inclined at different angles from the normal.
3. To measure and analyse the power loss of laser light through an epoxy layer between two glass microscope slides.
4. To measure the power loss of laser light through a fibre-optic cable and power loss per length.
5. To measure the power loss of laser light through a fibre-optic cable due to micro-bending.
6. To measure the power loss of laser light through a fibre cable with graded-index (GRIN) lens at the input and/or output.

3.4.1 TECHNIQUE AND APPARATUS

Appendix A presents the following details of the devices, components, tools and parts:

1. 2 × 2 ft. optical breadboard.
2. He–Ne laser light source and power supply.
3. Laser light sensor.
4. Laser light power metre.
5. Laser mount assembly.
6. Hardware assembly (clamps, posts, screw kits, screwdriver kits, sundry positioners, etc.).
7. Fibre-optic cable, 250 μm diameter, 500 m long.
8. 20X microscope objective lens.
9. Lens/fibre cable holder/positioner assembly.
10. Fibre cable holder/positioner assembly.
11. Standard glass microscope slides and slide holders/positioners.
12. Rotation stage.
13. Mode scrambler, as shown in Figure 3.7.
14. GRIN lenses.
15. GRIN lens/fibre cable holder/positioner assembly.
16. Denatured ethanol.
17. Tissue.
18. Swabs.
19. Epoxy.
20. Micro-spatula.
21. Timer.
22. Microscope.
23. Black/white card and cardholder.
24. Ruler.

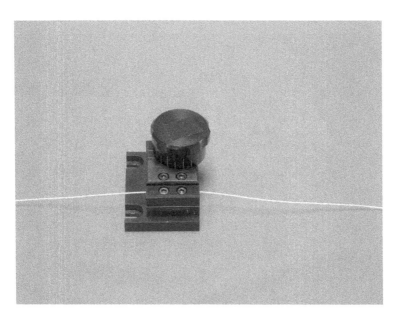

FIGURE 3.7 Mode scrambler.

3.4.2 PROCEDURE

Follow the laboratory procedures and instructions given by the instructor.

3.4.3 SAFETY PROCEDURE

Follow all safety procedures and regulations regarding the use of optical components, electrical and optical devices and test measurement instruments.

3.4.4 APPARATUS SET-UP

3.4.4.1 Laser Light Power Loss through One to Five Microscope Slides

This part of the experiment measures the power of the laser beam using a laser light sensor that is located at a fixed distance from the laser source. The measured laser light input power is the same value for all other cases in this experiment. The laser light passes through one glass microscope slide, then more slides are continually added, up to five slides in total. The output power will be measured for each slide when added. As slides are added, the output power will decrease, since the light is being refracted and absorbed more and more with increasing glass thickness. The distances between the laser source, the slides, and the laser power sensor will also have an effect on the measured losses. Therefore, these distances should be kept constant throughout the experiments.

The following steps illustrate the experimental lab set-up for Case (a):

1. Figure 3.8 shows the apparatus set-up.
2. Bolt the laser short rod to the breadboard.
3. Bolt the laser mount to the clamp using bolts from the screw kit.
4. Put the clamp on the short rod.
5. Place the He–Ne laser into the laser mount and tighten the screw. Follow the operation and safety procedures of the laser device in use.
6. Check the laser alignment with the line of bolt holes on the breadboard and adjust when necessary.
7. Place the laser sensor in front of the laser source at a fixed distance for all cases.

FIGURE 3.8   Laser light loss through one to five slides.

TABLE 3.1

Laser Power Loss through a Single to Five Slides

Power Input P_{in} (unit)		
Number of Glass Microscope Slides	**Power Output P_{out} (unit)**	**Loss (dB)**
One glass microscope slide		
Two glass microscope slides		
Three glass microscope slides		
Four glass microscope slides		
Five glass microscope slides		

8. Turn off the lights of the lab.
9. Turn on the laser power. Measure the laser input power (P_{in}). Fill out Table 3.1.
10. Prepare one standard glass microscope slide.
11. Clean the slide with a swab dampened with denatured ethanol. Follow the cleaning procedure for the optical components.
12. Mount a single slide into the slide holder/positioner.
13. Place the slide and slide holder/positioner in front of the laser source at a fixed distance between the laser source and the laser sensor, as shown in Figure 3.8. Make sure that the slide is perpendicular to the laser beam. Keep the same distance for all cases.
14. Measure the laser output power (P_{out}) from the slide. Fill out Table 3.1.
15. Add another slide, stacking the slides together. Place the slides and slide holder/positioner in front of the laser source at the same distance. Measure the laser output power from the slides. Fill out Table 3.1.
16. Repeat step 15 to add one slide at a time. Use up to five slides added together one at a time. Fill out Table 3.1.
17. Turn on the lights of the lab.

3.4.4.2 Laser Light Power Loss through a Single Slide Inclined at Different Angles

This case of the experiment deals with a single slide being rotated around the normal to the laser light source. The output power of the laser light depends on the angle that the slide is inclined. Most likely, there will be an angle for which the output power is the greatest; at all other angles, output power will be less. There is a specific inclination angle at which the laser beam will not pass through the slide. At this angle, the laser beam is reflected back to the incoming side of the laser beam.

The following steps illustrate the experimental lab set-up for Case (b):

1. Figure 3.9 shows the apparatus set-up.
2. Repeat steps 2–13 in Case (a) for installing the laser source and measuring the input laser power.
3. Measure the laser output power (P_{out}) from the slide. Fill out Table 3.2.
4. Mount the slide holder on top of the rotation stage.
5. Rotate the rotation stage at an angle of 10° from the normal of the light source. Measure the laser output power from the slide. Fill out Table 3.2.
6. Repeat step 5, rotating the rotation stage at an angle increment of 10° each time. Measure the laser light output. Fill out Table 3.2.
7. Turn on the lights of the lab.

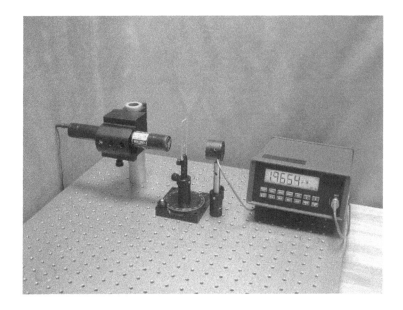

FIGURE 3.9   Laser light loss through a single slide inclined at different angles.

TABLE 3.2
Laser Power Loss through a Single Slide Inclined at Different Angles

Power Input P_{in} (unit)		
Laser Beam Incident Angle with the Normal (degrees)	Power Output P_{out} (unit)	Loss (dB)
0		
10		
20		
30		
40		
50		
60		
70		
80		
90		

3.4.4.3   Laser Light Power Loss through an Epoxy Layer between Two Slides

Having experienced the power loss of the laser light as it passes through a single slide in Case (a), add the following steps to measure the power loss of the laser light passing through two slides, as shown in Figure 3.10.

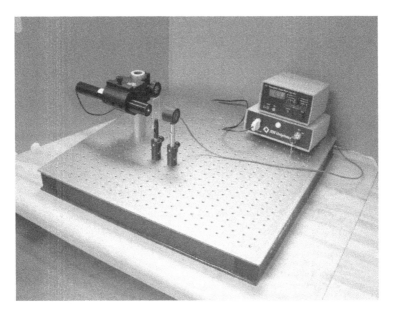

FIGURE 3.10 Laser light loss through air layer in a two-slide assembly.

TABLE 3.3
Power Loss of the Laser Light Passing through the Epoxy Layer in the Two-Slide Assembly

Power Input P_{in} (unit)			
Case		**Power Output P_{out} (unit)**	**Loss (dB)**
(a)	Single glass microscope slide		
(b)	Air gap between two glass microscope slides		
(c)	Epoxy layer between two glass microscope slides		

1. Repeat steps 2–13 in Case (a) for installing the laser source.
2. Measure the input laser power. Fill out Table 3.3 for laser power output for a single slide.
3. Strip a piece of fibre cable and cut two short lengths of the cladding to be used as a spacer between the two slides.
4. Prepare two standard slides.
5. Clean the two standard slides with a swab dampened with denaturated ethanol. Follow the optical components cleaning procedure.
6. Arrange the two slides with the spacers between them.
7. Mount the two-slide assembly into the slide holder/positioner.
8. Position the two-slide assembly in front of the laser source at a fixed distance, as shown in Figure 3.10.
9. Measure the laser output power (P_{out}) from the two-slide assembly with an air gap between the slides. Fill out Table 3.3.
10. Prepare the epoxy by mixing together the two-epoxy materials carefully.
11. Apply epoxy between the two slides. Use a micro-spatula to distribute the epoxy uniformly in the air gap between the slides. Follow the epoxy curing procedure for the epoxy that is

given in the lab. Note that a number of epoxy types used in manufacturing optical fibre devices cures in an oven set at 120°C temperature for 12 min.

12. Assemble the spacers and the two slides.
13. Mount the assembly into the slide holder/positioner, as shown in Figure 3.11.
14. Inspect the condition of the cured epoxy using a microscope, as shown in Figure 3.11. During the inspection, air bubbles, inhomogeneous epoxy mixing or the presence of uncured epoxy can be identified. Compare the colour of the cured epoxy with the manufacturer's specifications.
15. Position the assembly in front of the laser source at a fixed distance, as shown in Figure 3.12.
16. Measure the laser output power (P_{out}) from the assembly. Fill out Table 3.3.
17. Turn on the lights of the lab.

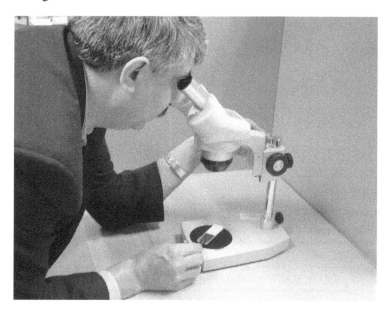

FIGURE 3.11 Epoxy-inspection process.

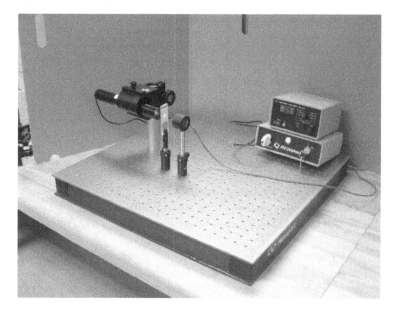

FIGURE 3.12 Laser light loss through the epoxy layer in the two-slide assembly.

3.4.4.4   Laser Light Power Loss through a Fibre-Optic Cable

This case of the experiment deals with the loss per length in a fibre-optic cable. Loss calculation in these optical components is similar to the above cases. The following steps illustrate the experimental lab set-up for Case (d):

1. Figure 3.13 shows the apparatus set-up.
2. Bolt the laser short rod to the breadboard.
3. Bolt the laser mount to the clamp using bolts from the screw kit.
4. Put the clamp on the short rod.
5. Place the He–Ne laser into the laser mount and tighten the screw. Turn on the laser device. Follow the operation and safety procedures of the laser device in use.
6. Check the laser alignment with the line of bolt holes of the breadboard and adjust when necessary.
7. Mount a lens/fibre cable holder/positioner to the breadboard so that the laser beam passes over its centre hole.
8. Add the 20X microscope objective lens to the lens/fibre cable holder/positioner.
9. Prepare a fibre cable with a good cleave at each end, as described in the end preparation section of the fibre-optic cable chapter.
10. Insert one end (input) of the fibre cable into the brass fibre cable holder and place it into the hole of the lens/fibre cable holder/positioner.
11. Mount the output end of the fibre cable into the brass fibre cable holder and place it in the hole of the fibre cable holder/positioner. Mount the assembly in front of the laser sensor.
12. Turn on the laser power.
13. Extend the end of the fibre cable until the fibre cable end is at the centre of the lens/fibre cable holder/positioner. This is a very important step for obtaining an accurate value of the laser beam intensity from the fibre cable output end.
14. Re-check the alignment of your light launching arrangement by making sure that the input end of the fibre cable remains at the centre of the laser beam.
15. Check to ensure that you have a circular output from the output end of the fibre cable. Point the output towards the centre of the laser sensor.

FIGURE 3.13   Fibre cable power loss apparatus set-up.

TABLE 3.4
Fibre Cable Power Loss Data

P_{in} (unit)	P_{out} (unit)	Loss (dB)	Cable Length (km)	Loss$_{cable}$ (dB/km)

16. Turn off the lights of the lab.
17. Measure the laser input power (P_{in}) and output power (P_{out}). Fill out Table 3.4.
18. Find the net length of the fibre-optic cable length that is being used in this experiment. Fill out Table 3.4.
19. Turn on the lights of the lab.

3.4.4.5 Laser Light Power Loss through a Fibre-Optic Cable due to Micro-Bending

Continue the following procedure as explained in Case (d) by adding a mode scrambler near the end of the fibre cable, as shown in Figure 3.14:

1. Place the mode scrambler at a convenient position on the fibre cable near the output end, as shown in Figure 3.14. Lay the fibre cable between the two corrugated surfaces of the mode scrambler, as shown in Figure 3.7. Rotate the knob clockwise until the corrugated surfaces just contact the fibre cable. Rotate the knob further clockwise and carefully observe the reduction in the power readout. Rotate the knob more until you break the fibre cable. In this case, no light will pass though the fibre cable.
2. Measure the output power while tightening the knob. Record the last power measurement before the fibre cable broke. Fill out Table 3.5.
3. Turn on the lights of the lab.

FIGURE 3.14 Mode scrambler apparatus set-up.

TABLE 3.5
Mode Scrambler Power Loss Data

P_{in} (unit)	P_{out} (unit)	Loss (dB)

3.4.4.6 Laser Light Power Loss through a Fibre-Optic Cable Coupled to a GRIN Lens at the Input and/or Output

Coupling light into a fibre cable is accomplished using a quarter-pitch GRIN lens that focuses the collimated beam from the laser source onto a small spot on the core of the fibre cable. The GRIN lens is coupled to the input and output ends of a fibre-optic cable. When the GRIN lens focuses the power of the collimated laser beam onto the core of the fibre cable, the input power increases. Similarly, inserting a GRIN lens at the output of the fibre cable increases the output power.

Follow the instructions from Case (a) regarding the use of a 20X lens. Figure 3.15 shows the experimental set-up for Case (f). Add the following steps:

1. Mount a GRIN lens/fibre cable holder/positioner to the breadboard at the input end of the fibre cable so that the laser beam passes through the centre hole.
2. Place a quarter-pitch GRIN lens into the groove of the GRIN lens/fibre cable holder/ positioner, as shown in Figure 3.15.
3. Insert the input end of the fibre cable into the brass fibre cable holder and place it into the hole of the GRIN lens/fibre cable holder/positioner.
4. Extend the end of the fibre cable so that the fibre cable end is at the centre of the GRIN lens/ fibre cable holder/positioner. This is a very important step for obtaining an accurate value of the laser beam intensity from the fibre cable output end.
5. Mount the output end of the fibre cable into the brass fibre cable holder and place it in the hole of the fibre cable holder/positioner. Mount the assembly in front of the laser sensor.
6. Verify the alignment of the input end of the fibre cable with the GRIN lens. Make sure that the focal point of the GRIN lens remains at front of the input end of the fibre cable.

FIGURE 3.15 Fibre cable power loss with a GRIN lens at the input and output ends.

TABLE 3.6

Laser Light Power Loss through a Fibre-Optic Cable Coupled to a GRIN Lens at the Input and/or Output

Situation	P_{in} (unit)	P_{out} (unit)	Loss (dB)	Cable Length (km)	$Loss_{cable}$ (dB/km)
GRIN lens at the fibre cable input end					
GRIN lens at the fibre cable output end					

7. Check to ensure that you have a circular output from the output end of the fibre cable. Point the fibre cable output end towards the centre of the laser sensor.

8. Optimise the coupling of the laser beam by finely adjusting the fibre cable holder/positioner.

9. Measure the laser output power (P_{out}). Fill out Table 3.6.

10. Continue the experiment by adding a GRIN lens to the output cable end. Make a very fine adjustment of the GRIN lens to the fibre cable output end position.

11. Measure the laser output power (P_{out}) with the GRIN lens at the output. Fill out Table 3.6.

12. Turn on the lights of the lab.

3.4.5 DATA COLLECTION

3.4.5.1 Laser Light Power Loss through One to Five Microscope Slides

1. Record the input power (P_{in}) without a slide in front of the laser light.
2. Measure the output power (P_{out}) for one to five slides.
3. Determine the losses for the slides.
4. Fill out Table 3.1 for Case (a).

3.4.5.2 Laser Light Power Loss through a Single Slide Inclined at Different Angles

1. Record the input power (P_{in}) without a slide in front of the laser light.
2. Measure the output power (P_{out}) for each inclination angle.
3. Determine the losses of the slide for each inclination angle.
4. Fill out Table 3.2 for Case (b).

3.4.5.3 Laser Light Power Loss through an Epoxy Layer between Two Slides

1. Record the input power (P_{in}) without a slide in front of the laser light.
2. Record the output power (P_{out}) of the laser light from the air gap filled between the two-slide assembly.
3. Record the output power (P_{out}) of the laser light from the epoxy-filled two-slide assembly.
4. Determine the loss for the slide and epoxy.
5. Fill out Table 3.3 for Case (c).

3.4.5.4 Laser Light Power Loss through a Fibre-Optic Cable

1. Measure the laser input power (P_{in}) and output power (P_{out}).
2. Find the length of the fibre-optic cable that is being used in this case.
3. Fill out Table 3.4 for fibre cable power loss measurements.

3.4.5.5 Laser Light Power Loss through a Fibre-Optic Cable due to Micro-Bending

1. Measure the laser output power (P_{out}) during the mode scrambler operation before the fibre cable breaks off. Measure power input (P_{in}) as in Case (d).
2. Fill out Table 3.5 for mode scrambler loss measurements.

3.4.5.6 Laser Light Power Loss through a Fibre-Optic Cable Coupled to a GRIN Lens at the Input and/or Output

1. Measure the laser output power (P_{out}) with the quarter-pitch GRIN lenses placed at the input and output ends of the fibre cable.
2. Find the length of the fibre-optic cable being used in this lab.
3. Fill out Table 3.6 loss measurements.

3.4.6 CALCULATIONS AND ANALYSIS

3.4.6.1 Laser Light Power Loss through One to Five Microscope Slides

1. Calculate the power loss (dB) of the laser light when it passes through a single and multiple of slides, using Equation 3.3.
2. Draw a curve for the power loss (dB) as a function of number of slides.

3.4.6.2 Laser Light Power Loss through a Single Slide Inclined at Different Angles

1. Calculate the power loss (dB) of the laser light when it passes through a single slide, using Equation 3.3.
2. Draw a curve for the power loss (dB) as a function of the angles with the normal.

3.4.6.3 Laser Light Power Loss through an Epoxy Layer between Two Slides

1. Calculate the power loss (dB) of the laser light when it passes through a single, two slides with an air gap, and two slides with an epoxy layer using Equation 3.3. Fill out Table 3.3 for Case (c).
2. Compare power loss (dB) among the three situations in this case.
3. Calculate the reflection coefficient (ρ) and the reflectance (R) at the air–glass interface or glass–epoxy interface, using Equations 3.1 and 3.2.

3.4.6.4 Laser Light Power Loss through a Fibre-Optic Cable

1. Calculate the cable power loss (dB) using Equation 3.3.
2. Calculate the cable power loss per unit length (dB/km) using Equation 3.4.
3. Fill out Table 3.4 for this case.

3.4.6.5 Laser Light Power Loss through a Fibre-Optic Cable due to Micro-Bending

1. Calculate the cable power loss with the mode scrambler, using Equation 3.3.
2. Fill out Table 3.5 for this case.
3. Compare the losses with and without the mode scrambler.

3.4.6.6 Laser Light Power Loss through a Fibre-Optic Cable Coupled to a GRIN Lens at the Input and/or Output

1. Calculate the cable power loss with the quarter-pitch GRIN lenses placed at the input and output ends of the fibre cable using Equations 3.3 and 3.4.
2. Fill out Table 3.6 for this case.

3.4.7   RESULTS AND DISCUSSIONS

3.4.7.1   Laser Light Power Loss through One to Five Microscope Slides

1. Report the power loss (dB) of the laser light.
2. Discuss the power loss as a function of number of slides.

3.4.7.2   Laser Light Power Loss through a Single Slide Inclined at Different Angles

1. Report the power loss (dB) of the laser light.
2. Discuss the power loss as a function of the angles with the normal.
3. Find the maximum and minimum losses that occur with the corresponding angles.

3.4.7.3   Laser Light Power Loss through an Epoxy Layer between Two Slides

1. Report the power loss (dB) of the laser light.
2. Report the percentage of the light reflected and transmitted at the air–glass interface or glass–epoxy interface.
3. Compare and discuss the power loss in each situation. Compare the power loss when air between the two slides is replaced by an epoxy layer.

3.4.7.4   Laser Light Power Loss through a Fibre-Optic Cable

1. Report the cable power loss (dB) and cable power loss per unit length (dB/km).
2. Compare the measured power loss values with the actual value of the power loss for the fibre cable, which is provided by the manufacturer.

3.4.7.5   Laser Light Power Loss through a Fibre-Optic Cable due to Micro-Bending

1. Report the calculated results for the fibre cable power loss (dB) with the mode scrambler.
2. Perform a comparison of the losses with and without the mode scrambler.

3.4.7.6   Laser Light Power Loss through a Fibre-Optic Cable Coupled to a GRIN Lens at the Input and/or Output

1. Report the calculated result for the cable power loss (dB) for this case.
2. Compare and discuss the calculated losses when using quarter-pitch GRIN lenses in this case, and the losses in Case (a).

3.4.8   CONCLUSION

Summarise the important observations and findings obtained in this lab experiment.

3.4.9   SUGGESTIONS FOR FUTURE LAB WORK

List any suggestions for improvements using different experimental equipment, procedures and techniques for any future lab work. These suggestions should be theoretically justified and technically feasible.

3.5   LIST OF REFERENCES

List any references that were used in the report. Use one format in writing the references. Never mix reference formats in a report.

3.6 APPENDIX

List all of the materials and information that are too detailed to be included in the body of the report.

FURTHER READING

Al-Azzawi, A. R., and Peter C., *Fiber Optics Principles and Practices,* Algonquin College Publishing Centre, Ottawa, Canada, 2002.

Chen, K. P., In-fiber light powers active fiber optical components, *Photon. Spectra,* April, 78–90, 2005.

Derickson, D., *Fiber Optic Test and Measurement,* Prentice Hall PTR, Englewood Cliffs, NJ, 1998.

Deveau, R. L., *Fiber Optic Lighting: A Guide for Specifiers,* 2nd ed., Prentice Hall PTR, Englewood Cliffs, NJ, 2000.

Edmund Industrial Optics, *Optics and Optical Instruments Catalog, 2004,* Edmund Industrial Optics, Barrington, NJ, 2002.

Hecht, J., *Understanding Fiber Optics,* 3rd ed., Prentice Hall, Inc., Englewood Cliffs, NJ, 1999.

Kao, C. K., *Optical Fiber Systems: Technology, Design, and Applications,* McGraw-Hill, New York, 1982.

Keiser, G., *Optical Fiber Communications,* 3rd ed., McGraw-Hill, New York, 2000.

Mynbaev, D. K., and Scheiner, L. L., *Fiber-Optic Communication Technology—Fiber Fabrication,* Prentice Hall PTR, Englewood Cliffs, NJ, 2005.

Okamoto, K., *Fundamentals of Optical Waveguides,* Academic Press, New York, 2000.

Palais, J. C., *Fiber Optic Communications,* 4th ed., Prentice Hall, Englewood Cliffs, NJ, 1998.

Salah, B. E. A., and Teich, M. C., *Fundamentals of Photonics,* Wiley, New York, 1991.

Senior, J. M., *Optical Fiber Communications, Principle and Practices*, 3rd ed., Personal Education Limited, Essex, England, 2009.

Shamir, J., *Optical Systems and Processes,* SPIE Optical Engineering Press, Washington, D.C., 1999.

Shotwell, R. A., *An Introduction to Fiber Optics,* Prentice Hall, Englewood Cliffs, NJ, 1997.

Sterling, D. J. Jr., *Technician's Guide to Fiber Optics,* 2nd ed., Delmar Publishers, New York, 1993.

Ungar, S., *Fiber Optics: Theory and Applications,* Wiley, New York, 1990.

Yeh, C., *Handbook of Fiber Optics: Theory and Applications,* Academic Press, San Diego, CA, 1990.

4 Fibre-Optic Cable Types and Installations

4.1 INTRODUCTION

Fibre-optic cables are significantly different from copper cables. Fibre-optic cables transmit data through very small cores over long distances at the speed of light. These cables come in a wide variety of configurations. Important considerations in any cable installation and operation are the following: bending radius, tensile strength, ruggedness, durability, flexibility, environmental conditions, such as temperature extremes and even appearance. Due to fibre-optic cables' light weight and extreme flexibility, fibre-optic cables are more easily installed than their copper counterparts. They are easy to handle and they can be pulled through conduit and piping systems over long distances using various installation techniques. The minimum bend radius and maximum tensile loading allowed on a fibre-optic cable are critical during and after installation. A tensile load causes attenuation and may ultimately crack the fibre-optic core. The tensile loading allowed during installation is higher than the permissible loads during operation. The minimum bend radius allowed during installation is likewise larger than the bend radius allowed during operation.

This chapter presents the types of fibre-optic cables in current use and methods of installation. Testing fibre cables after installation is an important aspect of any communication system. Some kinds of hardware used in telecommunications systems are presented in detail. This chapter also introduces the student to fibre cable installation and the testing of fibre cable after the installation is complete.

4.2 FIBRE-OPTIC CABLE TYPES AND APPLICATIONS

There are many fibre-optic cable types and designs available. Fibre cables can be classified into the following categories: fibre-optic cables for indoor applications, those for outdoor applications and for indoor/outdoor applications. The following sections explain each category in detail.

4.2.1 INDOOR FIBRE-OPTIC CABLE TYPES AND APPLICATIONS

Indoor cables (inside cable plant) are generally installed and operated in a controlled, stable environment. The cables must perform with minimal loss. Other factors that generate losses, such as environmental and mechanical stresses during and after installation, can cause failure. Outdoor cables (outside cable plant) have more factors that affect their performance. Indoor fibre cables are divided into the following types:

1. Simplex cables contain a single fibre cable, as shown in Figure 4.1. They are used for connections within equipment. They have a thicker outer jacket, which makes the cable easier to handle and adds mechanical protection.
2. Duplex cables contain two tight-buffered fibre cables inside a single jacket, as shown in Figure 4.2. They are used in equipment interconnections within workstations, test equipment, hubs and routers and so on.

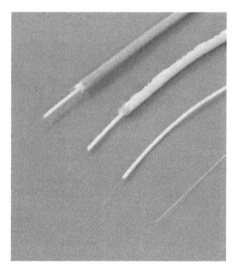

FIGURE 4.1   Simplex fibre cables.

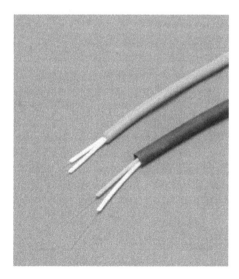

FIGURE 4.2   A duplex fibre cable.

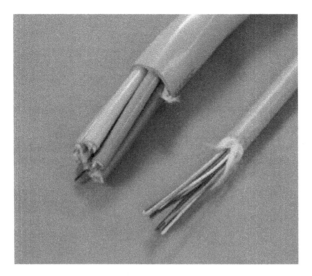

FIGURE 4.3 Multi-fibre cables.

3. Multi-fibre cables contain more than two fibre cables in one jacket, as shown in Figure 4.3. They have anywhere from three to several hundred fibre cables, all of which are colour-coded. They are used in vertical and horizontal fibre cable distributions between floors and telecommunication room.
4. Light cables, heavy cables and plenum cables are application-dependent.
5. Breakout cables have several individual simplex cables.

4.2.2 Outdoor Fibre-Optic Cable Types and Applications

Outdoor cables (outside cable plant) must withstand a variety of environment and mechanical stresses during and after installation. The cables must perform with minimal losses over a wide range of temperature and humidity changes. The cable must also have waterproof capabilities, have strength to endure the difficult installation conditions, provide protection against ultraviolet radiation and provide mechanical protection. There are many types and designs, depending on the manufacturer and application. Outdoor fibre cables are divided into the following types:

1. Overhead cables many be strung from telephone poles or along power lines.
2. Direct burial cables are placed directly in a trench dug in the ground and then covered by soil.
3. Indirect burial cables are placed inside a duct or conduit system.
4. Submarine cables are laid underwater.

4.2.3 Indoor/Outdoor Fibre-Optic Cable Types and Applications

There are many types of fibre cables that are used for indoor and outdoor applications. These cables use materials that enable them to pass the flame-retardant requirements of the indoor applications, and provide reliable waterproof performance for outdoor applications. They are able to withstand difficult installation conditions and temperature variations. They are widely used within buildings and between buildings in campus applications.

4.2.4 OTHER FIBRE-OPTIC CABLE TYPES AND APPLICATIONS

One of the fibre cables in common use is the ribbon cable. Ribbon cables are made of many fibres, which are embedded in a plastic or PVC material in parallel, forming a flat, ribbon like structure, as shown in Figure 4.4. The ribbon cable carries up to 12 fibre cables in a single ribbon. They are ideal for multi-fibre connector interconnect applications from equipment to a patch panel or as a patch cord.

Some ribbon cables consist up to 216 fibre cables, as shown in Figure 4.5. The cable consists of a single buffer tube that contains a stack of up to eighteen 12-fibre ribbons wrapped within a water-swellable foam tape and surrounded by a jacket. Dielectric or steel strength members located under the cable jacket provide tensile and mechanical strength. Some ribbon cables provide up to 864 fibre cables in a rugged, compact design to maximise the use of critical duct space. They are the ideal choice for maximising usage of duct or pipe space and getting the service running faster than other cable types. They are ideal for outdoor applications and upgraded communication systems.

FIGURE 4.4   A ribbon cable for indoor applications.

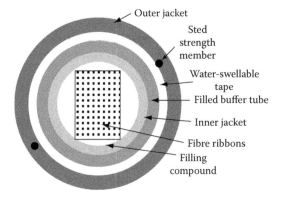

FIGURE 4.5   A cross-section of an outdoor ribbon cable.

4.3 FIBRE-OPTIC CABLE INSTALLATION METHODS

This section describes some of the common fibre-optic cable installation methods for inside and outside plants in local area networks, metropolitan area networks and wide area networks.

4.3.1 INDOOR FIBRE-OPTIC CABLE INSTALLATION

Generally, indoor cables are placed in conduits or trays. As standard fibre-optic cables are electrically non-conductive, they may be placed in trays alongside high-voltage cables without the special insulation required by copper cabling. Plenum fibre-optic cables can be placed without special restrictions in any plenum area within a building.

4.3.2 CABLE INSTALLATION IN TRAY AND DUCT SYSTEMS

The primary consideration in selecting a route for fibre-optic cable (through trays and ducts) is the avoidance of potential cutting edges and sharp bends. Areas where particular caution must be taken include the corners and exit slots in the sides of the trays, as shown in Figure 4.6.

If a fibre-optic cable is in the same tray or duct as a very large and heavy electrical cable, care must be taken to avoid placing an excessive weight on the fibre-optic cable. Figure 4.7 illustrates such a case.

Cables in ducts and trays are not subjected to tensile forces. The tensile load must be considered when determining the minimum bend radius at the top of a vertical run. Long vertical runs must be clamped at intermediate points to prevent excessive tensile loading on the fibre cable. The clamping force should not exceed that which is necessary to prevent the possibility of slippage. Clamping forces are often determined experimentally, because the force is dependent on the type of clamping and jacket materials. The clamping force should be applied over as long a length of the fibre-optic cable as practical. Clamping surfaces should be made of a soft material, such as rubber or plastic. Tensile load during vertical installation is reduced by installing the fibre cable in a top–down manner.

FIGURE 4.6 Fibre-optic cables in the tray system.

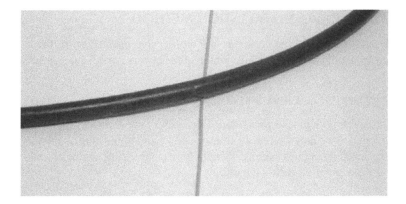

FIGURE 4.7   Heavy power cable crossing over the fibre-optic cable.

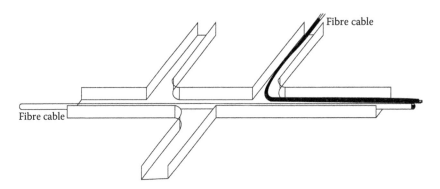

FIGURE 4.8   Wireway or tray with turn fittings.

4.3.3   Conduit Installation

Fibre-optic cables are pulled through a conduit (wireway or tray) by a wire or synthetic rope attached to the cable. Any pulling force must be applied to the cable strength member and not to the fibre-optic cable. In situations where the fibre-optic cable does not have connectors, the pull wire should be tied to the Kevlar strength member, or a pulling grip may be taped to the cable jacket or sheath. When determining the suitability of a conduit for a fibre-optic cable, the clearance between the conduit and any other cable is critical (described by a fill factor). Sufficient clearance must be provided, to allow the fibre-optic cable to be pulled through without excessive friction or binding. If the conduit makes a 90° turn, a fitting, such as that shown in Figure 4.8, must be used to allow the cable to be pulled in a straight line while avoiding any sharp bends in the cable.

4.3.4   Pulling Fibre-Optic Cable Installation

Fibre-optic cables are pulled using many of the same tools and techniques that apply to wire cable installations. Departures from standard methods are required, because connectors are usually pre-installed on the fibre-optic cable, smaller pulling forces are allowed, and there are minimum bend radius requirements. The pull tape must be attached to the optical cable in such a way that the pulling forces are applied to the outer Kevlar layer. Connectors should be protected to prevent damage. The recommended method for attaching a pulling tape to a simplex cable is the Kellems grip, as shown in Figure 4.9. The connector should be wrapped in a thin layer of foam rubber and inserted in a plastic sleeve for protection. The fibre-optic cable grip should be stretched and then wrapped tightly with electrical tape to provide a firm grip on the fibre-optic cable. The duplex fibre-optic

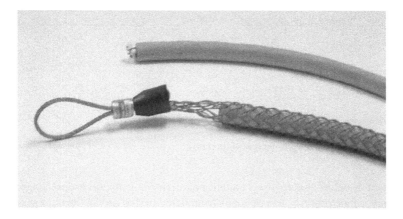

FIGURE 4.9 Kellems grip.

cable is supplied with Kevlar strength members extending beyond the outer jacket to provide a means of attaching the pulling tape. The Kevlar layer is attached with epoxy to the outer jacket and inner layers to prevent torsion while the cable is being pulled; the Kevlar is wrapped around the inner jacket in a helical pattern. The free ends of the Kevlar fibres are inserted into a loop at the end of the pulling tape and then epoxied back to themselves. The connectors are protected by foam rubber and a heat-shrink sleeving. The heat-shrink sleeving is clamped to the front of the steel ring in the pulling tape to prevent pushing the connectors along the rest of the fibre-optic cable. During an installation, the pulling force should be constantly monitored by a mechanical gauge. The pull tension can be monitored by a line tensiometer, a breakaway or by using a dynamometer and pulley arrangement. If a power winch is used to assist the pulling, a power capstan with adjustable slip clutch is recommended. The clutch, set for the maximum loading, will disengage if the set load is reached. The fibre-optic cable should be continuously lubricated using gel or powder if necessary.

4.3.5 FIBRE-OPTIC CABLES DIRECT BURIAL INSTALLATION

Fibre-optic cables can be buried directly in the ground using either ploughing or trenching methods, as shown in Figure 4.10. The ploughing method uses a cable-laying plough, which opens the ground, lays the cable and covers the cable in a single operation. In the trench method, a trench is dug with a machine (such as a backhoe), the cable is laid and the trench is filled. The trench method is more appropriate for short-distance installations. Buried fibre-optic cables must be protected against frost, water seepage, attack by burrowing and gnawing animals and mechanical stresses which could result from earth movements. Armoured fibre-optic cables specially designed for direct and indirect burial are available. Cables should be buried so they are below the frost line. Other buried cables should be enclosed in sturdy concrete ducts and polyurethane or PVC pipes, as shown in Figure 4.11. The duct holes and pipes should have an inside diameter several times larger than the outside

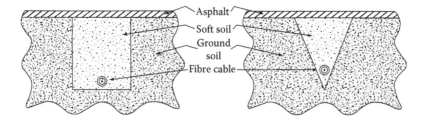

FIGURE 4.10 Fibre-optic cable direct burial methods.

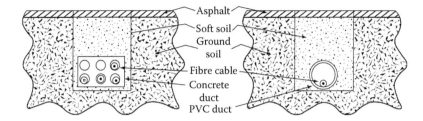

FIGURE 4.11 Fibre-optic cable indirect burial methods.

diameter of the fibre-optic cable to protect against earth movements. An excess length of fibre-optic cable in the duct or pipe prevents tensile loads from being placed on the fibre-optic cable.

4.3.6 FIBRE-OPTIC CABLE AERIAL INSTALLATION

Aerial installation includes stringing fibre-optic cables between telephone poles or along power lines, as shown in Figure 4.12. Unlike copper cables, fibre-optic cables may be run along power lines with no danger of inductive interference. Aerial fibre-optic cables must be able to withstand the forces of high winds, storms, ice loading and birds. Self-supporting aerial cables can be strung directly from pole to pole. Other cables must be attached to a high-strength steel wire, which provides additional support. The use of a separate support structure by lashing is the preferred method.

4.3.7 AIR-BLOWN FIBRE-CABLE INSTALLATION

There are many ways to upgrade or plan a new network system. Some network architects may choose a new and improved technology. The most recent technology is air-blown optical fibre invented by British Telecom (BT-London) and deployed in the early 1980s. This technology offers higher capacity installation and better security over other technologies. The air-blown fibre process involves the deployment of tube cable in place of traditional inner ductwork, as shown in Figure 4.13. These tube cables contain several individual tube cells inside a protective outer jacket. The individual tube cells can be in different diameters. Once the tube cable is in place, fibre cable is blown through the tubes to various locations and terminated or interconnected. Fibre in different types (single or multimode) is blown into the network. The air-blown fibre system can work with any panels and connectors of traditional fibres. The technology can be used to provide pathways between buildings and cities. It also allows upgrade, growth, change of high-speed voice, data and video local area networks.

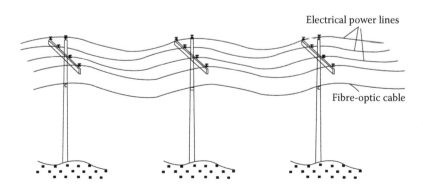

FIGURE 4.12 Fibre-optic cable aerial installation.

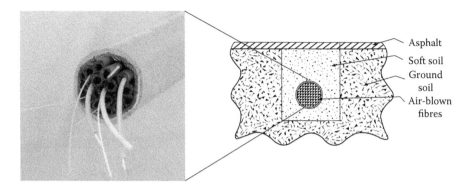

FIGURE 4.13 Air-blown fibre cable installation.

4.3.8 Other Fibre-Cable Installation Methods

In 2001, Toronto became the first major city in North America to test-drive a system to lay fibre-optic cable in sewer pipes deep below city streets using a remote-controlled robot. With this new technology, the city of Toronto managed to upgrade and extend the communication networks without digging up the streets. The project uses air-blown cables installed by a robot. The robot navigated cables and performed all-at-once infrastructure installation. Connections between the points were achieved quickly and efficiently. This technology is planned for use in other cities in Canada.

Other cities caught on quickly, and now several cities worldwide enjoy the benefits of having their own fibre networks upgraded. Vienna and Berlin applied this technology to upgrade their network systems.

In 2003, the city of San Francisco conducted a two-mile pilot project. The project enables San Francisco's Department of Telecommunications and Information Services to connect additional city facilities to E-Net. The project uses flexible cable, air-blown tubes and robots to install the tubes through sewer clean-outs. Each tube houses up to 19 individual fibre tubes. Distribution boxes housing the individual fibre tube connections are strategically placed to avoid unnecessary splice points along the network. The project demonstrates the importance of using new technology to advance network systems.

4.4 STANDARD HARDWARE FOR FIBRE-OPTIC CABLES

This section describes some types of common fibre-optic hardware that are used in all telecommunication systems.

4.4.1 Fibre-Splice Closures

There are different types of splice closures for fibre-optic cables. A splice closure is a standard piece of hardware in the telecommunication industry for protecting fibre-optic cable splices. Some small-closure types are for two to eight fibre mechanical splices and suitable for indoor/outdoor installation. Other closure types include splice trays and large splice closures for outdoor applications. Figure 4.14 shows a rack-mounted splice closure unit.

4.4.2 Rack with Panels

Consider building an application, such as a wiring centre, which will be used as a central distribution point, as shown in Figure 4.15. Fibre cables can be routed to their final destinations from the rack. An outdoor multi-fibre cable is brought (into the building) to the distribution rack, and then simplex or duplex fibre cables route the signals to different locations within the building.

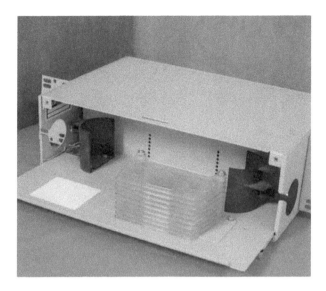

FIGURE 4.14   Rack-mounted splice closure unit.

4.4.3   CONNECTOR HOUSINGS

Connector housings are designed for main cross-connect, interconnect, intermediate cross-connects, for the local area network, and for data centre fibre distribution frames in telecommunication systems. Figure 4.15 shows a connector housing unit. They provide easy, open access to connectors for moves, adds and changes and for connector cleaning. They can be easily mounted on the racks or cabinets.

4.4.4   PATCH PANELS

Patch panels provide a convenient way to rearrange fibre-cable connections and circuits, as shown in Figure 4.15. A simple patch panel is a metal frame containing bushings in which fibre-optic cable connectors plug in on either side. One side of the panel is usually fixed, meaning that the fibre cables are not intended to be disconnected. On the other side of the panel, fibre cables can be connected and disconnected to arrange the circuits as required. Patch panels are widely used in the telecommunication industry to connect circuits to transmission equipment. They can also be used in the telecommunication room of a building to rearrange connections. The splice organiser and the patch panel serve similar functions in distribution. The primary difference is that the organiser is intended for fixed connections, whereas a patch panel is used for flexible connections.

4.4.5   SPLICE HOUSINGS

Splice housings storage and protection of fibre splices in individually accessible trays. The splice trays are explained in the fibre-optic connection chapter. They can easily be mounted on the racks for transition splicing between different cables at building entrance or pigtail splicing.

4.4.6   WALL OUTLETS

Fibre-optic wall outlets serve a similar function to electrical outlets, except that they allow connections to fibre cables carrying optical signals, as shown in Figure 4.16. In a building or office wired with fibre-optic cables, the outlet serves as a transition point between the horizontal cabling and the

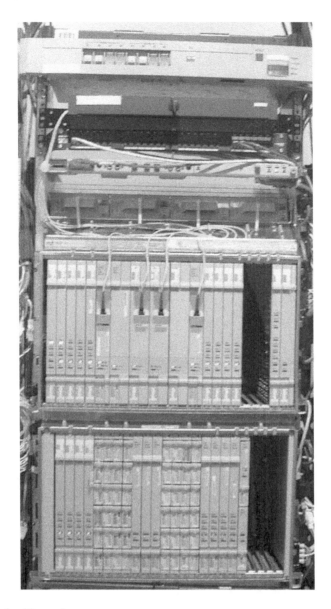

FIGURE 4.15   Rack with panels.

equipment. Wall outlets are designed to accommodate the fibre connectors at a 45 degree mating angle. This design avoids the fibre-optic cable bending at the back of the wall outlet. A short simplex or duplex cable, called a jumper or drop cable, typically runs from the wall outlet to the equipment being served.

4.4.7   FIBRE-OPTIC TESTING EQUIPMENT

Many types of fibre-optic testing equipment offer true multi-testing capability. They feature field-installable single-mode and multi-mode cables, optical time domain reflectometer modules, visual fault locator, optical return loss, fibre communicator, optical network simulator, built-in power meter, laser source and so on.

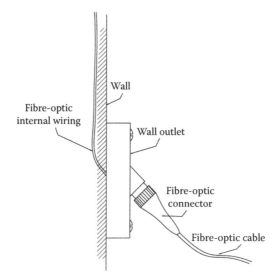

FIGURE 4.16   Wall outlet.

4.5   FIBRE-OPTIC CABLE TEST REQUIREMENTS

After the fibre cabling system is installed, it should be tested and certified that it meets performance specifications (TIA/EIA). Testing and certification have different implications when installing fibre-optic cables. Testing implies that certain values are measured, whereas certification, on the other hand, compares these measured values to standards or derived values to determine that the values are within specified limits. Testing is a quantitative procedure, whereas certification is both quantitative and qualitative.

Attenuation tests at different operating wavelengths should be performed on all fibre-optic cables. The test results should be documented, complied with the standards and should include the following information:

1. Fibre-optic cable ID.
2. Fibre-optic cable types.
3. Test equipment.
4. Testing and verification techniques.
5. Attenuation values.
6. Wavelengths.
7. Connector types.
8. Connector names.
9. Splice types and locations.
10. Reference setting at first wavelength.
11. Reference setting at second wavelength.
12. Building number and location.
13. Mapping and documentation requirements.
14. Relevant additional data.
15. Relevant additional comments.
16. Any special customer requirements.
17. Test results that must be supplied.
18. Date of test performed.

19. Contractors' names.
20. Technicians' names and signatures.
21. Any other requirements specified by the customer.

4.6 EXPERIMENTAL WORK

There are many types and applications of fibre-optic cable. The practical importance of the mechanical properties of fibre-optic cable is demonstrated here. The different types of common fibre-optic hardware, such as closures, organisers, rack boxes, distribution panels and wall outlets, play an important role in telecommunication systems. The student will also study and practice using the parts of the telecommunication system in the building. The student will become familiar with the correct manner of installing and testing a fibre-optic cable in a networking system.

4.6.1 Technique and Apparatus

Appendix A presents the details of the devices, components, tools and parts.

1. Figure 4.17 shows the tools that are required for this experiment.
2. Fibre-optic cable.
3. Fibre cable end preparation kit.
4. Fish wire.
5. Plastic tape/isolation tape.
6. Protective plastic spiral tube.
7. Vacuum handle for removing floor tiles.
8. Optical continuity checker.
9. Optical power metre and sensor.
10. Fibre-cable measurement set.
11. Connector assembly kit.
12. Mechanical optical splice kit.
13. Fusion splice.

Note: Tools and parts listed in items 10–13 are listed in case they are needed to build a connector or splice as an adapter between the fibre-optic cable and the testing devices.

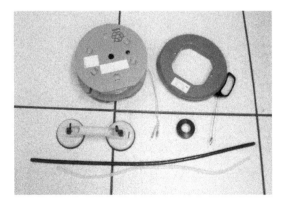

FIGURE 4.17   Fibre cable installation tools.

4.6.2 PROCEDURE

Follow the laboratory procedures and instructions given by the instructor.

4.6.3 SAFETY PROCEDURE

Follow all safety procedures and regulations regarding the use of fibre-optic cable, testing instrument and installation tools.

4.6.4 APPARATUS SET-UP

4.6.4.1 Fibre-Optic Cable Installation

Choose the area where the cable is to be installed. Try to choose two types of cable-installation procedures between two connections, such as an overhead cable tray system and an installation under raised floor tiles. The following steps will guide you throughout this experiment.

1. Locate the area and plan your fibre-optic cable installation in the telecommunications lab.
2. Determine the route that will be taken by the fibre-optic cable.
3. Ensure proper coiling of the fibre-optic cable. Arrange all of your cable in the correct entry position, using a spiral rubber tube to protect the fibre-optic cable from any sharp edges.
4. Remove every second tile from the floor, starting at the termination/distribution box along the fibre-optic cable raceways.
5. Properly attach the fibre-optic cable to the fish wire after passing it through the protective jacket in case of excessive bending.
6. Pull slowly and have someone assist with feeding the fibre-optic cable from the other end.
7. Proceed to pull the fibre-optic cable through the overhead cable tray using the fish wire.
8. Keep the fibre-optic cable neatly bundled, laying it on the floor away from the posts and other obstacles.
9. Carefully pull the fibre-optic cable through the tray and position it in the centre of the cable tray. Trim the excess fibre-optic cable, leaving sufficient service length of the cable for repair or maintenance.
10. Replace the tiles back into position, tidy up the site and store the tools.
11. Terminate each cable end and make the required end connections.
12. Carry out the fibre-optic cable continuity test.
13. Conduct cable end connection tests.
14. Measure the fibre-optic cable power loss.

4.6.5 DATA COLLECTION

Record the power and losses measurements of the fibre cable.

4.6.6 CALCULATIONS AND ANALYSIS

Verify the power and loss measurements and compare to cable specifications.

4.6.7 RESULTS AND DISCUSSIONS

Report the power and losses measurements with the cable specifications in charts or tables.

4.6.8 CONCLUSION

Summarise the important observations and findings obtained in this lab experiment.

4.6.9 SUGGESTIONS FOR FUTURE LAB WORK

List any suggestions for improvements using different experimental equipment, procedures and techniques for any future lab work. These suggestions should be theoretically justified and technically feasible.

4.7 LIST OF REFERENCES

List any references that were used in the report. Use one format in writing the references. Never mix reference formats in a report.

4.8 APPENDIX

List all of the materials and information that are too detailed to be included in the body of the report.

FURTHER READING

Boyd, W. T., *Fiber Optics Communications, Experiments and Projects*, 1st ed., Howard W. Sams & Co., Indianapolis, IN, 1987.

Chen, K. P., In-fiber light powers active fiber optical components, *Photon. Spectra*, April, 78–90, 2005.

Cole, M., *Telecommunications*, Prentice Hall, Englewood Cliffs, NJ, 1999.

Derfler, F. J. Jr., and Freed, L., *How Networks Work*, Millennium ed., Que Corporation, Indianapolis, IN, 2000.

Derickson, D., *Fiber Optic Test and Measurement*, Prentice Hall PTR, Upper Saddle River, Englewood Cliffs, NJ, 1998.

Dutton, H. J. R., *Understanding Optical Communications*, IBM, Prentice Hall, Englewood Cliffs, NJ, 1998.

Goff, D. R., and Hansen, K. S., *Fiber Optic Reference Guide: A Practical Guide to the Technology*, 2nd ed., Butterworth-Heinemann, London, 1999.

Golovchenko, E., Mamyshev, P. V., Pilipetskii, A. N., and Dianov, E. M., Mutual influence of the parametric effects and stimulated raman scattering in optical fibers., *IEEE J. Quantum Elect.*, 26, 1815–1820, 1990.

Groth, D., Barnett, D., and McBee, J., *Cabling—the Complete Guide to Network Wiring*, 2nd ed., Sams & Co., Indianapolis, IN, 2001.

Hecht, J., *Understanding Fiber Optics*, 3rd ed., Prentice Hall, Englewood Cliffs, NJ, 1999.

Herrick, C. N., and McKim, C. L., *Telecommunication Wiring*, 2nd ed., Prentice Hall, Englewood Cliffs, NJ, 1998.

Hioki, W., *Telecommunications*, 3rd ed., Prentice Hall, Englewood Cliffs, NJ, 1998.

Hitz, B., Photonic bandgap fiber eyed for telecommunications, *Photon. Spectra*, 122–124, April, 2004.

Hoss, R. J., *Fiber Optic Communications—Design Handbook*, Prentice Hall Publishing Co., Englewood Cliffs, NJ, 1990.

Kao, C. K., *Optical Fiber Systems: Technology, Design, and Applications*, McGraw-Hill, New York, 1982.

Keiser, G., *Optical Fiber Communications*, 3rd ed., McGraw-Hill, New York, 2000.

Keiser, G., *Optical Communications Essentials*, 1st ed., McGraw-Hill, New York, 2003.

Kolimbiris, H., *Fiber Optics Communications*, Prentice Hall, Englewood Cliffs, NJ, 2004.

Mynbaev, D. K., and Scheiner, L. L., *Fiber-Optic Communication Technology—Fiber Fabrication*, Prentice Hall PTR, Upper Saddle River, NJ, 2005.

Palais, J. C., *Fiber Optic Communications*, 4th ed., Prentice Hall, Englewood Cliffs, NJ, 1998.

Pease, B., When it comes to optical fiber installations, some companies are really "blowing it," *Fiber-Optic Product News*, 12, May, 2004.

Razavi, B., *Design of Integrated Circuits for Optical Communications*, McGraw-Hill, New York, 2003.

Shotwell, R. A., *An Introduction to Fiber Optics*, Prentice Hall, Englewood Cliffs, NJ, 1997.

Sterling, D. J. Jr., *Technician's Guide to Fiber Optics*, 2nd ed., Delmar Publishers, Albany, NY, 1993.

Sterling, D. J. Jr., *Premises Cabling*, Delmar Publishers, Albany, NY, 1996.

Vacca, J., *The Cabling Handbook*, Prentice Hall, Englewood Cliffs, NJ, 1998.

Yeh, C., *Handbook of Fiber Optics: Theory and Applications*, Academic Press, San Diego, CA, 1990.

5 Fibre-Optic Connectors

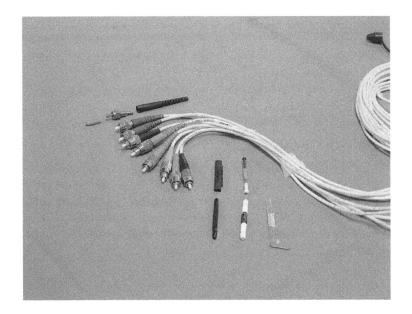

5.1 INTRODUCTION

The interconnection of optical components is a vital part of an optical system, having a major effect on performance. Interconnection between two fibre-optic cables is achieved by either connectors or splices that link the ends of the fibre cables optically and mechanically.

Connectors are devices used to connect a fibre-optic cable to an optical fibre device, such as a detector, optical amplifier, optical light power metre or link to another fibre cable. They are designed to be easily and reliably connected and disconnected. The connectors create an intimate contact between the mated halves to minimise the power loss across the junction. They are appropriate for indoor applications. Splices are used to permanently connect one fibre-optic cable to another. Splices are suitable for outdoor and indoor applications. Some types of splices are used to temporarily connect for quick testing purposes.

The key to a fibre-optic interconnection is precise alignment of the mated fibre cable cores so that the light couples from one fibre, across the junction, into the other fibre. This precise alignment creates a challenge for designers.

There is a difference between connection of two fibre cables and coupling of a light source into a fibre cable. Coupling techniques are explained in the fibre-optic cables chapter. This chapter presents the operating principles of the connectors and splices, and describes their types, properties and operations.

This chapter also presents four experimental cases for building a connection between two fibre-optic cables: linking two fibre cables by connectors and a fusion splice and testing the connection for losses.

5.2 APPLICATIONS OF CONNECTORS AND SPLICES

Connectors and splices make optical and mechanical connections between two fibre cables. It is easy to connect and disconnect a cable with a connector from another cable or a device. There are many applications for fibre connectors and splices in fibre systems, such as

- Connecting between a pair of fibre cables, using connectors or a splice, is an essential part of any fibre system.
- Interfacing devices to local area networks.
- Connecting and disconnecting fibre cables to patch panels where signals can be checked and routed in a fibre system.
- Connecting and splicing may be required on short fibre cables for wiring, testing devices, connecting instruments and devices and at other intermediate points between transmitters and receivers.
- Dividing a fibre system into subsystems, which simplifies the selection, installation, testing and maintenance of fibre systems.
- Temporarily connecting remote mobile systems and recording equipment in many fibre systems.

5.3   REQUIREMENTS OF CONNECTORS AND SPLICES

It can be very difficult to design a connector or a splice that meets all the requirements. A low-loss connector may be more expensive than a high-loss connector, or it may require relatively expensive application tooling. The lowest losses are desirable, but the other factors clearly influence the selection of the connector or splice as well.

The following is a list of the most desirable features for fibre connectors or splices required by customers and industry:

- Low loss (insertion and return): The connector or splice causes low loss of optical power across the junction between a pair of fibre cables.
- Easy installation and use: The connector or splice should be easily and rapidly installed without the need for special tools or extensive training.
- Repeatability: There should be no variation in power loss. Loss should be consistent whenever a connector is connected, disconnected and reconnected again, as many times as required.
- Economical: The connector, splice and special application tooling should be inexpensive.
- Compatibility with the environment: The connector or splice should be waterproof and not affected by temperature variations.
- Mechanical properties: The connector or splice should have high mechanical strength and durability to withstand the application and tension forces.
- Long life: The connector or splice should be built with a material that has a long life in various applications.

5.4   FIBRE CONNECTORS

Fibre connectors are designed to be easily connected and disconnected. Fibre-optic cable can be easily connected to a transmitter, receiver, power metres or another fibre cable. The key optical parameter for a fibre-optic connector is its attenuation. Signal attenuation in connectors is the sum of losses caused by several factors. The major factors are as follows:

- Overlap of fibre cable cores (also called lateral displacement)
- Alignment of fibre axes
- Fibre cable numerical aperture
- Reflection at the fibre cable junction/interface
- Connector-end polishing
- Fibre-cable spacing
- Connector end face profiles
- Insertion loss

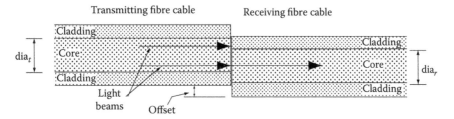

FIGURE 5.1   Overlap of fibre cable cores.

When the diameter of the transmitting fibre cable is greater than that of the receiving fibre cable, as shown in Figure 5.1, the diameter-mismatch loss (Loss_dia) is given by

$$\text{Loss}_\text{dia} 10 \log_{10} \frac{\left(\text{dia}_t^2 - \text{dia}_r^2\right)}{\text{dia}_t^2} \tag{5.1}$$

When the numerical aperture NA of the transmitting fibre cable is greater than that of the receiving fibre cable, as shown in Figure 5.2, the NA-mismatch loss (Loss_NA) is given by

$$\text{Loss}_\text{NA} = 10 \log_{10} \left(\frac{\text{NA}_r}{\text{NA}_t}\right)^2 \tag{5.2}$$

The formula for the loss due to end separation ($\text{Loss}_\text{Separation}$) between two fibre-optic cables (at separation distance) is rather involved. Assume that the transmitting and receiving fibres are identical. Figure 5.3 illustrates the separation (sometimes called air gap) between a pair of fibre cables. The formula for end separation loss ($\text{Loss}_\text{Separation}$) is given by

$$\text{Loss}_\text{Separation} = 10 \log_{10} \left(\frac{\dfrac{d}{2}}{\dfrac{d}{2} + S \tan(\arcsin)\left(\dfrac{\text{NA}}{n_0}\right)} \right) \tag{5.3}$$

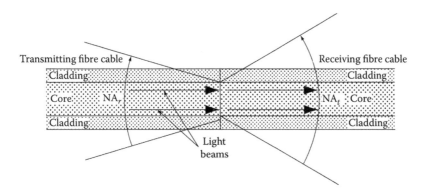

FIGURE 5.2   Numerical aperture (NA)-mismatch loss.

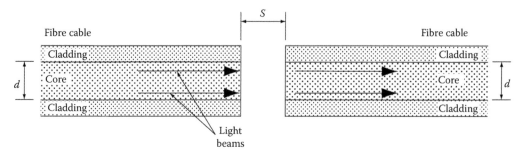

FIGURE 5.3   End separation loss.

where:
   d is core diameter
   S is the fibre spacing
   NA is the numerical aperture
   n_0 is the refractive index of the material between the two fibre cables.

A material known as index-matching fluid or gel applied between the two fibre cables reduces the reflection loss between the surfaces of the fibre cable ends. This loss, called Fresnel reflection loss, generally occurs between parallel optical surfaces. Most mechanical splices also use an index-matching gel to fill the gap between the connected fibre cable ends. An anti-reflection coating can also be applied to reduce this loss.

Additional losses may be experienced when two different types of fibre-cable connectors are connected using an adapter. You will find more details in the section dealing with adapters for different fibre connector types.

Insertion loss is a measure of the performance of a connector or splice. Insertion loss is calculated by

$$\text{Loss(dB)} = 10 \log_{10} \frac{P_2}{P_1}$$ (5.4)

where:
   P_1 is the initial power measured
   P_2 is the power measured after the connector has been mated.

5.5   MECHANICAL CONSIDERATIONS

The optical characteristics of the fibre connectors are significant. However, the mechanical characteristics are also important, and in some cases critical. Virtually all fibre connectors are designed to remain in place under working conditions. Connectors must withstand physical stresses, such as forces encountered during mating and unmating, and sudden stress induced by bending and tension. Connectors must also prevent contamination caused by dirt and moisture in the fibre cable ends.

5.5.1   DURABILITY

Durability is a concern with any type of connector. Repeated mating and unmating of the fibre connectors can cause wear in the mechanical components. Allowing dirt into the optics and straining the fibre cable will damage the exposed fibre cable ends.

5.5.2 ENVIRONMENTAL CONSIDERATIONS

Fibre connectors designed for indoor applications must be protected from environmental extremes to avoid excessive connector loss and poor system performance. Special hermetically-sealed connectors are required for outdoor use.

5.5.3 COMPATIBILITY

Compatibility refers to the need for the connector to be compatible with other connectors or with specifications. Specifications describe the type of connector to be used in specific applications. Compatibility exists on several levels and the most basic level being physical compatibility. The connector must meet certain dimensional requirements to allow it to mate with other connectors of the same style. The second level of compatibility involves connector performance, such as insertion loss, durability, operating temperature range and other requirements specified by the customers.

5.6 FIBRE-OPTIC CONNECTOR TYPES

Figure 5.4 shows the most common types of fibre-optic connectors. Fibre connectors are unique in that they must make both optical and mechanical connections. They must also allow the fibre cables to be precisely aligned to ensure that a connection is robust. Fibre connectors use various methods to achieve solid connections. Some of the types of fibre-optic connectors currently in use are listed below:

- Subscriber connectors (SC)
- SC/APC connectors
- FC/PC connectors
- FC/APC connectors
- LC connectors
- MU connectors

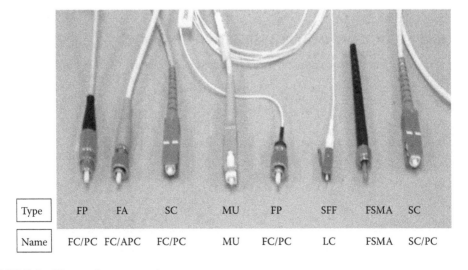

FIGURE 5.4 Fibre-optic connector types.

- Straight-tip connectors (ST)
- 568SC connectors (duplex SC)
- FDDI connectors (MIC)
- Biconic connectors
- SMA connectors
- Enterprise system connection (ESCON)
- Duplex connectors (ST)
- Polarising connectors
- MT multi-fibre connectors
- MT-RJ connectors
- D4-style connectors
- MFS/MPO connectors
- Plastic-fibre connectors
- E-2000 Diamond
- Fibre-optic connectors self-latch in push/pull system
- Special connectors for specific applications

5.7   ADAPTERS FOR DIFFERENT FIBRE-OPTIC CONNECTOR TYPES

An adapter is a passive device used to join two different types of connectors together. The type of adapter is identified by a nomenclature, such as SC, FC, ST or 568SC. Hybrid adapters join dissimilar connectors together, such as SC to FA. Figure 5.5 shows examples of some adapters.

5.8   FIBRE-OPTIC CONNECTOR STRUCTURES

Most fibre-optic connectors are built from a ferrule, a connector body, an epoxy material and a strain relief boot. Most connectors use a ferrule to hold the fibre and provide alignment. The most popular ferrule size is a 2.5-mm diameter, which is standard. Manufacturers offer a few types of ferrules made from different materials, such as ceramic, plastic and stainless steel.

Connectors may be attached to a device—outlet box or adapter—by direct connection, by coupling a threaded nut or by twisting a spring-loaded bayonet socket. The connector body is made from steel, ceramic or plastic. Epoxy is usually applied to secure the fibre cable end in the connector body. A strain relief boot made from plastic or rubber is used at the junction between the connector body and the fibre cable.

5.9   FIBRE-OPTIC CONNECTOR ASSEMBLY TECHNIQUES

The following sections present common assembly techniques that are used in building fibre-optic connectors.

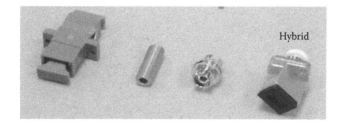

FIGURE 5.5   Fibre-optic adapter types.

5.9.1 COMMON FIBRE-CONNECTOR ASSEMBLY

The most common fibre connector assembly techniques use a fibre cable and a suitable connector. The fibre cable is most often epoxied into the connector. Epoxy provides good tensile strength to the connector to prevent the fibre cable from moving within the connector body, maintaining a good alignment. After the epoxy cures, the ferrule end is polished to a smooth finish by one of the many available procedures. Then the connector undergoes many inspections and test procedures to issue a data sheet for the customer.

5.9.2 HOT-MELT CONNECTOR

Hot-melt connectors are widely used in North American telecommunication systems. The hot-melt connectors use pre-loaded epoxy so that external mixing is not required. The prepared end of the fibre-optic cable is inserted into the connector ferrule, as shown in Figure 5.6. The cable (with the connector inserted) is loaded onto the connector holder and placed in an oven for a few minutes, which softens the epoxy around the fibre cable and cures the epoxy at the same time. The curing time is dependent on the type of epoxy. The end of the connector is then polished to a smooth finish. The polishing can be done by hand or by an industrial polishing machine. When such connectors are assembled in the field, a portable hand polisher is used.

5.9.3 EPOXYLESS CONNECTOR

Epoxyless connectors, also called crimp connectors, have been widely used for quick cable connections in telecommunication systems. When the connector is crimped, an insert compresses around the fibre cable. A front clamp on the bare fibre cable and a rear compression clamp add a higher clamping force on the fibre cable buffer coating to provide the necessary tensile strength. Special gripping tools are used in the assembly of the epoxyless connectors. The end of the fibre connector is polished to a smooth finish using a portable hand polisher before the connector is assembled in the field. The main advantage of an epoxyless connector is the speed of assembly. Some customers will tolerate a slightly higher loss to achieve a fast, easy termination. The epoxyless approach is a technology that is not limited to one connector type.

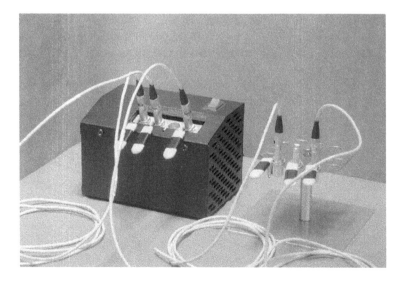

FIGURE 5.6 Hot-melt connectors assembly.

5.9.4   Automated Polishing

All fibre-optic connector-polishing machines are designed for accuracy, easy set-up and production efficiency. The polishing pressure, speed and duration can be adjusted to meet exact requirements. These machines precisely polish the ends of fibre-optic connectors in a repeatable and reliable manner. Polishing machines are available for dry or wet polishing process.

5.9.5   Fluid-Jet Polishing

Fluid-jet polishing (FJP) is another technique for shaping and polishing small surface areas of complicated optics made of brittle materials. This technique uses a fluid-jet system to guide pre-mixed slurry, at low pressures, onto the optical surface being machined. The surface is altered by the erosive effect of the abrasive particles in the stream.

5.9.6   Fibre-Optic Connector Cleaning

Contamination of connector ends can occur from something as simple as dust particles or fingerprints that can reduce light propagation through the fibre cable. This will degrade device performance, causing data error and loss. To avoid this, it is a common practice to clean fibre connectors prior to assembly and testing.

There are three major components of the fibre-optic connector system that users must consider when cleaning, mating and testing fibre-optic connectors: the adapter split sleeve, the outer diameter of the ferrule and the tip of the ferrule. There are many techniques for cleaning connectors, either wet or dry, by hand with recommended cleaning chemicals or with automated machines. Follow the cleaning procedure for each fibre connector type. Do not use the same procedure for other types of connectors. Cleaning standards for fibre-optic connectors promise savings in time and cost.

5.9.7   Connector Testing

There are many testing instruments available for testing connectors. Testing instruments range from a simple view scope to a sophisticated system. The condition of the end of the ferrule after the polishing process is usually inspected using simple instruments, as shown in Figure 5.7. This procedure is adequate for inspecting a connector built and polished in the field.

Handheld devices can measure the losses, optical powers, light sources and so on. The basic test measures the attenuation of the fibre cable with connectors by comparing the power through the fibre cable to that of a known reference fibre cable. The power through the fibre cable under test is measured in absolute units. The power through the reference fibre cable is also measured. Figure 5.8 illustrates a connector test set-up.

Using sophisticated systems for testing connectors saves time and cost in industrial production. These systems are very accurate. Each connector type refers to a standard test. The preferred test methods are compliant with the commercial building telecommunications cabling standard, TIA/EIA-568-B.1. The ANSI/TIA/EIA standards group developed intermateability standards for several connector types to ensure compatibility among manufacturers.

5.10   FIBRE SPLICING

The splicing process joins fibre-optic cable ends permanently. In general, a splice has a lower loss than a connector. Splices are typically used to join lengths of cable for outside applications. Splices may be incorporated into lengths of fibre-optic cable or housed in indoor/outdoor splice boxes, whereas connectors are typically found in patch panels or attached to equipment at fibre cable interfaces. The sources of loss, described in the section on fibre-cable connectors, are also applicable to splices. There are two types of splices: mechanical and fusion.

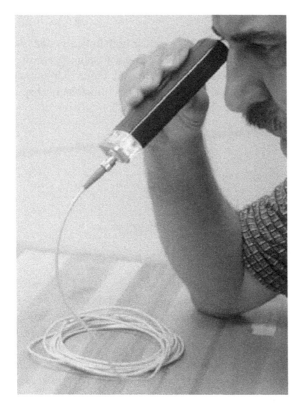

FIGURE 5.7 Connector end inspection with a view scope.

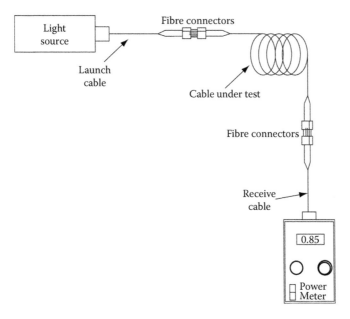

FIGURE 5.8 Connector testing set-up.

5.10.1 MECHANICAL SPLICING

Mechanical splices join two fibre cable ends together both optically and mechanically by clamping them within a common structure. In general, mechanical splicing requires less expensive equipment; however, higher consumable costs are experienced. Figure 5.9 shows the most common types of fibre-optic mechanical splices. A few important types of mechanical splices are listed below:

- Table-type splices
- Key-lock splices
- Fibre lock splices
- Twist lock splices
- Fastomeric splices
- Capillary splices
- Rotary- or polished-ferrule splices
- V-groove splices
- Elastomeric splices
- Finger splices
- Inner lock splices
- Many other types are available for special applications

Some mechanical and fusion splices are used with one type of splice closure. Figure 5.10 shows small and large splice closures. Splice closures are standard pieces of hardware in the telecommunication industry for protecting fibre-optic cable splices. Splices are protected mechanically and environmentally within the sealed closure. Splice closures are waterproof. Water is kept out by using non-flowing gel under permanent compression. They are suitable for indoor, outdoor and underground cable system installations. There are small and large closures available for different applications.

Splice trays are designed to hold fusion and mechanical splices, as shown in Figure 5.11. They are available in different sizes. Fusion and mechanical splices are held in a specially-designed splice organiser and splice holder, respectively. They are not sealed off environmentally. These trays are installed in a wall-mounted fibre splice cabinet in a communication system.

The following sections explain temporary mechanical splices that are used in different cases of the testing.

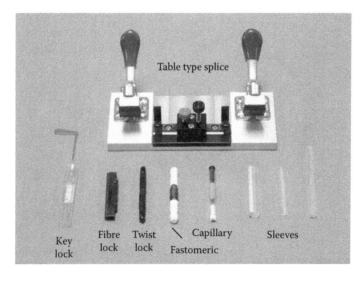

FIGURE 5.9   Some fibre-optic splice types.

(a)

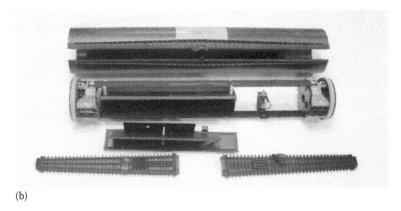

(b)

FIGURE 5.10   Splice closures: (a) a small closure and (b) a large closure.

FIGURE 5.11   Splice trays.

5.10.1.1   Key-Lock Mechanical Fibre-Optic Splices

Figure 5.12 shows the key-lock mechanical fibre-optic splice, commonly used to quickly mate and unmate fibre-optic cables. It is made from a U-shaped metal part covered by a transparent plastic body with two holes on each end. The prepared ends of the fibre cables are made longer than half of the length of the metal part. The fibre cable is inserted in the centre hole. When the key is inserted in the second hole towards the edge of the splice and turned by 90 degrees, the metal part opens and one fibre cable end can be easily inserted. This operation can be repeated on the other side to insert the second fibre cable. This type of splice provides a quick and easy way of joining two fibre cables with low signal loss. It may be used to temporarily or permanently connect fibre cables, wavelength division multiplexing (WDM) components and other fibre-optic elements.

5.10.1.2   Table-Type Mechanical Fibre-Optic Splices

Figure 5.13 shows a custom-made mechanical splice, used for quick mating and unmating of connections. This splice works like any other mechanical splice. The fibre cable ends are prepared and inserted into the mid-point of the block assembly. Screws are tightened to align the fibre cables on

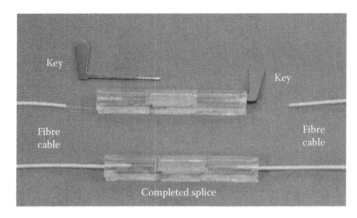

FIGURE 5.12   Key-lock mechanical fibre-optic splice.

FIGURE 5.13   Table-type mechanical fibre-optic splice.

both sides. L-clamps and K-clamps are placed in position to secure the fibre cables on both sides. Most fibre-optic companies use this kind of mechanical splice for quickly mating and unmating during manufacturing and testing processes. The splice loss associated with these instruments is acceptable by industry standards.

5.11   FUSION SPLICES

Fusion splicing is performed by placing the tips of two fibre cables together and heating them by a fast electrical-fusion process so that they melt into one piece. Fusion splices automatically align the two fibre cable ends and apply a spark across the tips to fuse them. They also include instrumentation to test the splice quality and display optical parameters pertaining to the join. A fusion splice is shown in Figure 5.20. When the fusion splice is completed, a cylindrical fusion protector is placed over the splice location. Fibre-fusion protectors are made from metal or polymer, and they are applied to ensure mechanical strength and environmental protection. Some types of fusion splice protectors (sleeves), as shown in Figure 5.9, are designed for use in place of the heat shrink method for fast, easy and reliable permanent installations. Fusion splices provide lower loss than mechanical splices.

5.11.1   SPLICE TESTING

Attenuation can be measured as the splice is being performed. Many fusion splicers come with instrumentation that measure the loss as the fibre cables are being aligned, and test the loss when the splice is completed.

TABLE 5.1
Connectors versus Splices

Connectors	Splices
Provide temporary connections	Provide permanent connections
Higher loss	Lower loss
Larger sizes	Smaller sizes
Immune, or not immune, to environmental effects (depends on the connector type)	Immune to environmental effects
It takes a long time to build a connector	It takes a very short time to build a splice
Diverse applications	Connection between a pair of fibre cables
Many types	Few types
New technology reduces installation time	Conventional technology keeps the same installation time
Building reasonable mechanical stability at the connection points	Building better mechanical stability at the connection points

5.12   CONNECTORS VERSUS SPLICES

There are definite differences between connectors and splices. Most companies make connectors and splices to satisfy customer requirements for smaller size and lower loss. As mentioned in this chapter, fibre-optic technology is moving forwards to create high-durability connectors and splices with small sizes and low cost in production and installation. Table 5.1 compares general important factors between the fibre connectors and splices.

5.13   EXPERIMENTAL WORK

The purpose of this experiment is to build a connector on a fibre cable end. Two fibre cables can be connected by two connectors with a sleeve or by a mechanical splice or by a fusion splice. The student will terminate the fibre-optic cable using an FSMA connector. The student will also be required to conduct mechanical and fusion splices and perform tests on the cable connectors and splices. Losses within the connectors, mechanical splices and fusion splices will be measured.

In this experiment the student will perform the following cases:

1. Building a connector
2. Connecting using two connectors
3. Connecting by a mechanical splice
4. Connecting by a fusion splice
5. Testing connection loss in two connectors
6. Testing connection loss in a mechanical splice
7. Testing connection loss in a fusion splice

5.13.1   TECHNIQUE AND APPARATUS

Appendix A presents the details of the devices, components, tools and parts.

1. 2 × 2 ft. optical breadboard.
2. He–Ne laser light source and power supply.
3. Laser light sensor.
4. Laser light power metre.
5. Laser mount assembly.

6. 20× microscope objective lens.
7. Lens/fibre cable holder/positioner assembly.
8. Fibre cable holder/positioner assembly.
9. Hardware assembly (clamps, posts, screw kits, screwdriver kits, lens/fibre cable holder/positioner, sundry positioners, brass fibre cable holders, fibre cable holder/positioner, etc.).
10. Fibre cable end preparation procedure and kit, and cleaning kit, as explained in the end preparation section of the fibre-optic cables chapter.
11. Connector holder/positioner assembly.
12. Black/white card and cardholder.
13. Water spray bottle.
14. 50%/50% mixing epoxy.
15. Fibre-optic cable, 900 micrometres diameter, 500 metres long.
16. Polishing disk.
17. FSMA connector kit, as shown in Figure 5.15.
18. Polishing pads, as shown in Figure 5.15.
    Size 63.0 µm GREY Colour
    Size 9.0 µm BLUE Colour
    Size 1.0 µm VIOLET Colour
    Size 0.3 µm WHITE Colour
19. Key-lock mechanical fibre-optic splicing unit, as shown in Figures 5.9 and 5.12.
20. Table-type mechanical fibre-optic splice unit, as shown in Figures 5.9 and 5.13.
21. Fusion splicing machine, as shown in Figure 5.20.
22. Ruler.

5.13.2 PROCEDURE

Follow the laboratory procedures and instructions given by the instructor.

5.13.3 SAFETY PROCEDURE

Follow all safety procedures and regulations regarding the use of fibre-optic cable. You must wear safety glasses and finger cots or gloves when working with and handling fibre-optic cables, optical components and optical cleaning chemicals.

5.13.4 APPARATUS SET-UP

5.13.4.1 Case (a): Building FSMA Connectors

The students will assemble a FSMA connector in this experiment. The structure of this connector is shown in Figure 5.14.

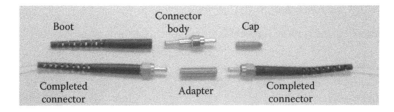

FIGURE 5.14 FSMA connectors.

5.13.4.1.1 Connector Body

The connector body is a continuation of the ferrule, as shown in Figure 5.14. The connector body accommodates the strain relief boot. The connector is attached to an adapter or a device by a threaded coupling nut.

5.13.4.1.2 Epoxy and Polish

A fibre cable is often epoxied to a connector. On account of its curing time, epoxy is generally considered to be an undesirable but necessary process in fibre-optic termination. After the epoxy cures, the ferrule end is polished to a smooth finish.

5.13.4.1.3 Strain Relief Boot

A black polymer strain relief boot shields the junction of the connector body and the fibre cable. Figure 5.15 shows the connector assembly kit. Building and testing a connector involves the following steps:

1. Cut two four-foot lengths of fibre cables from the spool.
2. Strip, cleave and clean the fibre cable ends so that there is about 1 to 11/4 inch of bare fibre extending beyond the jackets. Follow the fibre cable end preparation procedure as explained in the fibre-optic cables chapter.
3. Prepare a connector for the assembly process (the connector consists of two parts: ferrule-connector body and strain relief boot).
4. Insert the prepared end of the fibre cable into the connector until the fibre cable stops and the bare fibre emerges from the ferrule.
5. Prepare the epoxy according to the epoxy manufacturer's preparation procedure.
6. Pull the fibre cable ½ inch backwards; then, with a toothpick or small probe, apply the epoxy onto the fibre cable. Push and pull the cable (back and forth), until the epoxy adequately fills the gap between the fibre cable and the ferrule. The quantity of epoxy must be sufficient to support the fibre cable inside the connector body.
7. Insert the fibre cable into the ferrule completely.
8. Using a small probe, place a very small drop of epoxy onto the ferrule face where the fibre cable exits. This will seal off the space between the ferrule face hole and the fibre cable. The size of this drop is important; too large an epoxy drop will extend the polishing time.
9. Set the connector aside for the specified time to allow the epoxy to fully cure. Curing time depends on the type of epoxy.
10. Insert the fibre cable through the strain relief boot, positioning the boot on the connector body.

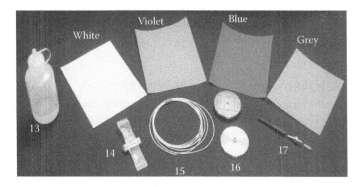

FIGURE 5.15 Connector assembly kit.

FIGURE 5.16   Polishing connector by hand.

11. Cleave the protruding fibre flush with the ferrule face. Remove the protruding piece of fibre and dispose of it in the designated container.
12. Place the size 63.0 μm GREY colour polishing pad on a flat and clean surface, such as a piece of glass sheet.
13. Screw the connector gently onto the polishing disk until finger-tight. Do not over-tighten.
14. Place several small drops of water on the polishing pad. Begin the polishing process by moving the polishing disk with your fingers in a figure 8 motion, as shown in Figure 5.16.
15. Polish until the epoxy bead and excess fibre cladding are removed and the fibre cable is flush with the ferrule face surface.
16. Apply only light pressure on the polishing disk, using enough water to keep the polishing pad and disk clean; apply constant motion.
17. About 20–30 motions should be sufficient to complete the rough polishing.
18. Rinse the polishing disk with water.
19. Inspect the ferrule face using an inspection microscope. Look at the quality of the polished ferrule face.
20. Repeat steps 14–19 using polishing pads of size 9.0 μm (BLUE colour), size 1.0 μm (VIOLET colour) and size 0.3 μm (WHITE colour).
21. About 20–30 motions should be enough to achieve a good quality finish on the ferrule face when using the white pad. Caution: do not over-polish.
22. Inspect the finished connector using the inspection microscope.
23. Once the two connectors are completed, test the connectors to measure the connection loss, which is calculated using Equation 5.4.

5.13.4.2   Case (b): Testing Connection Loss in Two Connectors

1. Figure 5.17 shows the experimental set-up for testing connection loss in two connectors.
2. Follow the instruction of the fibre-optic cable loss measurements and calculations explained in the fibre-optic cables chapter.
3. Measure the laser input power into the fibre cable. Fill out Table 5.2.
4. Couple the laser beam output to the fibre cable input.
5. Carefully align the laser with the lens/fibre cable holder/positioner so that the maximum amount of the laser beam is entering the core of the fibre cable.

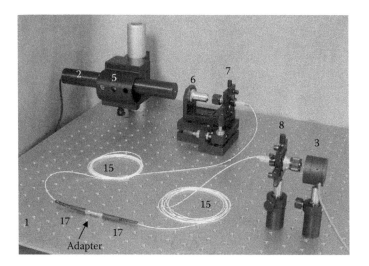

FIGURE 5.17   Mated connectors set-up test.

TABLE 5.2
Connectors and Splices Tests

		Cable 1			Cable 2	Cables-Connectors/
		P_{in} (unit)	$P_{out\,1}$ (unit)	Loss (dB)	$P_{out\,2}$ (unit)	Splices Loss (dB)
Case (b)						
Case (c)	(1)					
	(2)					
Case (d)						

6. Check to ensure that you have a circular output from the first optic connector. Direct the output towards the centre of the laser sensor.
7. Measure the laser output power from the first optic connector. Fill out Table 5.2.
8. Screw the two optic connectors into the in-line adapter, with one connector on each side, as shown in Figure 5.17.
9. Measure the laser output power from the second fibre cable end, as shown in Figure 5.17. Fill out Table 5.2.
10. Determine the connection assembly loss from the data. Fill out Table 5.2.
11. Mate and unmate the connectors three times to check the repeatability of the connection loss figures.

5.13.4.3   Case (c): Testing Connection Loss in a Mechanical Splice

5.13.4.3.1   Testing Connection Loss Using a Key-Lock Mechanical Splice

1. Figure 5.18 shows the experimental set-up for testing connection loss in a mechanical splice. Prepare a key-lock mechanical splice for the fibre cable connection process.
2. Repeat steps 2–7 from Case (b).
3. Insert the key fully into the edge hole and turn the key by 90 degrees to the open position.

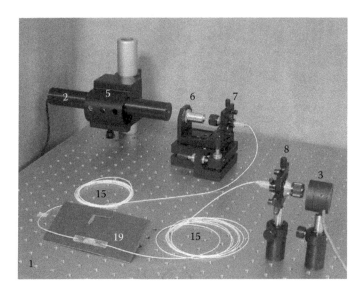

FIGURE 5.18   Key-lock mechanical splicing set-up test.

4. Insert the prepared first fibre cable output end into the centre hole on the side of the key-lock mechanical splice. The fibre cable end should be at the mid-point of the metal part of the mechanical splice. Then turn the key to the closed position to secure the fibre cable in the mechanical splice. Remove the key.
5. Repeat step 4 and gently insert the second fibre cable end into the other side of the mechanical splice. The two fibre cables should meet approximately at the centre of the metal part. Make sure that the fibre cable ends are as close (face-to-face) as possible without an air gap between them. If the mechanical splice shines during the test, the light is escaping because the two ends are not close enough.
6. Measure the laser output power from the second fibre cable, as shown in Figure 5.18. Fill out Table 5.2.
7. Determine the key-lock mechanical splicing loss from the collected data. Fill out Table 5.2.
8. Repeat the mechanical-splicing process three times to check the repeatability of the connection loss figures.

5.13.4.3.2   *Testing Connection Loss in a Table-Type Mechanical Splice*

1. Figure 5.19 shows the experimental set-up for testing connection loss in a table-type mechanical splice.
2. Prepare the table-type mechanical splice by cleaning and oiling it to be ready for the connection process.
3. Repeat the fibre cable ends as explained above.
4. Insert the first prepared fibre cable output end into the table-type mechanical splice to the mid-point of the black block assembly on one side.
5. Tighten the screw on the side where the first fibre cable was inserted to secure the first fibre cable in the splice.
6. Rotate the L-clamp and clamp down the K-clamp into position to secure the first fibre cable.
7. Repeat steps 4–6 to insert the second fibre cable into the other side of the splicer.
8. Measure the laser power at the second fibre cable output. Fill out Table 5.2.
9. Determine the connection loss from the data. Fill out Table 5.2.
10. Repeat the mechanical-splicing process three times to check the repeatability of the connection loss figures.

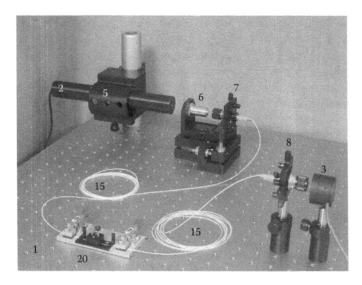

FIGURE 5.19 Table-type mechanical splicing set-up test.

5.13.4.4 Case (d): Testing Connection Loss in a Fusion Splice

1. Figure 5.20 shows the experimental set-up for testing connection loss in a fusion splicer machine.
2. Prepare a fusion splicer machine for the fibre cables connection process.
3. Repeat the steps in the mechanical splice for the fibre cables connection process.
4. Turn on the fusion splicer.
5. Insert one prepared end of each fibre cable into the bonder of the fusion splicer machine.
6. Make sure that the fibre cable ends are lined up straight, to ensure proper bonding.
7. Look through the window of the fusion splicer machine to inspect the positioning of the two fibre cables; ensure that they are lined up and that they are flush. Perform a fine alignment for the two fibre cable ends to be touched face-to-face.
8. Push the fusion key to start the fusion process.
9. The fusion splicer connects the two fibre cables permanently. When the splicing is complete, remove the fibre cable.
10. Add a fibre-optic fusion protector (sleeve a rubber or metal) to the fusion location on the fibre cable for mechanical protection.

FIGURE 5.20 Fusion splice set-up test.

11. The fusion splicer has a display panel and can display the splice test results and other optical parameters, as shown in Figure 5.20.
12. Set up the fusion splice test, as shown in Figure 5.20.
13. Measure the laser power output at the end of the second fibre cable. Fill out Table 5.2.
14. Determine the fusion splice connection loss from the data. Fill out Table 5.2.
15. Find the connection loss from the data that is provided by the fusion splicer machine.

5.13.5   DATA COLLECTION

5.13.5.1   Case (a): Building a Connector

No data collection is required for this case.

5.13.5.2   Case (b): Testing Connection Loss in Two Connectors

1. Measure the laser power at the first fibre cable input and output.
2. Measure the laser power from the second fibre cable output.
3. Fill out Table 5.2 for Case (b).

5.13.5.3   Case (c): Testing Connection Loss in a Mechanical Splice

Data collection is similar to Case (b). Fill out Table 5.2 for Case (c).

5.13.5.4   Case (d): Testing Connection Loss in a Fusion Splice

Data collection is similar to Case (b). Fill out Table 5.2 for Case (d).

5.13.6   CALCULATIONS AND ANALYSIS

5.13.6.1   Case (a): Building a Connector

No calculations or analysis are required for this case.

5.13.6.2   Case (b): Testing Connection Loss in Two Connectors

1. Calculate the power loss in the connectors, using Equation 5.4.
2. Fill out Table 5.2 for Case (b).

5.13.6.3   Case (c): Testing Connection Loss in a Mechanical Splice

1. Calculate the power loss in the mechanical splicer, using Equation 5.4.
2. Calculate the power loss in the table-type mechanical splice, using Equation 5.4.
3. Fill out Table 5.2 for Case (c).

5.13.6.4   Case (d): Testing Connection Loss in a Fusion Splice

1. Calculate the power loss in the fusion splicer using Equation 5.4. Read the fusion splice loss value from the fusion splicer once the fusion splice is complete.
2. Fill out Table 5.2 for Case (d).

5.13.7   RESULTS AND DISCUSSIONS

5.13.7.1   Case (a): Building a Connector

1. Report your observations of the connector assembly and of the connector inspection for the quality of polishing the end of the ferrule.
2. Compare the quality of the polished end of the ferrule with a similar industrial pre-assembled connector.

5.13.7.2   Case (b): Testing Connection Loss in Two Connectors
Report the calculated results for the power loss (dB) in the connector.

5.13.7.3   Case (c): Testing Connection Loss in a Mechanical Splice
1. Report the calculated results for the power loss (dB) in the mechanical splice.
2. Report the calculated results for the fibre cable power loss (dB) in the table-type mechanical splice.

5.13.7.4   Case (d): Testing Connection Loss in a Fusion Splice
1. Report the calculated results for the power loss (dB) in the fusion splicer.
2. Compare the results of the loss calculations in the connector, key-lock mechanical splice, table-type mechanical splice and fusion splice.

5.13.8   CONCLUSION
Summarise the important observations and findings obtained in this lab experiment.

5.13.9   SUGGESTIONS FOR FUTURE LAB WORK
List any suggestions for improvements using different experimental equipment, procedures and techniques for any future lab work. These suggestions should be theoretically justified and technically feasible.

5.14   LIST OF REFERENCES
List any references that were used in the report. Use one format in writing the references. Never mix reference formats in a report.

5.15   APPENDIX
List all of the materials and information that are too detailed to be included in the body of the report.

FURTHER READING
Al-Azzawi, A., and Casey, R. P., *Fiber Optics Principles and Practices*, Algonquin College Publishing Centre, Ottawa, Canada, 2002.
Agrawal, G. P., *Fiber-Optic Communication Systems*, 2nd ed., Wiley, New York, 1997.
Boyd, W. T., *Fiber Optics Communications, Experiments and Projects*, 1st ed., Howard W. Sams and Co., Indianapolis, IN, 1987.
Camperi-Ginestet, C., Kim, Y. W., Wilkinson, S., Allen, M., and Jokerst, N. M., Micro-opto-mechanical devices and systems using epitaxial lift off, in *JPL, Proceedings of the Workshop on Microtechnologies and Applications to Space Systems*, pp. 305–316, (SEE N94-29767 08-31), Category Solid-State Physics, Georgia Inst. of Tech., Atlanta, U.S.A., June, 1993.
Cole, M., *Telecommunications,* Prentice Hall, Englewood Cliffs, NJ, 1999.
Derickson, D., *Fiber Optic Test and Measurement,* Prentice Hall PTR, Upper Saddle River, NJ, 1998.
Dutton, H. J. R., *Understanding Optical Communications,* IBM/Prentice Hall, Research Triangle Park, NC/Englewood Cliffs, NJ, 1998.
Golovchenko, E., Mamyshev, P. V., Pilipetskii, A. N., and Dianov, E. M., Mutual influence of the parametric effects and stimulated Raman scattering in optical fibers, *IEEE J. Quantum Elect.,* 26, 1815–1820, 1990.
Green, P. E., *Fiber Optic Networks,* Prentice Hall Publishing Co., Englewood Cliffs, NJ, 1993.
Hecht, J., *Understanding Fiber Optics,* 3rd ed., Prentice Hall, Englewood Cliffs, NJ, 1999.
Herrick, C. N., and McKim, C. L., *Telecommunication Wiring,* 2nd ed., Prentice Hall, Englewood Cliffs, NJ, 1998.

Hoss, R. J., *Fiber Optic Communications—Design Handbook,* Prentice Hall Publishing Co., Englewood Cliffs, NJ, 1990.

Kao, C. K., *Optical Fiber Systems: Technology, Design, and Applications,* McGraw-Hill, New York, 1982.

Keiser, G., *Optical Fiber Communications,* 3rd ed., McGraw-Hill, New York, 2000.

Keiser, G., *Optical Communications Essentials,* 1st ed., McGraw-Hill Pub., New York, 2003.

Kolimbiris, H., *Fiber Optics Communications,* Prentice Hall, Englewood Cliffs, NJ, 2004.

L-com Connectivity Products, *Fiber Optic Connectors,* Master Catalog 2006, L-com Connectivity Products, U.S.A., June, 2006.

Mazzarese, D., Meeting the OM-3 challenge—Fabricating this new generation of multimode fiber requires innovative state of the art designs and testing processes, *Cabling Syst.,* 18–20, December 2002.

Mynbaev, D. K., and Scheiner, L. L., *Fiber-Optic Communication Technology—Fiber Fabrication,* Prentice Hall PTR, Englewood Cliffs, NJ, 2005.

Newport Corporation, *Projects in Fiber Optics Applications Handbook,* Newport Corporation, Fountain Valley, CA, 1986.

OZ Optics Limited, *Fiber Optic Components for Optoelectronic Packaging,* Catalog 2006, OZ Optics, Ottawa, Canada, 2006.

Pease, B., When it comes to optical fiber installations, some companies are really "blowing it," *Fiberoptic Product News,* 12, May 2004.

Salah, B. E. A., and Teich, M. C., *Fundamentals of Photonics,* Wiley, New York, 1991.

Senior, J. M., *Optical Fiber Communications, Principles and Practices,* 3rd ed., Personal Education Limited, Essex, England, 2009.

Shotwell, R. A., *An Introduction to Fiber Optics,* Prentice Hall, Englewood Cliffs NJ, 1997.

Sterling, D. J. Jr., *Technician's Guide to Fiber Optics,* 2nd ed., Delmar Publishers, Albany, NY, 1993.

Sterling, D. J. Jr., *Premises Cabling,* Delmar Publishers, Albany, NY, 1996.

Ungar, S., *Fiber Optics: Theory and Applications,* Wiley, New York, 1990.

Vacca, J., *The Cabling Handbook,* Prentice Hall, Englewood Cliffs, NJ, 1998.

Yariv, A., Universal relations for coupling of optical power between microresonators and dielectric waveguides, *Electron. Lett.,* 36, 321–322, 2000.

Yeh, C., *Handbook of Fiber Optics: Theory and Applications,* Academic Press, San Diego, CA, 1990.

Zhang, L., and Yang, C., Polarization selective coupling in the holey fiber with asymmetric dual cores, Paper ThP1, presented at LEOS2003, Tucson, AZ, October 26–30, 2003.

6 Passive Fibre-Optic Devices

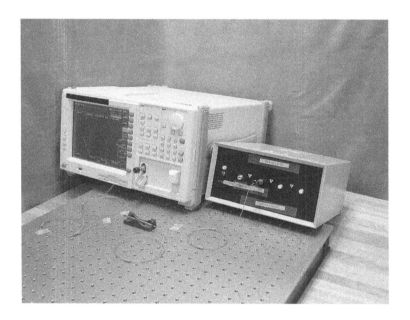

6.1 INTRODUCTION

A variety of passive fibre-optic devices are used in optical fibre communication systems to perform specific tasks. Passive devices work without using an external power supply, whereas active devices need external power to work. Passive devices split, redirect or combine light waves. This chapter describes in detail some common passive fibre-optic devices.

The simplest passive fibre-optic devices are couplers. Couplers direct multiple input light waves to multiple outputs. Normally, couplers split input signals into two or more outputs, or combine two or more inputs into one output. Couplers can have more than two inputs or outputs. Couplers work as power splitters, power taps and wavelength selectors.

Other passive devices, such as wavelength division multiplexers and de-multiplexers, filters, circulators and isolators, are also explained in detail in this chapter and Chapters 7 and 8. One section of this chapter allows the student to practice designing, manufacturing and testing a 50/50 Y-coupler in the lab. The student can plan the manufacturing process, and determine the process and handling times required to build one Y-coupler. The student will manufacture and test a Y-coupler in the lab, and test a 3 dB coupler, a 1×4 3 dB coupler, a Y-coupler, a 1×4 coupler, and a proximity sensor.

6.2 2 × 2 COUPLERS

In an optical network, there are many situations where it is necessary to combine and/or split light signals. Directional couplers form the basis of many data distribution networks. Figure 6.1 shows a four-port directional coupler with two inputs and two outputs. The arrows indicate the directions of power flow through the coupler. Power P_1 is incident on the input port, Port 1, of the coupler; the signal is divided between output ports, Ports 2 and 3, based on a set splitting ratio. Ideally, no power will reach Port 4, called the isolated port. By convention, the power P_2 emerging from Port 2 is equal to or greater than the power P_3 emerging from Port 3, depending on the designed splitting

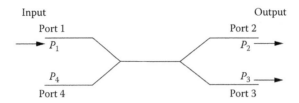

FIGURE 6.1 A four-port directional coupler.

ratio of the coupler. The splitting ratio is denoted by P_2:P_3 or else percentage of power, for example, 1:1 same as 50%/50% splits power in half.

The main parameters of couplers are optical power losses. There are several types of coupler losses, given in decibels (dB):

1. Throughput loss (L_{THP}) specifies the transmission loss between the input power P_1 at Port 1 and transmission power P_2 at Port 2. This loss is calculated by

$$L_{THP} = -10\log_{10}\left(\frac{P_2}{P_1}\right) \qquad (6.1)$$

2. Tap loss (L_{TAP}) specifies the transmission loss between the input power P_1 at Port 1 and the tap power P_3 at Port 3. This loss is calculated by

$$L_{TAP} = -10\log_{10}\left(\frac{P_3}{P_1}\right) \qquad (6.2)$$

3. Directional loss (L_D) specifies the loss between the input power P_1 at Port 1 and the isolated power P_4 at Port 4. Ideally, the isolated power loss is zero. This loss is calculated by

$$L_D = -10\log_{10}\left(\frac{P_4}{P_1}\right) \qquad (6.3)$$

4. Excess loss (L_E) specifies the power lost within the coupler. It includes radiation, scattering, absorption and coupling to the isolated port. This loss is calculated by

$$L_E = -10\log_{10}\left(\frac{P_2 - P_3}{P_1}\right) \qquad (6.4)$$

Table 6.1 lists splitting ratio values, throughput loss and tap loss for several common couplers.

TABLE 6.1

Characteristics of Several Common Couplers

Coupler Description (dB)	Splitting Ratio	L_{THP}(dB)	L_{TAP}(dB)
3	1:1	3	3
6	3:1	1.25	6
10	9:1	0.46	10
12	15:1	0.28	12

6.3   3 dB COUPLERS

A simple four-port coupler is often called a 3 dB coupler, if the input light splits into two equal portions at the output ports. The signal is split in half (3 dB = half). The 3 dB comes from the power loss formula: $[-10 \times \log_{10}(P_2/P_1) = -10 \times \log_{10}(0.5/1.0) = -10 \times (-0.3) = 3 \text{ dB}]$. Half of the light entering at Port 1 will exit at Port 2 and half at Port 3, as shown in Figure 6.2.

It is often useful to cascade many 3 dB couplers, as shown in Figure 6.3. The configuration shown is called a splitter. This splitter divides a single input into four equal outputs and is denoted as a 1 × 4 coupler. As might be expected, if the device is perfect, each output port will contain one fourth of the input power.

Figure 6.4 shows how a 1 × 8 coupler can be constructed by cascading several 3 dB couplers in a tree configuration. The signal input power will divide into eight equal outputs. Each output port will contain one eighth of the input power.

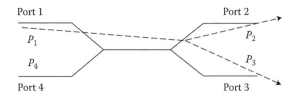

FIGURE 6.2   3 dB coupler.

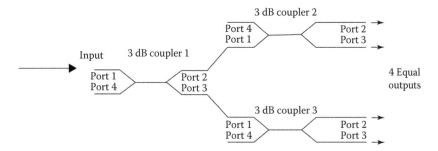

FIGURE 6.3   Cascaded 3 dB couplers to produce a 1 × 4 coupler.

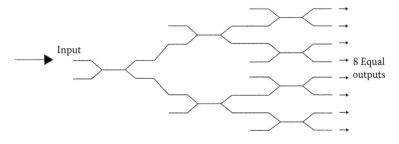

FIGURE 6.4   Cascaded 3 dB couplers to produce a 1 × 8 coupler.

6.4   Y-COUPLERS

Y-couplers or splitters, sometimes called 3 dB couplers, split the light equally. Y-couplers are 3 dB couplers in which Port 4 is not used. In the Y-coupler (splitter), as shown in Figure 6.5, the light entering Port 1 will be split equally between Ports 2 and 3 with almost no loss. They are extremely efficient at splitting light with little loss. Y-couplers are difficult to construct in fibre optics, but they are easy to construct in planar waveguide systems. The power loss in the Y-coupler system can be calculated by

$$\text{Loss (dB)} = -10 \log_{10} \left(\frac{P_{\text{out}}}{P_{\text{in}}} \right) \tag{6.5}$$

Y-couplers are very seldom built as separate planar devices; instead, they are manufactured on the same substrate as other devices. Connecting Y-couplers to a fibre-optic cable is expensive; and significant loss is experienced in the connections. However, Y-couplers of this kind are used extensively in complex planar devices.

It is often useful to cascade Y-couplers, as shown in Figure 6.6. The splitter configuration shown divides a single input into four equal outputs. If the device is perfect, each output port will contain one fourth of the input power. Figure 6.6 shows how a 1×4 Y-coupler can be constructed by cascading three 1×2 Y-couplers in a tree arrangement.

Figure 6.7 shows how a 1×8 Y-coupler can be constructed by cascading seven Y-couplers in a tree configuration. The power of the input signal will divide into eight equal outputs. Each output port will contain one eighth of the input power.

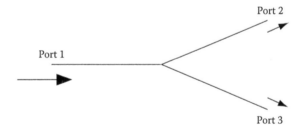

FIGURE 6.5   Y-coupler (splitter).

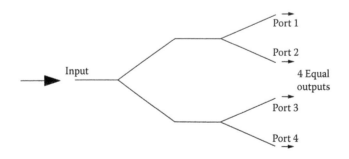

FIGURE 6.6   Cascaded 1×2 Y-couplers to produce a 1×4 coupler.

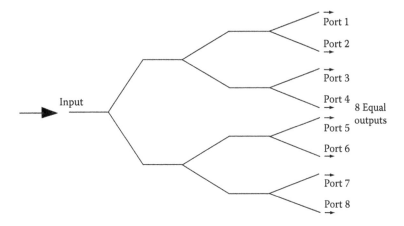

FIGURE 6.7   Cascaded Y-couplers to produce a 1 × 8 Y-coupler.

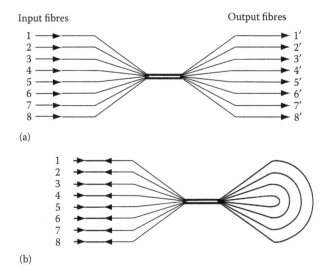

FIGURE 6.8   Star couplers: (a) transition star coupler and (b) reflection star coupler.

6.5   STAR COUPLERS

A star coupler is simply a multiple output coupler in which each input signal is made available on every output fibre. There are two star coupler designs, as shown in Figure 6.8. Figure 6.8a shows an 8 × 8 coupler. This coupler distributes the power from any input port to all of the output ports, splitting equally amongst the output ports. This type of coupler is called a transmission star coupler. Figure 6.8b shows a reflection star coupler; any input is split equally and is reflected back amongst all fibres. Star couplers are typically used in local area networks (LAN) and metropolitan area networks (MAN).

6.6   COUPLER CONSTRUCTION

In practice, many techniques can be used to construct couplers. Figure 6.9 shows the most common designs of manufactured couplers.

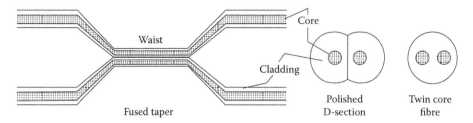

FIGURE 6.9   Some coupler configurations.

6.6.1   Fused Taper Couplers

Fused taper couplers consist of two regular single-mode fibre-optic cables that are in contact with one another, as shown in Figure 6.9. In this design, two fibre cables are heated, and then tension is applied to the ends of the fibre cables. Both the claddings and the cores of the fibre cables are drawn, and thus become thinner. During this process, the fibre cables fuse together, forming the waist. Sometimes the fibre cables are twisted tightly together before heating and stretching. Fused taper couplers are very common commercial devices.

6.6.2   Polishing D-Section Couplers

The design of polishing D-Section couplers involves embedding fibre cables into a solid material such as plastic. The cladding surface is polished along the length of the fibre cables until one side of the cladding is flat and removed to within 4 μm of the core. A D-shaped fibre cable section results, as shown in Figure 6.9. The plastic is then dissolved away. The two sections are joined along their flat surfaces, using index matching epoxy resin. Sometimes, the cladding on the fibre cables is thinned-out by etching with hydrofluoric acid, to reduce the amount of polishing. This is a relatively precise technique. However, it is more expensive to make than the fused taper method.

6.6.3   Twin Core Fibre Couplers

This design uses twin cores having a very small separation, as shown in Figure 6.9. The manufacturing cost of the twin core fibre cable is very low compared to the cost of other designs. However, it is more difficult to connect twin-core couplers to regular fibre cables than it is for other coupler designs.

6.7   THE PRINCIPLE OF RECIPROCITY

Reciprocity is a principle that applies to optical couplers and similar optical devices. The principle states that couplers work in both directions (forward and reverse) symmetrically. In Figure 6.10, a resonant coupler is set up to split input from Port 1 equally between Ports 2 and 3. The fused tapered directional coupler is designed to provide low-loss signal coupling with a range of splitting ratios. The construction of this coupler is described in Figure 6.9.

In multi-mode fibre cables, coupling occurs because higher-ordered modes no longer strike the core-cladding interface beyond the critical angle in the tapered regions. These modes are trapped by total reflection at the outer surface of the cladding; they have been converted into cladding modes. Rays from lower-ordered modes do not travel near the critical angle and are not converted. Power associated with these modes remains in the originating fibre. As the fused waveguides in Figure 6.10 share the same cladding, power from higher-ordered input modes is common to both fibre cables. The output tapers convert the cladding modes back into core-guided waves. The splitting ratio depends on the length of the taper (L) and the cladding thickness.

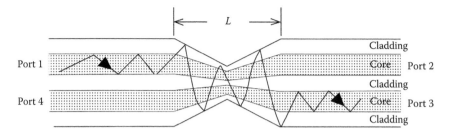

FIGURE 6.10   Fused biconically tapered directional coupler.

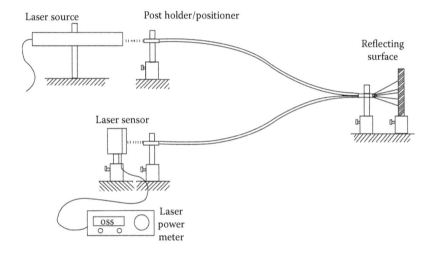

FIGURE 6.11   Proximity sensor using a fibre-optic cable bundle.

6.8   PROXIMITY SENSOR

Fibre optic cables can be used as sensors. An example of a proximity sensor that uses a bifurcated (Y-branched) fibre bundle is shown in Figure 6.11. Optical power from a light source is launched into one arm of the fibre bundle. It is reflected from a surface (reflecting mirror, white surface, black surface, mat colour surface, etc.) at the output end of the fibre cable bundle. A portion of the reflected light enters the second arm of the bundle where it is routed to the sensor. The amount of light sensed by the sensor depends on the distance between the end of the bundle and the reflecting surface. A fibre-optic bundle can be manufactured, tested and used in the experiment as a proximity sensor.

6.9   MACH–ZEHNDER INTERFEROMETER

Mach–Zehnder interferometers (MZIs) are used in a wide variety of applications within optics and fibre communication systems. Wavelength division multiplexers, opto-mechanical switches and modulators can be made using MZI techniques. Mach–Zehnder interferometers can be either passive or active devices.

   The basic structure of the MZI has the balanced configuration of a splitter and combiner, connected by a pair of matched waveguides or couplers. The MZI consists of three parts: (1) an input Y-coupler, which splits the signal into two beams; (2) the arms and (3) an output Y-coupler, which recombines the two components of the signal to regenerate the input signal. MZI operations and applications are generally presented in opto-mechanical switches.

6.10 OPTICAL ISOLATORS

Optical isolators are used in a wide variety of applications within fibre communication systems. Connectors, splices and other optical components are generally presented in fibre optic connectors. Return lights are generated from the various parallel optical surfaces and the end faces of these devices. Return lights are known as back reflections. Back reflections have a destabilising effect on oscillation of the laser source and on the operation of fibre-optic amplifiers, thus resulting in poor transmission performance in the data lines. Optical isolators are used to block the back reflections.

Figure 6.12 illustrates the isolator operation. An optical isolator is composed of a magnetic garnet crystal acting as a Faraday rotator; a permanent magnet for applying a designated magnetic field; and polarising elements that permit only forward light to pass, while shutting out backward light. For this reason, optical isolators are indispensable devices for eliminating the adverse effect of back reflection in fibre-optic systems. The optical isolator consists of two polarisation elements and a 45° Faraday rotator placed between the polarisation elements. The polariser and the analyser have a 45 degree difference in the direction of their light transmission axes. Forward light passing the optical isolator undergoes the following:

- When passing through the polariser, the incident light is transformed into linearly polarised light.
- When passing through the Faraday rotator, the polarisation plane of the linearly polarised light is rotated 45 degree.
- This light passes through the analyser without loss, because the light polarisation plane is now in the same direction as the light transmission axis of the analyser, which is tilted 45 degree from the polariser in the direction of Faraday rotation.

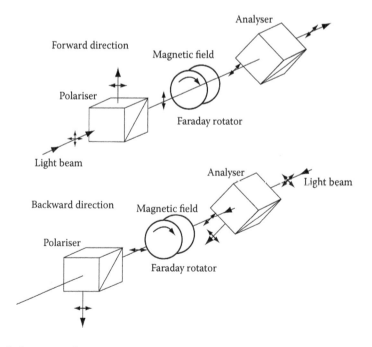

FIGURE 6.12 Isolator operation.

On the contrary, backward light undergoes a slightly different process:

- When passing through the analyser, the backward light is transformed into linearly polarised light with a 45 degree tilt in the transmission axis.
- When passing through the Faraday rotator, the polarisation plane of the backward light is rotated 45 degree in the same direction as the initial tilt.
- This light is completely shut out by the polariser because its polarisation plane is now tilted 90 degree from the light transmission axis of the polariser.

6.11 OPTICAL CIRCULATORS

Optical circulators are used in a wide variety of applications within fibre communication systems. In advanced optical communication systems, optical circulators are used for bi-directional transmissions, wavelength division multiplexing (WDM) networks, fibre amplifier systems, optical time domain reflectrometers (OTDR) and so on. Optical circulators are non-reciprocal devices that redirect a signal from port to port sequentially, in only one direction. The operation of an optical circulator is similar to that of an optical isolator; however, its construction is more complex. Figure 6.13a shows a three-port optical circulator. An input signal (λ_1) at Port 1 exits at Port 2, an input signal (λ_2) at Port 2 exits at Port 3 and an input signal (λ_2) at Port 3 exits at Port 1. Similarly, in a four-port optical circulator, as shown in Figure 6.13b, one could ideally have four inputs and four outputs. In practice, many applications do not need four inputs and four outputs. Therefore, in a four-port circulator, it is common to have three input ports and three output ports. This is done by making Port 1 an input-only port, Ports 2 and 3 input and output ports and Port 4 an output-only port. A typical application of the isolators is demonstrated in Figure 6.13. Input signals of different wavelengths are circulated to the next port in a clockwise direction.

There are many different circulator designs available for industrial applications. Circulators are commonly made of an assembly of the following optical components: polarising beam splitter, reflector prism, bi-refringent blocks, Faraday rotator and retardation plate.

Figure 6.14 shows the basic operation of such a three-port circulator for a light beam entering at Port 1 and exiting at Port 2. The light input at Port 1 is split into two separate orthogonal polarisation rays, which are called the ordinary (O) and extraordinary (E) components. The ordinary component passes through bi-refringent block 1 without refraction. The extraordinary component is refracted when it passes through bi-refringent block 1. Both O and E components pass through the Faraday rotator and phase retardation plates. As a result, both components are rotated 90 degree clockwise. The component that was the O in block 1 becomes the E component. The E component in block 1 becomes the O component. The two components meet bi-refringent block 2, which is similar to block 1. Then, the two components re-combine to form the original light ray when exiting block 2, which enters Port 2, as shown in Figure 6.14.

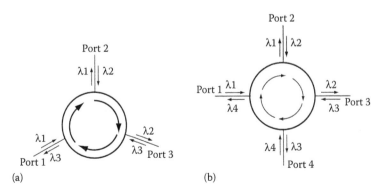

(a)   (b)

FIGURE 6.13 Circulator principle: (a) a three-port circulator and (b) a four-port circulator.

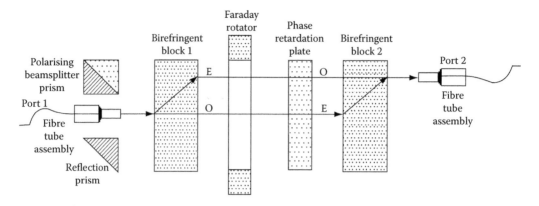

FIGURE 6.14 Optical circulator directs light beam from Port 1 to Port 2.

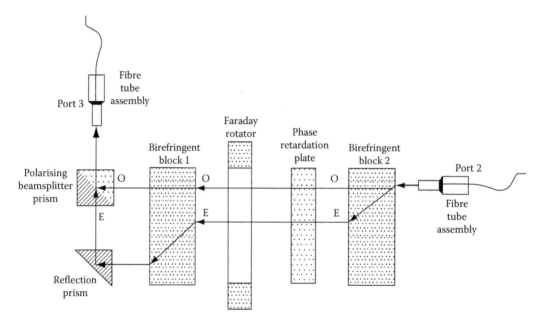

FIGURE 6.15 Optical circulator directs light beam from Port 2 to Port 3.

On the other hand, a light beam entering at Port 2 will exit from Port 3, as shown in Figure 6.15. The light entering Port 2 is split by bi-refringent block 2 into two separate rays, similar to the light passing through bi-refringent block 1. The ordinary component passes through bi-refringent block 2 without refraction. The E component is refracted when it passes through bi-refringent block 2. The two components meet bi-refringent block 1, the O component passes through to the polarising beam splitter, and the E component refracts towards the reflector prism. The two components recombine at the polarising beam splitter prism to form the original light ray, which exits from Port 3. Using the same principles, light can be traced when it enters at Port 3 and exits at Port 1.

6.12   OPTICAL FILTERS

Optical filters are used in a wide variety of applications within optics and optical fibre devices. They are also used in a wide variety of optical applications in the fields of microscopy, photometry, radiometry, imaging, instrumentation, displays, charge-coupled devices (CCDs), astronomy, aerospace and so on. Scientific, electronic, analytical, imaging and medical instrument companies are designing the next evolution of their products using a range of selected optical filters. There are many devices that are not called filters but have the same characteristics as a filter. Such devices are switches, modulators, array waveguide gratings, grating diffractions and grating multiplexers. These devices are presented in detail in different chapters throughout the book.

Optical filters are devices that allow specific wavelengths to pass while rejecting all other wavelengths. Optical filters can be divided into two categories: fixed filters and tunable filters.

6.12.1 Fixed Optical Filters

Fixed optical filters are commonly made from coloured glass (silica), thin metallic films or thin dielectric films, as shown in Figure 6.16. Some metallic films, such as inconel, chromium and nickel, are particularly insensitive to wavelength for absorption. On the other hand, the amount of absorption by coloured glass can vary as much as several orders of magnitude over only tens of nanometres of wavelength.

There is an extensive range of optical filter types. The following filters can be found in the market: visible filters, combination filters, infrared radiation cut-off filters, narrow band pass filters, calibration filters, laser application filters, ultraviolet transmitting filters, infrared transmitting filters, light balancing filters, skylight filters, sharp cut filters, contrast filters, colour temperature conversion filters, special application filters and so on.

Optical filter selection depends on a multitude of factors, including wavelength selection, shape and pass band width, blocking outside the pass band, transmittance colour matching, material and thickness, vibration and shock resistance, ordinary and advanced anti-reflection coatings, heat absorption and long-term stability.

FIGURE 6.16 Different types of filters.

6.12.2 Tunable Optical Filters

Tunable optical filters are versatile devices that are used in many photonic applications. They are essential in wavelength-flexible WDM systems, and they can also play a key role in wavelength-tunable lasers for WDMs.

There is an extensive range of optical tunable filter types. The following filters can be found in the market:

- Micron optic fibre Fabry–Perot (FFP) tunable filters.
- Digitally tunable optical filters based on dense wavelength division multiplexer (DWDM), thin film filters and semiconductor optical amplifiers.
- Narrowband tunable optical filters using fibre Bragg gratings.
- High-speed tunable optical filters using a semiconductor double-ring resonator.
- Wide-bandpass tunable optical filters.
- Micro-machined in-plane tunable filters using the thermo-optic effect of crystalline silicon.
- Acousto-optic tunable filters (AOTFs).
- Liquid crystal tunable filters (LCTFs).
- Others not listed here.

Some common types of tunable filters will be explained in detail in the following section.

Optical tunable filter selection depends on various factors, such as fast tuning speed, simple control mechanism, being scalable without additional insertion loss and long-term operation temperature stability.

6.12.2.1 Fibre Fabry–Perot Tunable Filters

The FFP tunable filter principle is based on Fabry–Perot etalon technology. An FFP tunable filter passes wavelengths that are equal to integer fractions of the cavity (etalon) length; all other wavelengths are attenuated or reflected back. The key to the design of the FFP tunable filter is its lensless fibre construction. There are no collimating optics or lenses; thus, the FFP tunable filter achieves high precision, maintains low loss and good transmission profile. Figure 6.17 shows a cross-section of an FFP tunable filter design. The design has two pieces of fibre, the ends of which are polished and silvered, so that each end acts like a mirror. The ends are placed precisely opposite to one another with a specific gap between them. The fibre assemblies are mounted on two piezoelectric crystals and packaged in a box. By applying a voltage across the crystals, the distance between the fibre ends changes, thus changing the resonant cavity length, and therefore, a change in the wavelength selection.

This design of tunable filter eliminates the pitfalls of other Fabry–Perot component technologies, including misalignment and environmental sensitivity. Fibre Fabry–Perot tunable filters have low

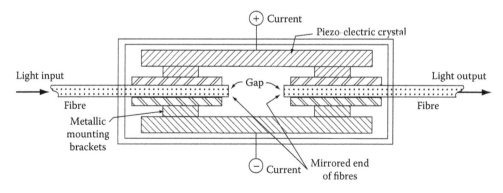

FIGURE 6.17 Fibre Fabry–Perot (FFP) tunable filter.

loss, high isolation, long-term alignment stability, high reliability and accurate power or wavelength measurements. Fibre Fabry–Perot tunable filters are used in optical performance monitoring, tunable optical noise filtering, dropping of a tunable channel for ultra dense WDM and so on.

The design of FFP tunable filter can be modified by putting a liquid crystal material into the gap between the ends of the two fibres. The index of refraction of the liquid crystal can be changed very quickly, by passing current through the liquid crystal. By changing the index of refraction of the crystal, a change in the wavelengths passing through the crystal can be achieved, thus eliminating unwanted wavelengths.

6.12.2.2   Mach–Zehnder Interferometer Tunable Filters

The Mach–Zehnder interferometer tunable filter has a ladder like structure, in which each section resembles a Mach–Zehnder interferometer, as shown in Figure 6.18. The output waveguide (across the top) and the input waveguide (across the bottom) are joined by regularly spaced linking waveguides, each longer than the previous one by ΔS, as illustrated in Figure 6.18. For constructive interference to occur at each coupler in the output waveguide, ΔS must be equal to an integral number of wavelengths of the input light. However, similar to a Mach–Zehnder that can be tuned by adjusting the refractive index of one or both arms, this filter can be tuned by adjusting the refractive index of the arms with an injected current. More details are presented on the Mach–Zehnder interferometer in Section 6.9 opto-mechanical switches.

6.12.2.3   Fibre Grating Tunable Filters

Fibre grating tunable filters are used in a wide variety of applications within optics and fibre communication systems. They are an important element in wavelength division multiplexer systems for combining and separating individual wavelengths. Fibre Bragg grating (FBG) transmits one wavelength and reflects all others. Basically, a grating is a periodic structure within an optical material. This variation in the structure of the optical material reflects or transmits light in a certain direction depending on the light wavelength. Therefore, gratings can be categorised as either transmitting or reflecting.

Figure 6.19 shows a design of an in-fibre Bragg grating tunable filter. The filter contains two Bragg gratings and a four-port circulator. The gratings have high reflectance in wavelength bands at specified wavelengths. The Bragg grating fibre is glued to the piezoelectric crystal contacts. Current can be applied on one or two Bragg gratings. The current deforms the crystal, stretching the gratings to match the wavelength of the signals. The wavelength filtered depends on the current level. This type of filter is used in aircraft or space borne differential absorption systems that measure water vapour in Earth's atmosphere. It is also used for a unique optical receiver that couples a laser radar signal from a telescope to the in-fibre Bragg grating filter.

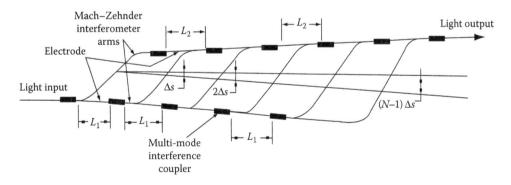

FIGURE 6.18   Mach–Zehnder interferometer tunable filter.

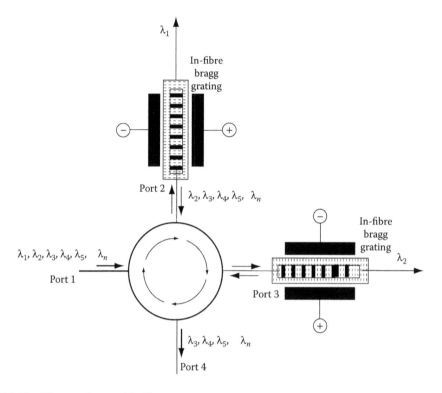

FIGURE 6.19   Fibre grating tunable filter.

6.12.2.4   Liquid Crystal Tunable Filters

Liquid crystal tunable filters (LCTFs) use electrically controlled liquid crystal elements, which select a specific visible wavelength of light for transmission through the filter. A typical wavelength-selective LCTFs is constructed from a stack of fixed filters that consist of interwoven bi-refringent crystal/liquid-crystal combinations and linear polarisers. The spectral region transmitted through an LCTF depends on the choice of polarisers, optical coatings and the liquid crystal characteristics (nematic, cholesteric, smectic, etc.). In general, visible-wavelength devices of this type usually perform in the range of 400–700 nm. This type of filter is ideal for use with electronic imaging devices, such as CCDs, because it offers excellent imaging quality with a simple optical pathway.

6.12.2.5   Acousto-Optic Tunable Filters

Acousto-optic tunable filters (ATOFs) apply the same technology used in acousto-mechanical switches. An incident beam of light impacts a dioxide crystal of an AOTF. The dioxide crystal is sandwiched between an acoustic transducer and absorber that can be regulated by the acoustic power and acoustic frequency sliders. Upon encountering the standing wave in the dioxide crystal, a portion of the incident light beam is diffracted into the output port, whereas the remainder of the beam passes through the crystal and is absorbed by a beam stop. As the slider moves, the amplitudes of the waves passing through the AOTF are increased or decreased. Wavelength selection is controlled by the acoustic frequency slide. ATOFs are employed to modulate the wavelength and amplitude of incident laser light.

6.12.2.6   Thermo-Optic Tunable Filters

Thermo-optic tunable filters apply the thermo-optic effect on an optical material, such as crystalline silicon. Changing the temperature results in a change in the index of refraction of the material.

For example, current applied to a resistive element creates heat that increases the temperature of the filter. Thus, by applying current, wavelength selection can be achieved. This type of filter is used for spectroscopy and in optical communication systems.

6.12.2.7 Other Types of Tunable Filters

There are many other types of tunable filters used in building various optical devices and systems. Many tunable filters perform specific functions. Tunable filters are also available in the market, such as wide-band-pass filters, gas tunable filters, active optical filters, volume holographic grating-based filters, digitally tunable filters based on DWDM thin film filters and high-speed tunable filters using a semiconductor double-ring resonator.

6.13 OPTICAL FIBRE RING RESONATORS

Optical fibre ring resonators are popular in communication systems. These devices are used in wavelength filtering, routing, switching, modulation and multiplexing/demultiplexing applications. They are sometimes called filters based on optical ring resonators. There are many arrangements for coupling ring resonators to achieve the desired function. Ring resonators are either passive or active (when power is applied).

Optical ring resonators consist of a waveguide, such as a fibre, in a closed loop coupled to one or more input/output directional couplers for incoupling and outcoupling, as shown in Figure 6.20. When light of the appropriate wavelength is coupled into the loop via the input waveguide, the light builds up in intensity over multiple round-trips due to constructive interference at the coupler. As only specific wavelengths resonate in the loop, it functions as a filter. Wavelength selection depends on the ring loop length being an integral number of wavelengths.

6.14 OPTICAL MODULATORS

There are many devices that are not called optical modulators but which have the characteristics of modulators. Such devices include switches, filters, array waveguide gratings, grating diffractions and grating multiplexers. These devices are presented in detail in different chapters throughout the book.

Optical modulators also use materials that change their optical properties under the influence of an electric, acoustic or magnetic field; electro-absorption; vibration; light or heat. Optical modulators also use the DWDM and in-fibre Bragg grating techniques.

An optical fibre modulator, which operates by using magnetic fluids, is a common example. The modulator consists of a bare fibre core surrounded by magnetic fluids, instead of a SiO_2 cladding layer. When applying a magnetic field, the index of refraction of the magnetic fluid changes. When light propagates along the fibre, controlling the occurrence of total reflection at the interface between the fibre core and the magnetic fluid is very rapid. As a result, the intensity of the output light is modulated by varying the magnetic field strength. This allows only specific wavelengths to pass through the fibre, reflecting the rest.

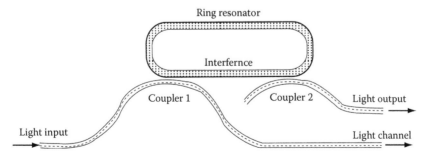

FIGURE 6.20 Optical fibre ring resonator.

6.15   OPTICAL ATTENUATORS

Optical attenuators are devices that decrease the optical power in a fibre by a fixed or adjustable amount. Optical attenuators can be divided into two categories: fixed and variable. They are widely used in fibre optic telecommunication systems. The basic principle of operation of a common attenuator is controlled by the offset of the input and output fibres, or an air gap between the input and output fibres. Other techniques are used in the operation of advanced optical attenuators; they use temperature, twisting, tension or current acting on some type of optical material. Some of these techniques are explained throughout the book.

Optical attenuator selection depends on a multitude of factors, including wavelength, attenuation levels and temperature stability.

6.15.1   Fixed Attenuators

Once the desired level of attenuation is determined, fixed attenuators can be set to deliver a precise power output. Fixed attenuators are used to reduce the optical power transmitted through fibre-optic cables. The most common uses include equalising power between fibres and multi-fibre systems and reducing receiver saturation. They are available in plug and in-line styles.

6.15.2   Variable Attenuators

Variable attenuators allow a range of adjustability, delivering a precise power output at multiple decibel loss levels. Variable attenuators produce precise levels of attenuation, with flexible adjustment. By using simple adjustment controls, the attenuation can be easily modified to any level, as shown in Figure 6.21. They are available with other connector and in-line styles.

6.16   OPTICAL SPECTROMETERS

Optical spectrometers are devices that disperse the incoming light into spectrum by a prism or a grating, as shown in Figure 6.22. Isaac Newton's first scientific paper, published in 1672, described

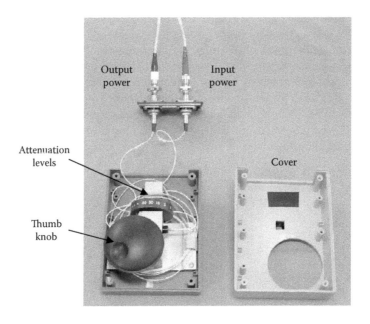

FIGURE 6.21   Inside a variable attenuator.

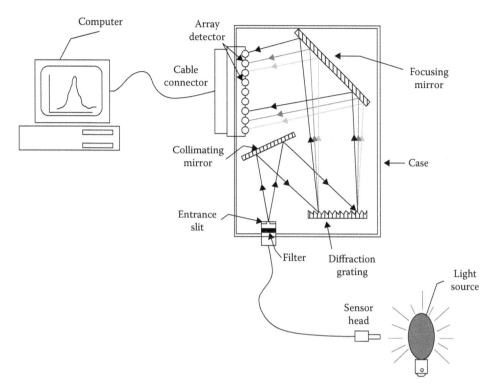

FIGURE 6.22 Optical secptrometer analyser.

his experiments with light and colour. Newton passed a beam of sunlight through a prism, spreading the light into a spectrum of colours. Newton found that when white light is passed through a prism, a spectrum of colours is observed. This separation of white light into its constituent spectral wavelengths is called chromatic dispersion. In real life, dispersion can be observed in rainbows, on the surface of CD's and an oil layer on water.

Spectrometers are used in many fields. For example, they are used in astronomy to analyse the radiation from astronomical objects and deduce chemical composition. This allows astronomers to detect many of the chemical elements by their characteristic spectral fingerprints. If the object is glowing by itself, it will show spectral lines caused by the glowing gas itself. Chemical compounds may also be identified by absorption. Typically these are dark bands in specific locations in the spectrum caused by energy being absorbed as light from other objects passes through a gas cloud. Much of our knowledge of the chemical makeup of the universe comes from spectra. Optical spectrum devices are also widely used by fire brigades. It detects and identifies a light wave coming from a fire source. Then the brigades can easily fight the fire by a suitable fire distinguisher. Also, it is used for air and water quality analysis, agricultural research, metallurgical analysis, polymer analysis and biotechnology applications.

The following steps explain manufacturing processes of an optical spectrometer device, as shown in Figure 6.23:

1. Prepare quantity of the required components to manufacture the device.
2. Check and clean the base and the lid.
3. Fix an entrance slit to the base.
4. Apply an epoxy and mount a focusing mirror in the slot in base.
5. Apply an epoxy and mount a collimating mirror in the slot in base.
6. Leave the device in the oven to cure the epoxy for a certain time, depends on the type of epoxy.

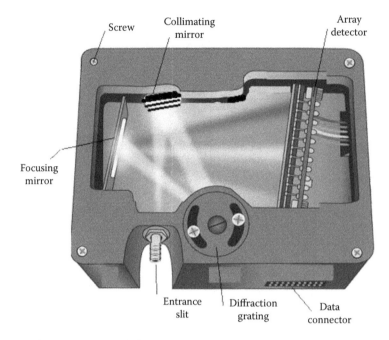

FIGURE 6.23 Optical spectrometer device.

7. Mount a diffraction grating in the hole in the base.
8. Mount an array detector to the base.
9. Connect the array cables to data connector.
10. Connect the data connector to a computer.
11. Connect an external light source with known spectrum.
12. Put on the computer.
13. Test the device by adjusting the diffraction grating to get the required spectrum on the computer screen.
14. Keep fine adjustments of the diffraction grating until the device spectrum matches the standard spectrum on the computer screen.
15. Add black silicon rubber to the base lip.
16. Mount the cove and fastened by four screws.
17. Clean the outside of the device by liquid cleaner matching the black silicon rubber.
18. Test the device for accuracy and consistency. Check the results with the first test before the cover packaging.
19. The device can be checked by quality control to match the specifications and check the paper attachments for shipping. Quality control process can be applied at any sections in the manufacturing processes.
20. Packaging the device with customer requirements.

6.17   EXPERIMENTAL WORK

In this experiment, the student will measure power levels at the input and output ports, calculate the power losses and determine the splitting ratio for optical couplers in the following cases:

1. Testing a 3 dB coupler
2. Testing a 1×4 3 dB coupler
3. Manufacturing a Y-coupler

4. Testing a Y-coupler
5. Testing a 1 × 4 Y-coupler
6. Testing a proximity sensor

6.17.1 TECHNIQUE AND APPARATUS

Appendix A presents the details of the devices, components, tools and parts.

1. 2 × 2 ft optical breadboard.
2. He–Ne laser light source and power supply.
3. Laser light sensor.
4. Laser mount assembly.
5. 20X microscope objective lens.
6. Lens/fibre cable holder/positioner.
7. Hardware assembly (clamps, posts, screw kits, screwdriver kits, sundry positioners, etc.)
8. Fibre-optic cable.
9. Fibre cable end preparation procedure and kit, and cleaning kit, as explained in the Fibre-Optic Cables Chapter, at the end preparation section.
10. Fibre cable holder/positioner assembly.
11. 3 dB coupler, as shown in Figure 6.24.
12. Y-Coupler, as shown in Figure 6.25. [Note: for Case (f), a Y-coupler (splitter) is used as a proximity sensor, as shown in Figure 6.30].
13. Reflecting surfaces (mirror, white card and black card).
14. Ruler.

6.17.2 PROCEDURE

Follow the laboratory procedures and instructions given by the professor and/or instructor.

6.17.3 SAFETY PROCEDURE

Follow all safety procedures and regulations regarding the use of fibre-optic cable. You must wear safety glasses and finger cots or gloves when working and handling fibre-optic cables, optical components, epoxy and optical cleaning chemicals.

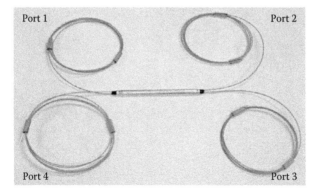

FIGURE 6.24 3 dB coupler.

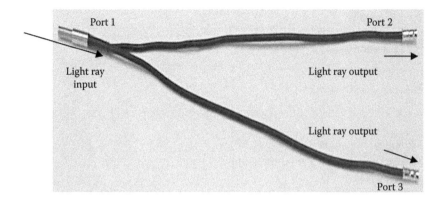

FIGURE 6.25   Y-coupler.

6.17.4   Apparatus Set-Up

6.17.4.1   Testing a 3 dB Coupler

1. Figure 6.26 shows the experimental set-up.
2. Bolt the laser short rod to the breadboard.
3. Bolt the laser mount to the clamp using bolts from the screw kit.
4. Put the clamp on the short rod.
5. Place the He–Ne laser into the laser mount and tighten the screw. Turn on the laser device. Follow the operation and safety procedures of the laser device in use.
6. Check the laser alignment with the line of bolt holes and adjust when necessary.
7. Mount a lens/fibre cable holder/positioner to the breadboard, so that the laser beam passes through its centre hole.

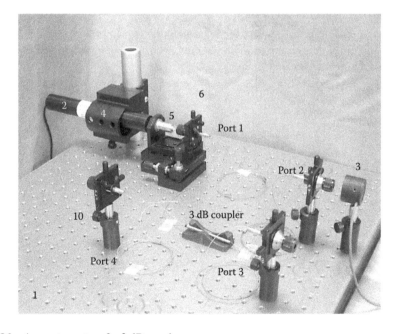

FIGURE 6.26   Apparatus set-up for 3 dB coupler.

TABLE 6.2

3 dB Coupler Test

	Input Power (unit)	Output Power (unit)			Calculated Loss (dB)			
	Port 1	Port 2	Port 3	Port 4	L_{THP}	L_{TAP}	L_D	L_E
Splitting Ratio								

8. Add the 20× microscope objective lens to the lens/fibre cable holder/positioner. It is better to use a 20× microscope objective lens to focus the collimated laser beam onto the input port.

9. Prepare the end of each port with a good cleave, as described in the end preparation procedure of the Fibre-Optic Cables chapter.

10. Mount Port 1 into the lens/fibre cable holder/positioner, as explained in the Fibre Optic Cables Chapter.

11. Mount Ports 2–4 of the 3 dB coupler into a fibre cable holder/positioner, one for each port.

12. Align the laser source with Port 1 of the 3 dB coupler.

13. Verify the alignment of your laser beam launching arrangement, by making sure that the 3 dB coupler input port remains at the centre of the laser beam.

14. Measure the input power of the laser before it enters Port 1 of the 3 dB coupler.

15. Place the laser sensor in front of Ports 2, 3 or 4 of the 3 dB coupler.

16. Measure the output power from Ports 2–4 of the 3 dB coupler. Fill out Table 6.2.

6.17.4.2 Testing a 1 × 4 3 dB Coupler

A 1 × 4 coupler can be built by cascading three 3 dB couplers, as shown in Figures 6.3 and 6.27. To connect the 3 dB coupler ports, a mechanical splice can be used, as shown in Figure 6.27. Fibre connectors and splices technique is presented in the Fibre Connectors Chapter. Continue the procedure explained in Section 6.17.4, adding the following steps:

1. Figure 6.27 shows the experimental set-up.

2. Add two fibre cable holder/positioners to the breadboard.

3. Mount the output ports of 3 dB coupler 2 into a fibre cable holder/positioner one for each port.

4. Use mechanical splices to connect Port 2 of 3 dB coupler 1 to Port 1 of 3 dB coupler 2, as shown in Figure 6.27.

5. Place the laser sensor in front of Ports 2, 3 or 4 of 3 dB coupler 2.

6. Measure the power at each output port of cascaded 3 dB coupler 2. Fill out Table 6.3.

7. Repeat steps 4–6, by adding 3 dB coupler 3 to the set-up, to form a 1 × 4 coupler. This can be constructed by connecting Port 3 of 3 dB coupler 1 to Port 1 of 3 dB coupler 3. Fill out Table 6.3.

6.17.4.3 Manufacturing a Y-Coupler in the Lab

This case enables students to practice designing, manufacturing, and testing a 50/50 Y-coupler in the lab. Students can plan the manufacturing process and determine the process and handling times required to build one Y-coupler, as shown in Figure 6.25. The students can also list tools,

---

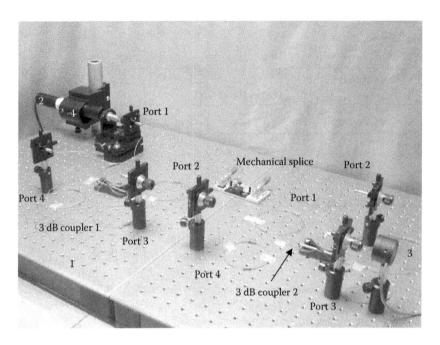

FIGURE 6.27 Apparatus set-up for 1 × 4 coupler.

TABLE 6.3
1 × 4 3 dB Coupler Test

		3 dB Coupler 1						
	Input Power (unit)	Output Power (unit)			Calculated Loss (dB)			
	Port 1	Port 2	Port 3	Port 4	L_{THP}	L_{TAP}	L_D	L_E
Splitting Ratio								

		3 dB Coupler 2						
	Input Power (unit)	Output Power (unit)			Calculated Loss (dB)			
	Port 1	Port 2	Port 3	Port 4	L_{THP}	L_{TAP}	L_D	L_E
Splitting Ratio								

(*Continued*)

TABLE 6.3 (*Continued*)
1 × 4 3 dB Coupler Test

	3 dB Coupler 3							
	Input Power (unit)	Output Power (unit)			Calculated Loss (dB)			
	Port 1	Port 2	Port 3	Port 4	L_{THP}	L_{TAP}	L_D	L_E
Splitting Ratio								

instruments, and parts requirements and design a floor plan for a production line for such a product. The manufactured Y-coupler can be used in the lab both as a self-sufficient product and a testing tool. The following steps explain the manufacturing process for a Y-coupler of 30 cm long (the length of the end is 5 cm and the arems are 25 cm each):

1. Prepare fibre cable, heat shrink tubing, epoxy, polishing pads, and tools.
2. Cut ten lengths of fibre-optic cable, 34 cm long (add 4 mm to the required length, to give allowance for the polishing steps). Make sure that all lengths are equal.
3. Cut two lengths 25 cm and one length 5 cm of heat shrink tubing. Choose a suitable heat shrink tubing diameter to accommodate the ten and five fibre cables.
4. Insert the five fibre cables through the 25 cm long heat shrink tube. Make sure that all the fibre cables are even at the ends and they are slightly sticking out by 2 mm of the heat shrink tube.
5. Apply heat, using an air heat gun, to shrink the tube tidily around the fibre cables.
6. Repeat steps 4 and 5 to make another Y-coupler arm.
7. Polish the end of the Y-coupler arms at the heat shrink tube side. Follow the polishing process, as explained in the fibre-optic cables and fibre-optic connector chapters.
8. Rinse the Y-coupler arm ends with water.
9. Inspect the Y-coupler arm ends using an inspection microscope. Look at the quality of the polished end of the Y-coupler arms.
10. Combine the two sides of the Y-coupler arms (5 cm).
11. Insert the ten fibre cables (two sides of the Y-coupler arms [5 cm]) through the 5 cm long heat shrink tube. Make sure that all the fibre cables are even at the ends and they are slightly sticking out by 2 mm of the heat shrink tube.
12. Apply heat, using an air heat gun, to shrink the tube tidily around the ten fibre cables.
13. Polish the 5 cm end of the Y-coupler arm. Follow the polishing process, as explained in the fibre-optic cables and fibre optic connector chapters.
14. Inspect the Y-coupler end using an inspection microscope. Look at the quality of the polished end of the Y-coupler.
15. Add a metal tube on each end of the Y-coupler (head and arms) for coupling purposes. Fix the metal tubes using an epoxy. Follow the epoxy instructions for the curing time.
16. Polish each end of the Y-coupler arms for a final accurate length of the Y-coupler.
17. Use the microscope to inspect the polishing quality of the ends. The polishing process may need repeating, to get a mirror-finished surface on each end.
18. Test the Y-coupler to verify the splitting ratio and calculate power loss.

6.17.4.4 Testing a Y-Coupler

As explained in Section 6.4, a Y-coupler can be tested, as shown in Figure 6.28.

1. Figure 6.28 shows the apparatus set-up.
2. Install the laser source as explained Case (a).
3. Mount one post holder/positioner in front of the laser source. Check the laser alignment, so that the laser beam passes over the post's centre hole.
4. Insert input Port 1 of the Y-coupler into the post holder/positioner.
5. Mount the two output ports of the Y-coupler into the posts.
6. Verify the alignment of the laser light launching arrangement, by making sure that input Port 1 of the Y-coupler remains at the centre of the laser beam.
7. Measure the laser input power before it enters the Y-coupler. Fill out Table 6.4.
8. Place the laser sensor in front of the output ports of the Y-coupler.
9. Measure the laser output power from the output ports of the Y-coupler. Fill out Table 6.4.

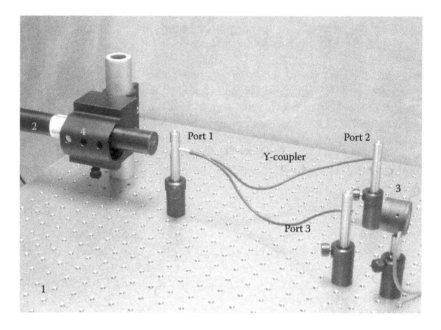

FIGURE 6.28 Apparatus set-up for Y-coupler test.

TABLE 6.4
Y-Coupler Test

	Input Power (unit)	Output Power (unit)		Calculated Loss (dB)	
	Port 1	Port 2	Port 3	LOSS$_{1-2}$	LOSS$_{1-3}$
Splitting Ratio					

6.17.4.5   Testing a 1 × 4 Y-Coupler

Continue the procedure explained in Case (d), adding the following steps, to create a 1 × 4 coupler.

1. Figure 6.29 shows the experimental set-up.
2. Add four post holder/positioners to the breadboard.
3. Couple two new Y-coupler inputs to the first Y-coupler outputs. Be sure that the coupling is good, with minimum light loss.
4. Mount the two output ports of the two new Y-couplers into the posts, as shown in Figure 6.29.
5. Measure the input power of the laser beam before it enters the cascaded Y-coupler. Fill out Table 6.5.
6. Measure the power at each output port of the cascaded Y-coupler. Fill out Table 6.5.

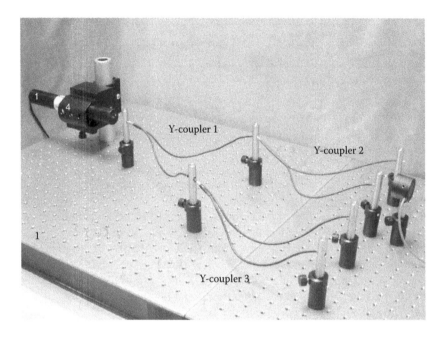

FIGURE 6.29   Cascaded Y-couplers to produce a 1 × 4 couple apparatus set-up.

TABLE 6.5

1 × 4 Y-Coupler Test

	Y-Coupler 1				
	Input Power (unit)	Output Power (unit)		Calculated Loss (dB)	
	Port 1	Port 2	Port 3	Loss$_{1-2}$	Loss$_{1-3}$
Splitting Ratio					

(Continued)

TABLE 6.5 (*Continued*)
1 × 4 Y-Coupler Test

	Y-Coupler 2				
	Input Power (unit)	Output Power (unit)		Calculated Loss (dB)	
	Port 1	Port 2	Port 3	Loss$_{1-2}$	Loss$_{1-3}$
Splitting Ratio					

	Y-Coupler 3				
	Input Power (unit)	Output Power (unit)		Calculated Loss (dB)	
	Port 1	Port 2	Port 3	LOSS$_{1-2}$	LOSS$_{1-3}$
Splitting Ratio					

6.17.4.6 Testing a Proximity Sensor

For the purpose of the lab experiment set-up, a proximity sensor can be made from a Y-coupler (splitter). Continue the procedure explained in Case (d), adding the following steps, to use a proximity sensor.

1. Rotate the Y-coupler 180° so that it works as a proximity sensor, as shown in Figure 6.30.
2. Figure 6.31 shows the experimental set-up.
3. Mount Port 1 of the Y-coupler into the post, so that Port 1 is facing the laser beam.
4. Align the laser beam with Port 1 of the proximity sensor.
5. Mount Ports 2 and 3 of the proximity sensor into a post for each port, as shown in Figure 6.31.
6. Measure the input power of the laser beam before it enters Port 1 of the proximity sensor.
7. Place a mirror and mirror holder at a distance (*d*) from Port 2 of the proximity sensor, such that the mirror reflects the light back into the proximity sensor.
8. Measure the output power at Port 3. Fill out Table 6.6.
9. Repeat steps 7 and 8 for at least five distances, measuring the output power for each distance.
10. Repeat steps 7–9 using white and black reflecting surfaces.

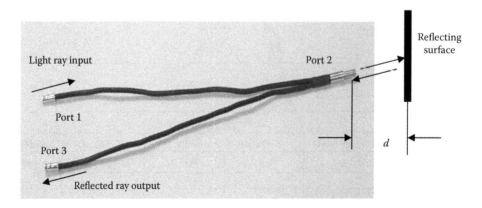

FIGURE 6.30   A proximity sensor.

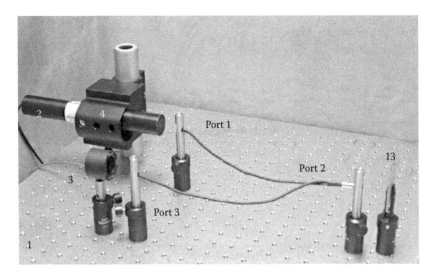

FIGURE 6.31 Proximity sensor apparatus set-up.

TABLE 6.6
Proximity Sensor Test

(I) Mirror

Input Power Port 1 (unit) =		
Output Power Port 2 (unit) =		
Calculated Loss Loss$_{1-2}$ (dB) =		
	Output Power (unit)	**Calculated Loss (dB)**
Distance (d) (unit)	**Port 3**	**Loss$_{1-3}$**

(II) White Surface

Input Power Port 1 (unit) =		
Output Power Port 2 (unit) =		
Calculated Loss Loss$_{1-2}$ (dB) =		
	Output Power (unit)	**Calculated Loss (dB)**
Distance (d) (unit)	**Port 3**	**Loss$_{1-3}$**

(Continued)

TABLE 6.6 (*Continued*)
Proximity Sensor Test

(III) Black Surface

Input Power Port 1 (unit) =		
Output Power Port 2 (unit) =		
Calculated Loss Loss$_{1-2}$ (dB) =		
	Output Power (unit)	Calculated Loss (dB)
Distance (d) (unit)	Port 3	Loss$_{1-3}$

6.17.5 DATA COLLECTION

6.17.5.1 Testing a 3 dB Coupler

1. Measure the input power of the laser beam before it enters the 3 dB coupler. Then measure the output power at Ports 2–4 of the 3 dB coupler.
2. Fill out Table 6.2 with the collected data.

6.17.5.2 Testing a 1 × 4 3 dB Coupler

1. Measure the input power and output power of the laser beam at each port of the cascaded 3 dB couplers that produced the 1 × 4 3 dB coupler.
2. Fill out Table 6.3 with the collected data.

6.17.5.3 Manufacturing a Y-Coupler in the Lab

1. List the tools, instruments and parts that are required. Design a floor plan for a production line to manufacture a Y-coupler in the lab.
2. Estimate the manufacturing, testing and handling times required to manufacture a Y-coupler in the lab.

6.17.5.4 Testing a Y-Coupler

1. Measure the input power of the laser beam before it enters the Y-coupler. Measure the output power at Ports 2 and 3 of the Y-coupler.
2. Fill out Table 6.4 with the collected data.

6.17.5.5 Testing a 1 × 4 Y-Coupler

1. Measure the input power of the laser beam before it enters the cascaded Y-couplers. Measure the output power at each port of the cascaded Y-couplers.
2. Fill out Table 6.5 with the collected data.

6.17.5.6 Testing a Proximity Sensor

1. Measure the input power of the laser beam before it enters the sensor.

2. Measure the output, which is caused by backscattering reflection from the mirror, white surface or black surface. Repeat this for five distances, measuring the output power.

3. Fill out Table 6.6(I) for the mirror, (II) the white surface and (III) the black surface with the collected data.

6.17.6 CALCULATIONS AND ANALYSIS

6.17.6.1 Testing a 3 dB Coupler

1. Calculate the power loss in each output port of the 3 dB coupler, using Equations 6.1 through 6.4.
2. Calculate the splitting ratio of each output port.
3. Fill out Table 6.2.

6.17.6.2 Testing a 1 × 4 3 dB Coupler

1. Calculate the power loss in each output port of the 3 dB couplers, using Equations 6.1 through 6.4.
2. Calculate the splitting ratio of each output port.
3. Fill out Table 6.3.

6.17.6.3 Manufacturing a Y-Coupler in the Lab

Estimate the manufacturing, testing and handling times that are required to manufacture a Y-coupler in the lab.

6.17.6.4 Testing a Y-Coupler

1. Calculate the power loss in each output port of the Y-coupler, using Equation 6.5.
2. Calculate the splitting ratio of each output port.
3. Fill out Table 6.4.

6.17.6.5 Testing a 1 × 4 Y-Coupler

1. Calculate the power loss in each output port of the Y-coupler, using Equation 6.5.
2. Calculate the splitting ratio of each output port.
3. Fill out Table 6.5.

6.17.6.6 Testing a Proximity Sensor

1. Calculate the power loss in each output port of the proximity sensor, using Equation 6.5.
2. Fill out Table 6.6 (I)–(III).
3. Plot the output power as a function of $(1/d^2)$, when using the mirror, white surface and black surface.
4. Find the point where a linear $(1/d^2)$ dependence begins. This is the distance at which the finite diameter of the fibre core is no longer significant in determining the amount of reflected light accepted by the proximity sensor. Explain why the amount of reflected power accepted by the proximity sensor drops from a maximum when approaching the reflecting surface being used (mirror, white surface and black surface).

6.17.7 RESULTS AND DISCUSSIONS

6.17.7.1 Testing a 3 dB Coupler

1. Report the measurements of the input power into Port 1 and output power from Ports 2–4 of the 3 dB coupler.
2. Report the loss calculation for the 3 dB coupler.

3. Determine the splitting ratio of the 3 dB coupler.

4. Compare the results to the manufacturer's specifications.

6.17.7.2   Testing a 1 × 4 3 dB Coupler

1. Report the measurements of the input power of Port 1 and the output power from Ports 2–4 of the 3 dB couplers.

2. Report the loss calculation results for the 3 dB couplers.

3. Report the splitting ratio of the 1 × 4 coupler.

4. Compare the results to the manufacturer's specifications.

6.17.7.3   Manufacturing a Y-Coupler in the Lab

1. Report the manufacturing and testing process, and floor plan. Propose any other manufacturing process, which would reduce manufacturing time and manpower and using different materials.

2. Report the manufacturing, testing and handling times required to manufacture a Y-coupler in the lab.

6.17.7.4   Testing a Y-Coupler

1. Report the input power and output power of the Y-coupler.

2. Report the loss calculation for each output port.

3. Report the splitting ratio of the Y-coupler.

4. Compare the results to the theoretical ratio.

6.17.7.5   Testing a 1 × 4 Y-Coupler

1. Report the input power and output power of the cascaded Y-couplers.

2. Report the loss calculation for each output port.

3. Report the splitting ratio of the 1 × 4 Y-coupler.

4. Compare the results to the theoretical ratio.

6.17.7.6   Testing a Proximity Sensor

1. Report the input power and output power of the proximity sensor.

2. Report the loss calculation for each output port.

3. Report the calculation results for the output power, when using the mirror, white surface and black surface.

4. Compare the results of using the three different reflective surfaces.

6.17.8   CONCLUSION

Summarise the important observations and findings obtained in this lab experiment.

6.17.9   SUGGESTIONS FOR FUTURE LAB WORK

List any suggestions for improvements using different experimental equipment, procedures and techniques for any future lab work. These suggestions should be theoretically justified and technically feasible.

6.18   LIST OF REFERENCES

List any references that were used in the report. Use one format in writing the references. Never mix reference formats in a report.

6.19 APPENDIX

List all of the materials and information that are too detailed to be included in the body of the report.

FURTHER READING

Agrawal, G. P., *Fiber-Optic Communication Systems*, 2nd ed., John Wiley & Sons, New York, 1997.

Bise, R. T., Windeler, R. S., Kranz, K. S., Kerbage, C., Eggleton, B. J., and Trevor, D.J., Tunable photonic bandgap fiber, *Proceedings of Optical Fiber Communication Conference and Exhibit*, pp. 466–468, OFC, Anaheim, CA, 2002.

Camperi-Ginestet, C., Kim, Y. W., Wilkinson, S., Allen, M., and Jokerst, N. M., Micro-opto-mechanical devices and systems using epitaxial lift off, *JPL Proceedings of the Workshop on Microtechnologies and Applications to Space Systems*, pp. 305–316, (SEE N94-29767 08-31), Category Solid-State Physics, Georgia Institute of Technology, Atlanta, GA, June 1993.

Cantore, F. and Della Corte, F. G., 1.55-µm silicon-based reflection-type waveguide-integrated thermo-optic switch, *Opt. Eng.*, 42 (10), 2835–2840, 2003.

Dutton, H. J. R., *Understanding Optical Communications*, IBM/Prentice Hall, Upper Saddle River, NJ, 1998.

Eggleton, B. J., Kerbage, C., Westbrook, P., Windeler, R., and Hale, A., Microstructured optical fiber devices, *Opt. Express*, 9, 698–713, 2001.

Fedder, G. K., and Howe, R. T., Multimode digital control of a suspended polysilicon microstructure, *IEEE J. MEMS*, 5 (4), 283–297, 1996.

Fedder, G. K., Iyer, S., and Mukherjee, T, Automated optimal synthesis of microresonators, *Technical Digest of the IEEE International Conference on Solid-State Sensors and Actuators (Transducers'97)*, Chicago, IL, June 16–19, Vol. 2, pp. 1109–1112, 1997.

Fedder, G. K., Santhanam, S., Reed, M. L., Eagle, S. C., Guillou, D. F., Lu, M. S.-C., and Carley, L. R., Laminated high-aspect-ratio microstructures in a conventional CMOS process, *Sens. Actuators*, A, 57 (2), 103–110, 1996.

Francon, M., *Optical Interferometry*, Academic Press, New York, pp. 97–99, 1966.

Gebhard, B. and Knowles, C. P., Design and adjustment of a 20 cm Mach–Zehnder interferometer, *Rev. Sci. Instrum.*, 37 (1), 12–15, 1996.

Goff, D. R., and Hansen, K. S., *Fiber Optic Reference Guide: A Practical Guide to the Technology*, 2nd ed., Butterworth-Heinemann, Boston, MA, 1999.

Hariharan, P., Modified Mach–Zehnder interferometer, *Appl. Opt.*, 8 (9), 1925–1926, 1969.

Hecht, J., *Understanding Fiber Optics*, 3rd ed., Prentice Hall, Englewood Cliffs, NJ, 1999.

Heidrich, H., Albrecht, P., Hamacher, M., Nolting, H.-P., Schroeter-Jannsen, H., and Weinert, C. M., Passive mode converter with a periodically tilted InP/GaInAsP rib waveguide, *IEEE Photon. Technol. Lett.*, 4, 34–36, 1992.

Hibino, K., Error-compensating phase measuring algorithms in a Fizeau interferometer, *Opt. Rev.*, 6 (6), 529–538, 1999.

Him, F. C., Effects of passivation of GaN transistors, *Paper presented to Epitaxy Group of the IMS at NRC*, Ottawa, Canada, 2005.

Hitz, B., A new tunable optical filter, *Photonics Spectra*, Technology World, 10 (9), pp. 32, October 2003.

Ishida, O., Takahashi, H., and Noue, Y. I., Digitally tunable optical filters using arrayed-waveguide grating (AWG) multiplexers and optical switches, *J. Lightwave Technol.*, 15, 321–327, 1997.

Jackel, J., Goodman, M. S., Baran, J. E., Tomlinson, W. J., Chang, G.-K., Iqbal, M. Z., Song, G. H. et al., Acoustooptic tunable filters (AOTF's) for multiwavelength optical cross-connects: crosstalk considerations, *J. Lightwave Technol.*, 14, 1056–1066, 1996.

Kao, C. K., *Optical Fiber Systems: Technology, Design, and Applications*, McGraw-Hill, New York, 1982.

Koga, M., Compact quartzless optical quasi-circulator, *Electron. Lett.*, 30, 1438–1440, 1994.

Kolimbiris, H., *Fiber Optics Communications*, Prentice Hall, Englewood Cliffs, NJ, 2004.

Lequine, M., Parmentier, R., Lemarchand, F., and Amra, C., Toward tunable thin-film filters for wavelength division multiplexing applications, *Appl. Opt.*, 41 (16), 3277–3284, 2002.

Li, X., Chen, J., Wu, G., and Ye, A., Digitally tunable optical filter based on DWDM thin film filters and semiconductor optical amplifiers, *Opt. Express*, 13 (4), 1346–1350, 2005.

Litchinitser, N., Dunn, S., Steinvurzel, P., Eggleton, B., White, T., McPhedran, R., and de Sterke, C., Application of an arrow model for designing tunable photonic devices, *Opt. Express*, 12 (8), 1540–1550, 2004.

Loreggia, D., Gardiol, D., Gai, M., Lattanzi, M. G., and Busonero, D., Fizeau interferometer for global astrometry in space, *Appl. Opt.*, 43 (4), 721–728, 2004.

Mynbaev, D. K., and Scheiner, L. L., *Fiber-Optic Communication Technology—Fiber Fabrication*, Prentice Hall PTR, Englewood Cliffs, NJ, 2005.

OFS Leading Optical Innovations, Nonzero dispersion optical fiber, *True Wave Reach Fiber Product Catalog*, OFS Laboratories, Somerset, NJ, 2003.

Okamoto, K., *Fundamentals of Optical Waveguides*, Academic Press, San Diego, CA, 2000.

OZ Optics Limited, *Fiber Optic Components for Optoelectronic Packaging*, Catalog 2006, OZ Optics, Ottawa, Canada, 2006.

Page, D., and Routledge, I., Using interferometer of quality monitoring, *Photon. Spectra*, 147–153, November 2001.

Palais, J. C., *Fiber Optic Communications*, 4th ed., Prentice Hall, Englewood Cliffs, NJ, 1998.

Sadot, D. and Boimovich, E., Tunable optical filters for dense WDM networks, *IEEE Commun. Mag.*, 36, 50–55, 1998.

Shamir, J., *Optical Systems and Processes*, SPIE Optical Engineering Press, Bellingham, WA, 1999.

Stone, J., and Stulz, L. W., Pigtailed high finesse tunable FP interferometer with large, medium and small FSR, *Electron. Lett.*, 23, 781–783, 1987.

Sugimoto, N., Shintaku, T., Tate, A., Kubota, E., Terui, H., Shimokozono, M., Ishii, M., and Inoue, Y., Waveguide polarization-independent optical circulator, *IEEE Photon. Technol. Lett.*, 11, 355–357, 1999.

Tolansky, S., *An Introduction to Interferometry*, Longmans/Green and Co., London, pp. 115–116, 1955.

Ungar, S., *Fiber Optics: Theory and Applications*, Wiley, New York, 1990.

Vail, E., Wu, M. S., Li, G. S., Eng, L., and Chang-Hasnain, C. J., GaAs micromachined widely tunable Fabry–Perot filters, *Electron. Lett.*, 31 (3), 228–229, 1995.

Yeh, C., *Applied Photonics*, Academic Press, San Diego, CA, 1994.

Yun, S.-S., and Lee, J.-H., A micromachined in-phase tunable optical filter using the thermo-optic effect of crystalline silicon, *J. Micromech. Microeng.*, 13, 721–725, 2003.

7 Wavelength Division Multiplexer

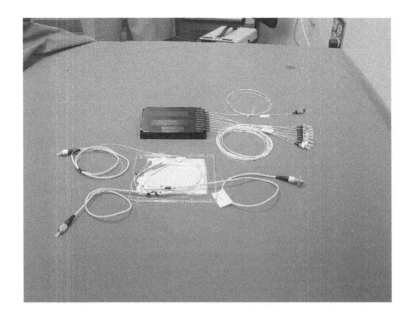

7.1 INTRODUCTION

A powerful aspect of fibre communication systems is that signals of many different wavelengths can be sent along a single optical fibre cable simultaneously, without signal interference. The technology that combines a number of wavelengths into the same fibre cable is known as wavelength-division multiplexing (WDM). Similarly, separating and distributing the signals is known as de-multiplexing. Multiplexing and de-multiplexing techniques are used in fibre communication systems to increase link capacity, flexibility and speed and to reduce cost.

This chapter presents the operating principles of WDM and describes the components needed for its utilisation in systems. These components range from simple devices to sophisticated tunable optical sources and wavelength filters.

Multiplexer and de-multiplexer devices can be either passive or active in design. Passive designs are based on the use of a prism, diffraction grating or filter. Active designs combine passive devices with tunable filters.

This chapter presents two experiments in both multiplexing/de-multiplexing of light signals from many light sources, using different techniques.

7.2 WAVELENGTH DIVISION MULTIPLEXING

WDM is an optical technology that permits several wavelengths to be coupled into the same fibre cable, effectively increasing the aggregation bandwidth per fibre cable. Figure 7.1 illustrates the components of a basic communication system. The de-multiplexer (de-mux)

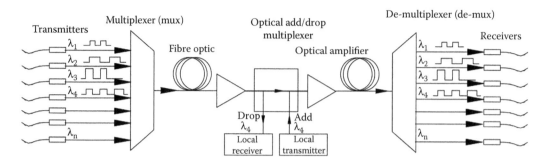

FIGURE 7.1 A schematic diagram of basic communication system.

decouples what the multiplexer has coupled. The de-mux separates several wavelengths in a single fibre cable, and directs them individually onto many fibre cables, which are connected to the receiver channels.

WDM systems are based on the ability of a fibre cable to carry many different wavelengths without mutual interference. Each wavelength represents an optical channel within the fibre cable. Several optical methods are available to combine individual channels within a fibre cable, and to extract them at appropriate points along a network. WDM technology has evolved to the point that channel wavelength separations can be as small as a few nanometres, giving rise to dense wavelength division multiplexing (DWDM) systems.

The output of each laser transmitter in a WDM system is set to one of the channel frequencies. These signals of various frequencies must then be multiplexed (superimposed or combined) and then inserted into a single fibre-optic cable. These signals then travel through the cable from the multiplexer to an add/drop multiplexer. The add/drop multiplexer routes one wavelength, λ_4, to a point and picks up another signal at the same wavelength, also λ_4. Note that this is a different signal, as shown in Figure 7.1. A de-multiplexer is used to extract the multiplexed channels at the receiver end. Multiplexing and de-multiplexing devices employ narrowband filters. Multiplexer and de-multiplexer devices can be cascaded and combined to achieve the desired results. Several devices exist to perform such filtering, including thin-film fibre Bragg gratings, optic gratings, tapered fibres, liquid crystal filters and integrated optical devices. These multiplexing technologies are also used for other optical fibre applications. These applications include telephone and data communications, SONET/SDH networks, inter-exchange networks and in links for trunk exchange and local exchange hubs.

7.3 TIME-DIVISION MULTIPLEXING

Time-division multiplexing (TDM) is a technique for transmitting multiple digitised data, voice and video signals simultaneously over one fibre cable. This is accomplished by interleaving pulses representing bits from different channels or time slots. The public-switched telephony network (PSTN) is based on the TDM technologies and is often called a TDM access network.

The time-division multiplexer is a device that uses TDM techniques to combine several slower speed data streams into a single high-speed data stream, as shown in Figure 7.2. Data from multiple sources is broken into portions (bits or bit groups); these portions are transmitted in a defined sequence. The transmission order must be maintained so that the input streams can be reassembled at the destination. Typically, using the same TDM techniques, the same device can also perform the reverse process: de-compose the high-speed data streams into multiple low speed data streams, a process called de-multiplexing. Therefore, a time-division multiplexer and de-multiplexer are very often packaged in the same box.

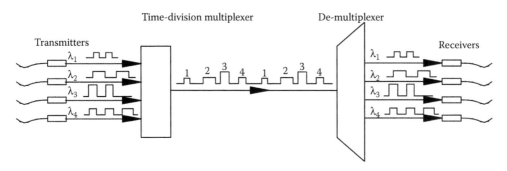

FIGURE 7.2 A schematic diagram of time-division multiplexing.

7.4 FREQUENCY-DIVISION MULTIPLEXING

Frequency-division multiplexing (FDM) is a scheme in which numerous analogue signals are combined for transmission on a single communications line or channel. Each signal is assigned a different frequency (sub-channel) within the main channel. This technology is used in broadcast radio, television and cable television. Home local area networks (LAN) use this technology to ensure compatibility between the different services sharing the same telephone wire, specifically voice and the home network. To eliminate interference, each service has a frequency spectrum that is different from the others. Traditionally, FDM is used for analogue signals, but it also can be used for digital signals.

When FDM is used in a communication network, each input signal is sent and received at maximum speed at all times. However, if many signals must be sent along a single long-distance line, the necessary bandwidth is large, and careful design is required to ensure that the system will perform properly. In some systems, TDM is used instead.

7.5 DENSE WAVELENGTH DIVISION MULTIPLEXING

DWDM is an acronym for dense wavelength division multiplexing, an optical technology used to increase bandwidth over existing fibre-optic backbones. DWDM is a fibre-optic transmission technique that employs light wavelengths to transmit data as parallel bits or a serial string of characters. Using DWDM, up to 80 (and theoretically more) separate wavelengths or channels of data can be multiplexed into a single light stream, and then transmitted on a single fibre-optic cable. Each channel carries a TDM signal. In a system with each channel carrying data, billions of bits per second, can be delivered by the fibre-optic cable. DWDM is also sometimes called WDM. As each channel is de-multiplexed at the end of the transmission back into the original source, different data formats can be transmitted together, at different rates. Specifically, the internet data, synchronous optical network (SONET) data and asynchronous transfer mode (ATM) data can all be transmitted at the same time within the same optical fibre. Utilising DWDM technology is a suitable solution for high-speed data transmission, without the addition of more fibre cables.

7.6 COARSE WAVELENGTH DIVISION MULTIPLEXING

CWDM is an acronym for coarse wavelength division multiplexing, which is a technology that combines up to 16 wavelengths onto a single fibre. When there are just a few channels (up to 16 channels) and they are spaced more widely (10 nm or more) apart, the system is called CWDM. The coarse wavelength division multiplexer and de-multiplexer (CWDM) are designed to multiplex and de-multiplex wavelength signals in metropolitan, access and enterprise networks, and for cable television applications. They are a low-cost approach for systems that use un-cooled laser sources, and are an alternative to more expensive DWDM components based on 100 or 200 GHz channel spacing. CWDMs are used to isolate a specific wavelength channel, whereas CWDM channel splitters are used to isolate a band of channels.

7.7 TECHNIQUES FOR MULTIPLEXING AND DE-MULTIPLEXING

The following presents some of the techniques that are used in the multiplexing and de-multiplexing of many wavelengths.

7.7.1 MULTIPLEXING AND DE-MULTIPLEXING USING A PRISM

A simple way of multiplexing or de-multiplexing wavelengths can be done using a prism. Figure 7.3 shows the de-multiplexing of multiple wavelengths exiting from the fibre cable. The first lens makes the diverging light beam become parallel and incident on the prism surface. Each component of the light is refracted differently when it exits from the prism. This spreading of the wavelengths produces the rainbow effect. Each wavelength is refracted by a different angle from the next wavelength. This angle is called the angle of refraction. The angle of refraction depends on the wavelength, the apex angle and the refraction index of the prism. The second lens focuses each wavelength to the designated output receiver via a fibre-optic assembly. The same components can be used in reverse to multiplex different wavelengths onto one fibre-tube assembly. Therefore, this device is bi-directional.

7.7.2 MULTIPLEXING AND DE-MULTIPLEXING USING A DIFFRACTION GRATING

Another technology based on the principles of diffraction uses a diffraction grating. When a light source is incident on a diffraction grating, each wavelength is diffracted at a different angle, and therefore, to a different point in space. It is necessary to use a lens to focus the wavelengths onto individual fibres, as shown in Figure 7.4. Separate wavelengths can be combined onto the same output port, or a single mixed input may be split into multiple outputs, one per wavelength. This device is bi-directional.

7.7.3 OPTICAL ADD/DROP MULTIPLEXERS/DE-MULTIPLEXERS

Figure 7.5 illustrates a schematic representation of a design for an optical add/drop multiplexer/ de-multiplexer (OADM), which is widely used in communication systems. Between multiplexing

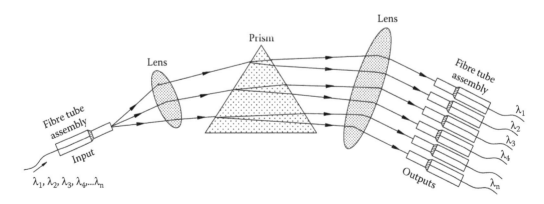

FIGURE 7.3   Multiplexing and de-multiplexing of wavelengths using a prism.

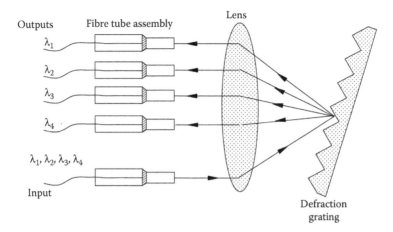

FIGURE 7.4   Multiplexing and de-multiplexing of wavelengths using a diffraction grating.

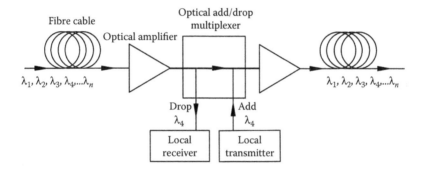

FIGURE 7.5   An optical add/drop multiplexer.

and de-multiplexing points in a DWDM system, there is a span where multiple wavelengths exist. An OADM removes or inserts one or more wavelengths at some point along this span. Rather than combining or separating all wavelengths, the OADM can remove some, while passing on others. OADMs are a key part of moving towards the goal of all-optical networks. The design shown in Figure 7.5 includes both pre- and post-amplification components that may or may not be present in an OADM design.

There are two general types of OADMs. The first generation is a fixed device, physically configured to drop specific pre-set wavelengths while adding others. The second generation is reconfigurable and capable of dynamically selecting which wavelengths are added and dropped. Thin-film filters are used for OADMs in metropolitan DWDM systems, because of their low loss, low cost and high stability. The new third generation of OADMs involves other technologies, such as tunable fibre gratings, FBGs and circulators.

7.7.4   ARRAYED WAVEGUIDE GRATINGS

Arrayed waveguide gratings (AWGs) are also based on the principles of diffraction. An AWG device is sometimes called an optical waveguide, a waveguide grating router, a phase array or a phasar. An AWG device consists of an array of curved-channel waveguides (W_1, W_2, W_3,...W_n) with a fixed difference in the length of optical path between the adjacent channels, as shown in Figure 7.6. The waveguides are connected to cavities at the input (S_1) and output (S_2). When light enters the input cavity, it is diffracted and enters the waveguide array. There the optical path length difference

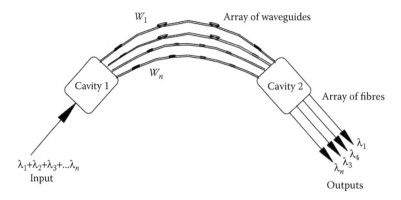

FIGURE 7.6   An arrayed waveguide grating device.

of each waveguide creates phase delays in the output cavity, where an array of fibres is coupled. The process results in different wavelengths having constructive interference at different locations, where the output ports are aligned.

7.7.5   FIBRE BRAGG GRATING

A Bragg grating is made of a small section of fibre cable, which is modified by exposure to ultra-violet radiation to create periodic changes in the refractive index of the core of the fibre cable. Figure 7.7 shows a FBG fibre cable. The Bragg grating reflects some of the light waves when travelling through it. The reflected waves usually occur at one particular wavelength. The reflected wavelength, known as the Bragg resonance wavelength, depends on the change in refractive index that is applied to the Bragg grating fibre. This also depends on the basic parameters of the grating (the grating period, the grating length and the modulation depth).

7.7.6   THIN FILM FILTERS OR MULTI-LAYER INTERFERENCE FILTERS

Figure 7.8 shows one multiplexing technique that uses interference filters in devices called thin film filters, or multi-layer interference filters. By positioning the thin filters in the optical path, wavelengths can be distributed. The property of each filter is such that it transmits one wavelength while reflecting others. By arranging the thin filters in a device, a de-multiplexer is created, and many wavelengths can be de-multiplexed.

There are several designs that use spectral filters positioned in the optical path to sort out wavelengths. These designs can be used as de-multiplexers.

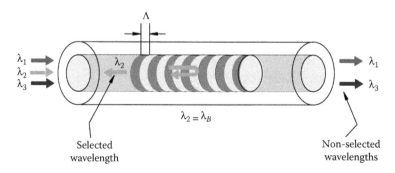

FIGURE 7.7   A fibre Bragg grating.

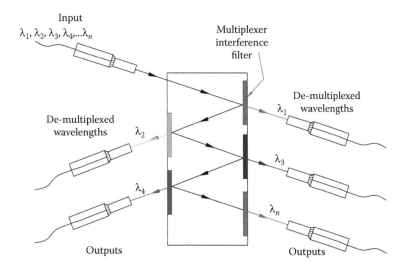

FIGURE 7.8 Multi-layer interference filters.

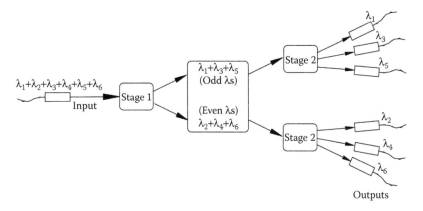

FIGURE 7.9 Periodic filters, frequency slicers and interleavers multiplexing.

7.7.7 PERIODIC FILTERS, FREQUENCY SLICERS AND INTERLEAVERS MULTIPLEXING

Figure 7.9 is a schematic diagram of periodic filters, frequency slicers and interleaver components that share the same functions, and are usually used together to make a multiplexer device. Stage 1 is a type of periodic filter, called an AWG. Stage 2 represents a frequency slicer. In this instance, stage 2 is another AWG; an interleaver function on the output is provided by six Bragg gratings. Six wavelengths are received at the input to the AWG at stage 1, which then breaks the wavelengths down into odd and even wavelengths. The odd and even wavelengths go to their respective stage 2 frequency slicer, and then are delivered by the interleaver in the form of six discrete, interference-free optical channels.

7.7.8 MACH–ZEHNDER INTERFEROMETER

Interferometers are based on the interferometric properties of light and the principles of Mach–Zehnder interferometer. Mach–Zehnder interferometers can be used to direct a specific wavelength to a specific output port, as shown in Figure 7.10. The interferometer consists of parallel titanium-diffused waveguides on a lithium niobate substrate. The incoming signal is split evenly along the

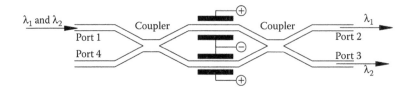

FIGURE 7.10   Mach–Zehnder interferometer.

arms of the Mach–Zehnder interferometer, and then recombined at the output. Electrodes are fixed on the arms of the Mach–Zehnder interferometer, and two couplers are connected to the interferometer. Although voltage is applied to the electrodes, a specific wavelength can be directed to either Port 2 or 3. More detail about the principle and operation of the Mach–Zehnder interferometer is presented in the optical fibre devices chapter.

7.8   WAVELENGTH DIVISION MULTIPLEXERS AND DE-MULTIPLEXERS

There are many types of wavelength division multiplexers and de-multiplexers available in the market to suit the requirements of any communication system. The following sections present the most widely used devices.

7.8.1   2-CHANNEL WDM DEVICES

Figure 7.11 shows single-mode wavelength division multiplexer devices (1 × 2 WDM 1310/1550, 1480/1550 nanometre). These two examples combine or separate light at different wavelengths and from different directions. The devices work to multiplex or de-multiplex optical signals within the 1310 and 1550 nanometre windows. These devices utilise a micro-optic filter-based technology to provide high performance, low insertion loss, low polarisation dependence, high isolation and environmental stability. These devices are extensively used in EDFA, CATV, WDM networks and in fibre-optics instrumentation.

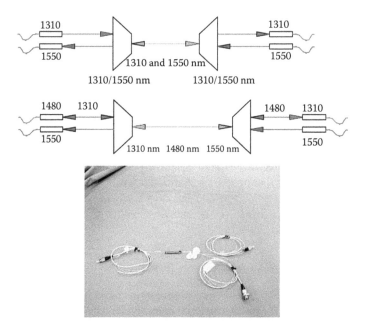

FIGURE 7.11   Two-channel WDM device.

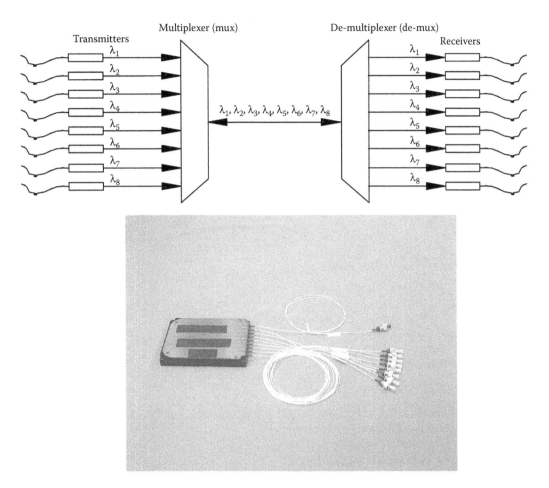

FIGURE 7.12 Eight-channel WDM device.

7.8.2 8-Channel WDM Devices

Figure 7.12 shows a wavelength division multiplexer device, which is used for increasing fibre-optic network signal capacity. The capacity is increased by enabling the simultaneous transmission of several wavelengths over the same common fibre. This device combines or separates light of eight different wavelengths. The optical channel wavelengths cover the 1480–1625 nm range, with 20 nm channel spacing. These devices utilise a micro-optic filter based technology to provide high performance, low insertion loss, low polarization dependence, high isolation and environmental stability. These devices are extensively used in EDFA, CATV fibre links, long-haul loops, subscriber loops, WDM networks and fibre-optics instrumentation.

7.9 EXPERIMENTAL WORK

This experiment is designed to demonstrate the concept and practicality of WDM and wavelength division de-mux devices. These devices are the key components in optical telecommunications systems. WDM is a process that combines n different signals for simultaneous transmission over the same fibre cable. Similarly, de-mux is a process that separates n different signals from the same fibre cable. WDM and de-mux devices use the concept of passive optical diffraction to combine and separate signals.

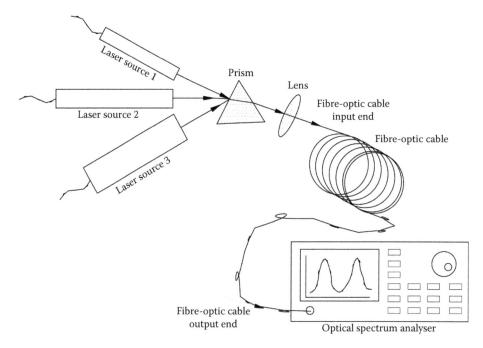

FIGURE 7.13   A schematic diagram of the WDM set-up.

Students will measure the power, wavelength and insertion loss of the individual signals from these devices. The following cases will be examined:

7.9.1   Wavelength Division Multiplexer

Figure 7.13 shows a schematic diagram of the set-up used in this experiment. Three laser beams are launched into a fibre cable through a prism. The prism acts as part of a diffraction grating, combining the beams onto a fibre cable. The power, wavelength and losses of the laser beams can be measured at different locations of the experimental set-up. The index of refraction of the prism can be calculated, by measuring the angles of the prism, and the angles of refraction and deviation of the beam.

7.9.2   Wavelength Division De-Multiplexer

Figure 7.14 shows a schematic diagram of white light dispersed into a coloured spectrum by a glass prism. The wavelengths of such a spectrum are examined in this experiment. White light is launched into a lens, and later through a prism. The prism acts as part of a diffraction grating that separates the light into a coloured spectrum, and directs each colour onto a fibre-optic cable. Each colour of the spectrum has a specific wavelength that can be measured. The index of refraction of the prism can be calculated by measuring the angles of the prism, and the angles of refraction and deviation of the beams of colour.

7.9.3   Technique and Apparatus

Appendix A presents the details of the devices, components, tools and parts.

1. 2 × 2 ft optical breadboard
2. Three laser light sources and power supplies

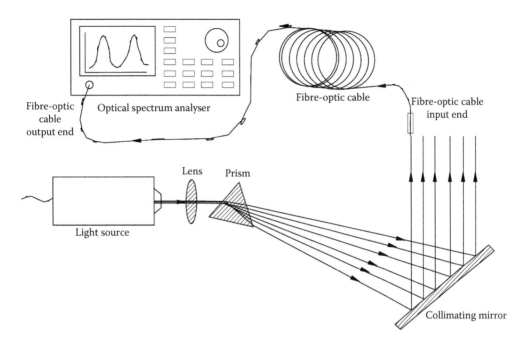

FIGURE 7.14 A schematic diagram of the de-mux set-up.

3. Three laser clamps
4. Laser light power detector
5. Laser light power meter
6. Light source and jack
7. Optical spectrum analyser
8. Multi-axis translation stages, as shown in Figure 7.15
9. Magnification lens, lens holder and rotation stages, in Figure 7.16
10. Collimating mirror and mirror holder, as shown Figure 7.17
11. White card and cardholder, as shown in Figure 7.18
12. Hardware assembly (clamps, posts, screw kits, screwdriver kits, lens/fibre cable holder/positioner, positioners, post holders, brass fibre cable holders, fibre cable holder/positioner, laser holder/clamp, etc.)
13. Prism and prism holder
14. Fibre-optic cable end preparation procedure and kit, and cleaning kit, as presented in the Fibre-Optic Cables Chapter, in the end preparation section
15. Connector assembly
16. Building connectors kit
17. Fibre-optic cable, 250 micrometres diameter, 500 metres long
18. Fibre cable end holder/positioner assembly
19. 20X, 30X or 40X microscope objective lens
20. Lens/fibre cable holder/positioner
21. Rotation stages
22. Black/white card and cardholder
23. Protractor
24. Ruler

FIGURE 7.15   Prism, prism holder and multi-axis translation stage.

FIGURE 7.16   Magnification lens and lens holder.

7.9.4   PROCEDURE

Follow the laboratory procedures and instructions given by the professor and/or instructor.

7.9.5   SAFETY PROCEDURE

Follow all safety procedures and regulations regarding the use of electric and optical devices, optical components, fibre-optic cable, laser devices, measurement instruments and optical cleaning chemicals.

FIGURE 7.17   Collimating mirror and mirror holder.

FIGURE 7.18   White card and cardholder.

7.9.6   APPARATUS SET-UP

7.9.6.1   Wavelength Division Multiplexer

1. Figures 7.13 and 7.19 show the apparatus set-up.
2. Mount three multi-axis translation stages on the breadboard.
3. Place laser source into the laser mount clamp and tighten the screw.
4. Bolt three laser sources to the multi-axis translation stages, one to each stage.
5. Align each laser to point at a central location on the breadboard.
6. Turn the lasers on individually. Follow the operation and safety procedures of the laser device in use. Use the optical spectrum analyser (or the laser power detector for quick

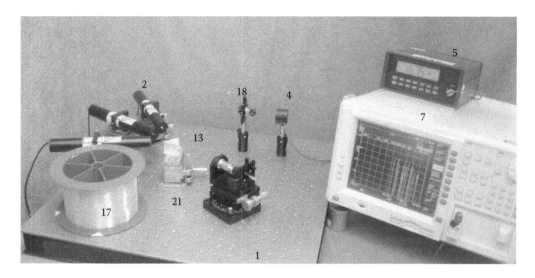

FIGURE 7.19 Apparatus set-up for a WDM.

power out results) to measure the power and wavelength of each laser source. Ask the
supervisor/instructor how to operate the optical spectrum analyser and reference the laser.
Refer to manufacturer's operation manual for the optical spectrum analyser in use.

7. Measure the laser power and wavelength of each laser beam, after it exits from the laser
sources. Figure 7.20 shows a sample of data collected by the optical spectrum analyser and
printed on a paper chart. Fill out Table 7.1.

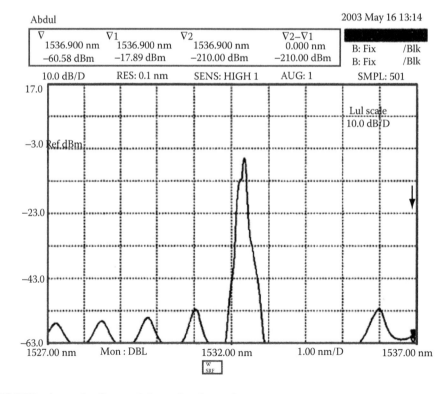

FIGURE 7.20 A sample of spectral data printed on a chart.

TABLE 7.1
Laser Powers and Wavelengths Measurements

	At Laser Source		Exiting from Prism			Exiting from Lens			Exiting from Lens		
	Laser Power P_{laser} (unit)	Wavelength λ_{laser} (unit)	Laser Power P_{prism} (unit)	Wavelength λ_{prism} (unit)	Prism Insertion Loss (unit)	Laser Power P_{lens} (unit)	Wavelength λ_{lens} (unit)	Prism + Lens Insertion Loss (unit)	Laser Power P_{cable} (unit)	Wavelength λ_{cable} (unit)	Prism + Lens + Cable Insertion Loss (unit)
Laser Source #1											
Laser Source #2											
Laser Source #3											
Laser Sources Turnd on at the Same Time											

8. Mount a multi-axis translation stage at a central location on the breadboard.

9. Mount and clamp a prism on top of the stage platform, and ensure that one side of the prism is facing the three laser beams.

10. Make a rough alignment of the three laser beams that pass through the prism and exit from one spot on the other side of the prism, as shown in Figure 7.21. Check for output from all laser beams, separately, with a piece of black/white card.

11. Finely align the three laser beams passing through the prism. Ensure that the output of all laser beams is all on one spot on the black/white card, as shown in Figure 7.21.

12. Figure 7.22 shows another way to align beams, using a glass rod. The beams exit from the glass rod in concentric rings. One beam hits a point on the black/white card. This point will be the centre for the second and third beams, as shown in Figure 7.22. The glass rod can only be used for quick beam alignment purposes. When the alignment is completed as explained above, use the first laser beam to continue the lab set-up.

13. Measure the laser power, wavelength and insertion loss of each laser beam after exiting from the prism. Figure 7.20 shows a sample of collected data printed on a chart. Fill out Table 7.1.

14. Mount a lens/fibre cable holder/positioner to the breadboard, so that the laser beam passes over the centre hole.

15. Add the 20X, 30X or 40X microscope objective lens to the lens/fibre cable holder/positioner.

16. Measure the laser power, wavelength and insertion loss of each laser source, after it exits from the lens. Fill out Table 7.1.

17. Prepare a fibre cable to have a good cleave at each end.

18. Insert one end (input) of the fibre cable into the brass fibre cable holder, and put it into the hole of the lens/fibre-cable holder/positioner.

19. Locate the input of the fibre cable at the focal point of the microscope objective lens.

20. Mount the output end of the fibre cable into the brass fibre cable holder, and place it in the hole of the fibre-cable holder/positioner.

21. Finely align the input end of the fibre-cable holder, so that the input end is at the centre of the lens/fibre cable holder/positioner, and is positioned at the focal point of the microscope objective lens. This is a very important step for obtaining an accurate value of the laser beam power from the fibre cable output end.

(a)   (b)

FIGURE 7.21   Red laser beam spot exiting from the prism. (a) Three laser-beams alignment and (b) One spot of the three laser beams on the target board.

FIGURE 7.22 Laser beams exiting from the glass tube falling on the black/white card.

FIGURE 7.23 Red laser beam spot on the black/white card.

22. Check to make sure that you have a proper circular spot from the output end of the fibre cable, as shown in Figure 7.23.
23. Repeat steps 21 and 22 to check the alignment of the second and third laser beams through the fibre cable.
24. Connect the output end of the fibre cable to the optical spectrum analyser. There are many ways to achieve the connection, either by using a fibre cable holder/connector to connect the fibre cable directly to the input of the optical spectrum analyser, or by building a connector on the output end of the fibre cable to suit the optical spectrum analyser input connection point.

TABLE 7.2

Prism Angles, Incidence and Reflected Angles of the Laser Beams

Prism Angles Apex Angle (°), Side Angle (°), and Side Angle (°)			
	Laser Beam Incidence Angle on Prism (°)	Laser Beam Refracted Angle from Prism (°)	Deviation Angle (°)
Laser Source # 1			
Laser Source # 2			
Laser Source # 3			

25. Measure the laser power, wavelength and insertion loss of each laser beam after exiting from the fibre cable. Fill out Table 7.1.
26. Find the prism angles. Fill out Table 7.2.
27. Use a protractor (or ruler and geometry) to measure the angle of incidence of each laser beam incident on the prism surface, and the refracted angle of the laser beams exiting from the prism. Fill out Table 7.2.
28. Now, turn on the three laser sources at the same time. Measure the laser power outputs, wavelengths and insertion loss for two laser sources exiting from the prism, the lens and the fibre cable. Fill out Table 7.1.
29. Turn on two laser sources at the same time, by blocking either laser source # 1, # 2 or #3. Measure the laser power outputs, wavelengths and insertion loss for the other laser beams exiting from the prism, the lens and the fibre cable. Fill out Table 7.3.

7.9.6.2   Wavelength Division De-Multiplexer

1. Figures 7.14 and 7.24 show the apparatus set-up.
2. Mount a light source on a jack and plug the light power supply into a 110V wall outlet.
3. Place the input end of the fibre cable into the brass fibre-cable holder, and place it in the hole of the fibre cable holder/positioner assembly.
4. Locate the input end of the fibre cable in front of the light source.
5. Connect the output end of the fibre-optic cable to the optical spectrum analyser. There are many ways to achieve the connection to the optical spectrum analyser. Either use a fibre-cable holder/connector to connect the fibre cable directly to the input of the optical spectrum analyser, or build a connector on the output end of the fibre cable to suit the optical spectrum analyser input connection point.
6. Measure the power and wavelength of the light source using the optical spectrum analyser (or the light power detector for quick power out results). Ask the supervisor/instructor how to operate the optical spectrum analyser and reference the laser. Refer to manufacturer's operation manual for the optical spectrum analyser in use.
7. Figure 7.20 shows a sample of data collected by the optical spectrum analyser printed on a paper chart. Fill out Table 7.4.
8. Mount a magnification lens into a lens holder/positioner and onto the breadboard, so that the light beam passes over the centre of the lens.
9. Mount a multi-axis translation stage on the breadboard at a location facing the lens.

TABLE 7.3

Laser Powers and Wavelengths of Two Laser Sources Turned on at the Same Time

	Exiting from Prism		Exiting from Lens			Exiting from Fibre Cable		
	Laser Power P_{prism} (unit)	Wavelength λ_{prism} (unit)	Prism Insertion Loss (unit)	Laser Power P_{lens} (unit)	Wavelength λ_{lens} (unit)	Prism + Lens Insertion Loss (unit)	Laser Power P_{cable} (unit)	Wavelength λ_{cable} (unit)	Prism + Lens + Cable Insertion Loss (unit)
Laser Source # 1 + Laser Source # 2									
Laser Source # 1 + Laser Source # 3									
Laser Source # 2 + Laser Source # 3									

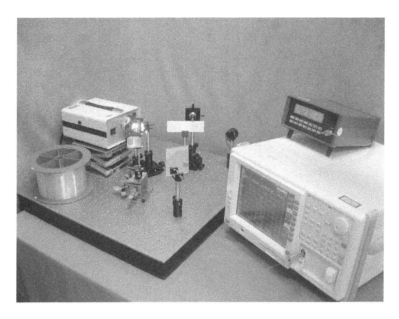

FIGURE 7.24   Apparatus set-up for a de-mux.

TABLE 7.4
Light Source and Colours Powers and Wavelengths

	Light Source Input to Prism	Spectrum Colours						
		Colour 1	Colour 2	Colour 3	Colour 4	Colour 5	Colour 6	Colour 7
Power (unit)								
Wavelength (unit)								

10. Mount and clamp a prism on top of the stage platform, and ensure that one side of the prism is facing the light beam exiting from the lens.
11. Use the prism multi-axis translation stage to make a rough alignment of the light beam passing through the prism and exiting from the other side of the prism.
12. Mount the white/black card and cardholder to intercept the light exiting the prism. This light will disperse into a spectrum (rainbow), as shown in Figure 7.25.
13. Use the prism multi-axis translation stage to finely align the spectrum exiting the prism. A clear coloured spectrum (rainbow) can be captured on the white/black card.
14. Replace the white/black card and cardholder with the collimating mirror and mirror holder, to intercept the spectrum (rainbow) of light exiting the prism.
15. Mount the white card and cardholder, to intercept the spectrum (rainbow) of the light that reflects from the mirror.
16. Locate the fibre cable input end holder/positioner assembly behind the white card, to capture the first colour of the spectrum.

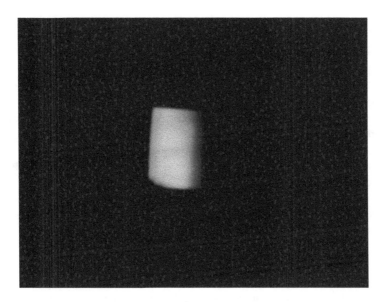

FIGURE 7.25 Spectrum (rainbow) of the light from the prism falling on the black/white card.

17. Measure the power and wavelength of each colour of the spectrum. Fill out Table 7.4.
18. Find the prism angles. Fill out Table 7.5.
19. Use a protractor (or ruler and geometry) to measure the angle of incidence of the light source incident on the prism surface, and the angles of refraction and deviation of each colour of the spectrum. Fill out Table 7.5.

TABLE 7.5
Prism, Light Source and Colour Angles

Prism Angles Apex Angle (º), Side Angle (º) and Side Angle (º)		
Light Source Incidence Angle on Prism (º)		
Colours	**Angle of Refraction of the Colours Exiting from Prism (º)**	**Angle of Deviation of the Colours (º)**
Colour 1		
Colour 2		
Colour 3		
Colour 4		
Colour 5		
Colour 6		
Colour 7		

7.9.7 Data Collection

7.9.7.1 Wavelength Division Multiplexer

1. For each laser source, measure the power and wavelength of the beams at the laser sources. Measure the power, wavelength and insertion loss of the laser beams, after exiting the prism, the lens and the fibre cable. Fill out Table 7.1.
2. Find the prism angles. Measure the angle of incidence of each laser beam incident on the prism, and the refracted angle of the laser beams leaving the prism. Find the deviation angle of the laser beams. Fill out Table 7.2.
3. Measure the power, wavelength and insertion loss of the laser beams when all laser sources are turned on together. Fill out Table 7.1.
4. Measure the power, wavelength and insertion loss of the laser beams, after exiting the prism, the lens and the fibre cable, when two laser sources (#1 and #2, #1 and #3, and #2 and #3) are turned on. Fill out Table 7.3.

7.9.7.2 Wavelength Division De-Multiplexer

1. Measure the power and wavelength of the light source input to the prism. Fill out Table 7.4.
2. Measure the power and wavelength of each colour of the spectrum exiting the prism. Fill out Table 7.4.
3. Find the prism angles. Measure the angle of incidence of light on the prism, and the angles of refraction and deviation for each colour of the spectrum exiting the prism. Fill out Table 7.5.

7.9.8 Calculations and Analysis

7.9.8.1 Wavelength Division Multiplexer

1. Report the power, wavelength and insertion loss of the laser beams exiting from the prism, the lens and the fibre cable. Report this for the laser beams turned on individually, three together, or two at a time.
2. Calculate the index of refraction (n) of the prism, using Equation 7.1.

$$n = \frac{\sin(\frac{A+\delta}{2})}{\sin \frac{A}{2}} \qquad (7.1)$$

where:
  A is the apex angle of the prism
  δ is the deviation angle of the light exiting from the prism

7.9.8.2 Wavelength Division De-Multiplexer

1. Report the power, wavelength and insertion loss of the laser beams exiting from the prism, the lens and the fibre cable.
2. Calculate the index of refraction (n) of the prism, using Equation 7.1.

7.9.9 Results and Discussions

7.9.9.1 Wavelength Division Multiplexer

1. Compare the power and wavelength of the laser beams from the laser sources (input) to the laser beams exiting from the fibre cable (output).

2. Compare the powers, wavelengths and insertion loss, when the laser sources are turned on individually, three together or two at a time.

3. Compare the insertion loss at the prism, the lens and the fibre cable, amongst the above set-ups.

7.9.9.2 Wavelength Division De-Multiplexer

1. Report the power and wavelength of the light source, and each colour of the spectrum.
2. Report the angles of refraction and deviation of each colour of the spectrum.
3. Compare the power and wavelength of the light source, and each colour of the spectrum.

7.9.10 CONCLUSION

Summarise the important observations and findings obtained in this lab experiment.

7.9.11 SUGGESTIONS FOR FUTURE LAB WORK

List any suggestions for improvements using different experimental equipment, procedures and techniques for any future lab work. These suggestions should be theoretically justified and technically feasible.

7.10 LIST OF REFERENCES

List any references that were used in the report. Use one format in writing the references. Never mix reference formats in a report.

7.11 APPENDIX

List all of the materials and information that are too detailed to be included in the body of the report.

FURTHER READING

Agrawal, G. P., *Fiber-Optic Communication Systems*, 2nd ed., John Wiley & Sons, New York, 1997.

Cole, M., *Telecommunications*, Prentice Hall, Engelwood Cliffs, NJ, 1999.

Derfler, F. J., Jr. and Freed, L., *How Networks Work*, Millennium Edition, Que Corporation, Indianapolis, IN, 2000.

Dutton, H. J. R., *Understanding Optical Communications*, IBM, Prentice Hall, Engelwood Cliffs, NJ, 1998.

EXFO, *Guide to WDM Technology and Testing*, EXFO Electro-Optical Engineering, Quebec City, Canada, 2000.

Glance, B., Wavelength-tunable add/drop optical filter, *IEEE Photon. Techn. Lett.*, 8, 245–247, 1996.

Goff, D. R., and Hansen, K. S., *Fiber Optic Reference Guide: A Practical Guide to the Technology*, 2nd ed., Butterworth-Heinemann, London, 1999.

Hecht, J., *Understanding Fiber Optics*, 3rd ed., Prentice Hall., Englewood Cliffs, NJ, 1999.

Hioki, W., *Telecommunications*, 3rd ed., Prentice Hall, Englewood Cliffs, NJ, 1998.

Jackel, J., Goodman, M. S., Baran, J. E., Tomlinson, W. J., Chang, G.-K., Iqbal, M. Z., Song, G. H. et al., Acousto-optic tunable filters (AOTF's) for multiwavelength optical cross-connects: Crosstalk considerations, *J. Lightwave Technol.*, 14, 1056–1066, 1996.

Kao, C. K., *Optical Fiber Systems: Technology, Design, and Applications*, McGraw-Hill, New York, 1982.

Kartalopoulos, S. V., *Introduction to DWDM Technology—Data in a Rainbow*, SPIE Optical Engineering Press, Bellingham, WA, 1999.

Keiser, G., *Optical Fiber Communications*, 3rd ed., McGraw-Hill, New York, 2000.

Keiser, G., *Optical Communications Essentials*, 1st ed., McGraw-Hill, New York, 2003.

Kolimbiris, H., *Fiber Optics Communications*, Prentice Hall, Engelwood Cliffs, NJ, 1998.

Laude, J. P., *DWDM Fundamentals, Components, and Applications*, Artech House, Boston, MA, 2002.

Li, X., Chen, J., Wu, G., and Ye, A., Digitally tunable optical filter based on DWDM thin film filters and semi-conductor optical amplifiers, *Opt. Express.*, 13 (4), 1346–1350, 2005.

Malacara, D., *Geometrical and Instrumental Optics*, Academic Press, Boston, MA, 1988.

Mazzarese, D., Meeting the OM-3 challenge—Fabricating this new generation of multimode fiber requires innovative state of the art designs and testing processes, *Cabling Systems* (December 2002), pp. 18–20, 2002.

Palais, J. C., *Fiber Optic Communications*, 4th ed., Prentice Hall, Englewood Cliffs, NJ, 1998.

Razavi, B., *Design of Integrated Circuits for Optical Communications*, McGraw-Hill, New York, 2003.

Senior, J. M., *Optical Fiber Communications Principle and Practice*, 2nd ed., Prentice Hall, Englewood Cliffs, NJ, 1992.

Shamir, J., *Optical Systems and Processes*, SPIE Optical Engineering Press, Bellingham, WA, 1999.

Watanabe, T., Inoue, Y., Kaneko, A., Ooba, N., and Kurihara, T., Polymeric arrayed-waveguide grating multi-plexer with wide tuning range, *Electron. Lett.*, 33 (18), 1547–1548, 1997.

Yeh, C., *Handbook of Fiber Optics: Theory and Applications*, Academic Press, San Diego, CA, 1990.

8 Optical Amplifiers

8.1 INTRODUCTION

The development of the optical amplifier (OA) in the late 1980s revolutionised communication systems. OAs had an important impact similar to the invention of the laser in the early 1960s. Both devices contributed to the development of communication systems and other applications, such as lower pump optical power, single pixel multi-colour displays and light emitting devices. An OA is a device that amplifies the optical signal directly, without converting it to an electrical signal and then to an optical signal again. OAs are used for amplifying a weak optical signal in order to increase the distance the signal can be transmitted down the transmission lines. In comparison, repeaters and generators convert the signal to electrical form, regenerate or amplify the signal and then convert it to optical form again. The conversion of the signal from one form to another is a complex process, subject to high losses, slow speed and more costly than simpler OAs.

This chapter presents the fundamentals of OAs, their characteristics and classifications of common ones; it also describes their applications. The three basic technologies are: erbium-doped fibre optical amplifiers (EDFAs), semiconductor optical amplifiers (SOAs) and Raman OAs. These technologies and applications are presented in detail in this chapter.

8.2 BASIC APPLICATIONS OF OPTICAL AMPLIFIERS

OAs are used to boost signals transmitted over long distances in network systems. Figure 8.1 is a schematic diagram illustrating the components of a basic communication system.

The characteristics and advantages of the OAs have led to many applications. OAs can be used to boost signal power after multiplexing, or before de-multiplexing or at any point in modern optical networks. OAs are ideal for Merto and Long–Haul dense wavelength multiplexing (DWDM) as well as single wavelength applications. The optical design, coupled with sophisticated control circuitry, allows these OAs to provide constant gain even with signals being added to the network,

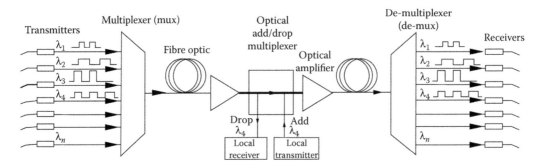

FIGURE 8.1   A schematic diagram of basic communication system.

such as λ4 in Figure 8.1. OA operation is independent of the signal data rate. OAs have different designs and packaging shapes. The following sections present the general applications of OAs in optical networks:

8.2.1   IN-LINE OPTICAL AMPLIFIERS

An OA is used as an in-line amplifier allowing signals to be amplified within the optical signal path, as shown in Figure 8.2. OAs can compensate for transmission loss and thus increase the distance between the transmitters and receivers. This application enables the signal to travel through lines hundreds of kilometres long.

8.2.2   POST-AMPLIFIER

An OA is used as a post-amplifier when placed immediately after the transmitter, as shown in Figure 8.3. An OA will boost the light signal to the required power level at the beginning of a fibre line. This arrangement enables the signal to travel hundreds of kilometres down the fibre cable. A common application of the post-amplification technique, together with an optical pre-amplifier at the receiving end, enables continuous underwater transmission distances up to a few kilometres.

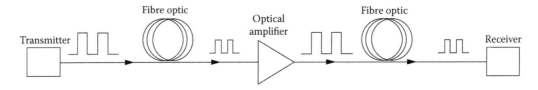

FIGURE 8.2   In-line optical amplifier.

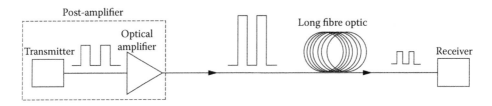

FIGURE 8.3   Post-amplifier application.

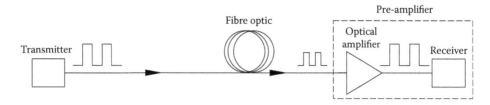

FIGURE 8.4   Pre-amplifier application.

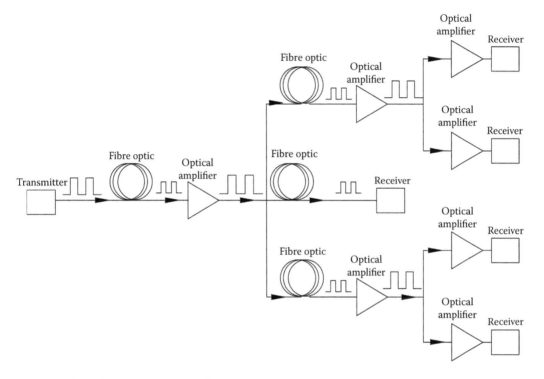

FIGURE 8.5   In local area network application.

8.2.3   PRE-AMPLIFIER

An OA is used as a pre-amplifier when placed immediately before the receiver, as shown in Figure 8.4. An OA will boost the light signal to the required power level before being received by an optical receiver. The pre-amplifier enables the signal to be processed directly by the receiver. The pre-amplification technique is commonly used before optical heterodyne detectors or avalanche photodiodes.

8.2.4   IN LOCAL AREA NETWORKS

OAs are used to boost signals in local area networks, when placed in sub-centres within the transmission lines and branches, as shown in Figure 8.5. This type of arrangement also enables OAs to have the characteristics suitable for analogue transmission. The video content multiplexed by head-end equipment is converted from optical/electrical transmission into the 1550 nm band-wavelength optical video signal that is suitable for OA. The optical video signals are optically amplified to compensate the losses of splitting and transmission. The signals repeatedly undergo amplification, splitting and transmission.

8.3 TYPES OF OPTICAL AMPLIFIERS

Fundamental operation of each type of OA will be presented in the classifications of the following sections.

8.3.1 DOPED FIBRE OPTICAL AMPLIFIERS

Doped fibre OAs are made from optical fibre whose core is doped with atoms of an element that can be excited by an external pump light to a state where stimulated emission can occur. Pump light from an external laser source is steadily pumped in one end or both ends of the fibre. The pump light is guided along the fibre length where it excites the doped atoms of the core. The core also guides the input light signal and the amplified light resulting from stimulated emission. The doped material types and doping concentrations in the fibre core depend on the wavelengths of light to be amplified. This is a general description of the operation of the doped fibre OAs. The following sections present details of each type.

8.3.1.1 Erbium-Doped Fibre Optical Amplifiers

In the late 1980s, a group of researchers at the University of Southampton in the United Kingdom successfully developed the erbium-doped fibre amplifier (EDFA). The EDFA then became the dominant type of OA. EDFAs combined with wavelength division multiplexing (WDM) technology are widely used in long-distance optical communications, networks and signal modulation. EDFAs operate in the 1540–1570 nm range, called the C-band by convention. Erbium ions have quantum levels that allow them to be stimulated to emit light in the C-band, which is the wavelength band having the least power loss in most silica-based fibre, where high-quality amplifiers are most needed.

Figure 8.6 shows the basic operation of the EDFA. An erbium-doped fibre amplifier consists of a 10–30 metre length of optical fibre, the core of which is doped with a rare-earth element of erbium (Er^{3+}). The fibre is pumped with a laser light at 980 or 1480 nm to raise the erbium ions to a high energy state. When an erbium ion is in a high energy state, an incident photon of input light signal can stimulate the erbium ions to give up some of its energy in the form of light and return to a lower energy state, which is more stable. This operation is called stimulated emission, and is generally presented in light production and laser theory.

The pump laser power supplies the optical energy for the amplifier. The pumped laser light is mixed with the input light signal via a coupler at the input fibre cable. The mixed light is guided into the fibre section with erbium ions included in the core. A photon of the laser excites the erbium ion to its higher-energy state. When the photon of the input light signal meets the exited erbium atom, the erbium atom gives up some energy in the form of a photon and returns to its lower-energy state. The new photon is in exactly the same phase and direction as the light signal that is being

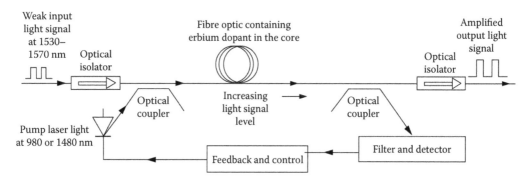

FIGURE 8.6 Erbium-doped fibre optical amplifier basic operation.

amplified. Thus, the light signal is amplified along the fibre core in a forward direction of travel only. Figure 8.6 also shows the need to have a pump laser beam along the length of a fibre to provide the energy for EDFAs. This design requires power and optics, such as couplers and filters.

EDFAs also have gain that varies with a signal's wavelength, which creates problems in many WDM applications. This can be solved by using special optical passive filters that are designed to compensate for the gain variation of the EDFA.

Pumping power can be applied in a forward direction, as shown in Figure 8.6, backward from the output end, or in both directions. Optical isolators are commonly used at the output end or both ends of the EDFA, to prevent the pump power signal and light signal from returning back down the fibre, or unwanted reflections that may affect laser stability.

As explained above, the pump laser light is supplied using a coupler at the inlet of the fibre cable, as shown in Figure 8.6. Pump laser light can be pumped in the direction and/or opposite direction of the light signal. The pump signal can be coupled in various locations in the amplification system, as shown in Figure 8.7; Figure 8.7 also shows that the pump power can be (a) at the end of the fibre cable using a coupler, or (b) coupled on both ends. Figure 8.7c shows a design for the remote pumping of the power, used where the pump laser is at a long distance from the amplifier, such as in undersea systems.

EDFAs have a number of main technical characteristics, such as efficient pumping, wavelength selection, minimal polarisation sensitivity, low insertion loss, low distortion and inter-channel crosstalk, high power output, low noise, very high sensitivity, low power consumption and low cost.

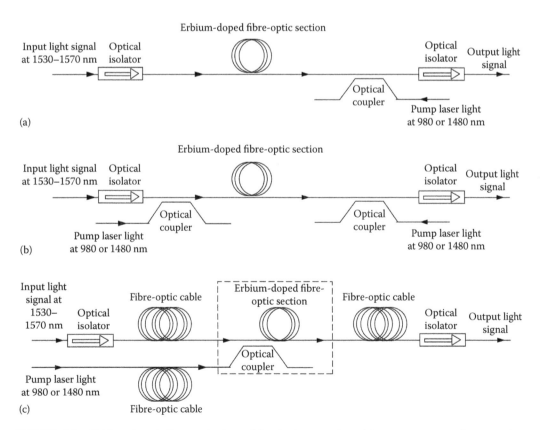

FIGURE 8.7 Erbium-doped fibre optical amplifier basic operation: (a) counter-propagating pumping, (b) bi-directional pumping and (c) remote pumping.

8.3.1.2 Praseodymium-Doped Fluoride Optical Amplifiers

EDFAs have shifted the optical telecommunication emphasis towards the third transmission window, called long-wavelength band in the 1510–1600 nm range. There is still great interest in 1300 nm in O-band amplifiers. This is mainly because a substantial part of the fibre-optic network worldwide is designed for operation in the second transmission window of about 1310 nm. Praseodymium-doped fluoride fibre amplifiers (PDFFAs) can provide substantial gain in this region. However, to compete with EDFAs, the quantum efficiency of the 1310 nm transition of Pr3 + should be increased. Low-phonon-energy glass hosts are needed for this purpose. Other alternatives are directed towards Gallium–Lanthanum–Sulfide (GLS) and Gallium–Sulfide–Iodide (GSI) glasses.

8.3.1.3 Neodymium-Doped Optical Amplifiers

Neodymium (Nd)-doped OAs amplify in the 1310 nm band. Nd will amplify over the 1310–1360 nm range when doped into Fluorozirconate (ZBLAN) glass and over the 1360–1400 nm range when doped into silica. The most efficient pump wavelengths are at 795 nm and 810 nm.

8.3.1.4 Telluride-Based, Erbium-Doped Fibre Optical Amplifiers

Telluride-based, erbium-doped OAs offer the potential optical bandwidth of more than 76 nm in the 1532–1608 nm band, thus increasing the potential bandwidth of an erbium-doped OA from 30 nm to more than 110 nm.

8.3.1.5 Thulium-Doped Optical Amplifiers

Thulium-doped OAs amplify between approximately 1450–1500 nm in the S-band.

8.3.1.6 Other Doped Fibre Optical Amplifiers

Other fibre OAs use doping materials, such as ytterbium (Yb). The host fibre material can be silica, a fluoride-based glass or a multi-component glass. Some plastic fibre amplifiers are under research and development. Modern plastics have characteristics similar to doped glass.

8.3.2 Semiconductor Optical Amplifiers

The functional applications of semiconductor optical amplifiers (SOAs) were first studied in the early 1990s. Since then, the diversity and scope of such applications have been steadily growing. SOAs are another common type of in-line amplifiers that are developed to support DWDM and to expand to the other wavelength bands supported by fibre optics. They have many applications in optical fibre communication, switching and signal processing systems.

SOAs are based on the same technology as basic semiconductor Fabry–Perot laser diodes, but they have anti-reflection (AR) coating at the end faces. Fabry–Perot laser diodes are generally presented in the laser theory. The structure of the SOAs is much the same as the diode, with two stacked slabs of specially designed semiconductor material, with another material between them that forms the active layer, as shown in Figure 8.8. An electrical current is passed through the device in order to excite electrons to high-state level. The electrons then fall back to the non-excited ground state, emitting photons by stimulated emission. An incoming optical signal stimulates emission of photons at its own wavelength. This is accomplished by blocking the cavity reflectors using an AR coating on both end faces. Fibre-optic cables are attached to both ends. As explained in EDFAs, optical isolators are commonly used at both ends of the SOA to prevent light signals from returning back. Depending on the material of the active layer, they operate from 1310 nm to 1550 nm in telecommunication systems.

SOAs are typically constructed in a small package. In addition, they transmit bi-directionally, making the reduced size of the device an advantage over EDFAs. They can be integrated with optical devices, such as semiconductor lasers, modulators and DWDM. But the actual

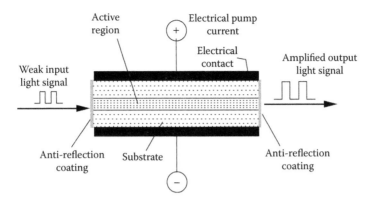

FIGURE 8.8   Semiconductor optical amplifier.

performance is still not compatible with EDFAs. They have high noise, less gain, medium polarisation dependence and high optical gain non-linearity with fast transient time. High non-linearity makes the SOAs attractive for optical signal processing, such as all-optic switching, wavelength conversion and regeneration, time de-multiplexing, clock recovery and pattern recognition. A number of SOA chips can be integrated on the same substrate to create high-density switching matrices.

SOAs are classified into two groups: Fabry–Perot cavity amplifiers (FPA) and travelling wave amplifiers (TWA). The difference depends on the efficiency of the reflection value of the AR coating material used.

8.3.3   RAMAN FIBRE OPTICAL AMPLIFIERS

Raman OAs differ in principle from EDFAs. They utilise stimulated Raman scattering (SRS) to create optical gain. SRS occurs when light waves interact with molecular vibrations in a material having a solid lattice structure. In Raman scattering, the molecule absorbs the light, then quickly re-emits a photon with energy equal to the original photon, plus or minus the energy of a molecular vibration mode. This has the effect of both scattering light and shifting its wavelength.

When a fibre transmits two suitably spaced wavelengths, SRS can transfer energy from one wavelength to the other. In this case, one wavelength excites the molecular vibration; then light of the second wavelength stimulates the molecule to emit energy at the second wavelength.

Figure 8.9 shows the topology of a typical Raman OA. The pump laser and circulator comprise the two key elements of the Raman OA. The pump laser, in this case, has a wavelength of 1535 nm. Raman amplifiers work in the 1550 nm window. The circulator provides a convenient means of injecting light backwards into the transmission fibre with minimal optical loss. The pump laser is coupled into the transmission fibre either in the same direction as the transmission signal, which is called *co-directional pumping*, or is coupled into the transmission fibre in the opposite direction, which is called *contra-directional pumping*. Contra-directional pumping is more common because co-directional pumping has the problem of optical non-linearity (non-linear amplification). In contra-directional pumping, the attenuation of the pump light is so small that it travels a great distance, several kilometres along the transmission fibre. It also keeps pump photons from reaching the receiver, where they could interfere with reception of the desired signal.

Optical isolators are commonly used at both ends of the Raman amplifier to prevent pump power and light signal from returning back down the fibre or unwanted reflections that may affect laser stability. Raman amplifiers are used as pre-amplifiers, power amplifiers and distributed amplifiers in digital and analogue transmissions in communication systems.

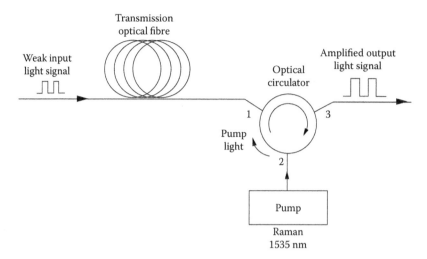

FIGURE 8.9   Raman optical amplifier.

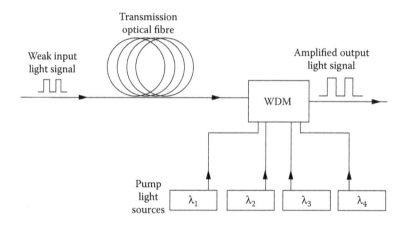

FIGURE 8.10   Raman optical amplifier with multiple pump light sources.

Figure 8.10 shows another technique for pumping light into Raman amplifiers. The amplification bandwidth can be extended, using multiple pump light sources along with a WDM. This can be done by using more than two pump light sources, producing a broadband amplifier over bands of more than 100 nm, for example, in the range of 1500–1600 nm.

Raman amplifiers have a number of main technical characteristics, such as efficient pumping, simplicity, wider wavelength coverage, minimal polarisation sensitivity, low insertion loss, high gain, low noise, fast reaction to changes of the pump power, low power consumption and low cost. These amplifiers are used in long-haul and ultra-haul DWDM transmission systems.

8.3.4   PLANER WAVEGUIDE OPTICAL AMPLIFIERS

Rare-earth-doped planar waveguide OAs are becoming increasingly important and provide compact and inexpensive alternatives to fibre amplifiers. In addition, planar technology is quite suitable for optical integration and will be essential to the development of fully integrated advanced optical devices.

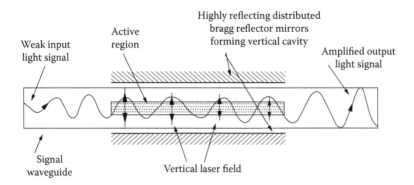

FIGURE 8.11 Linear optical amplifier.

8.3.5 LINEAR OPTICAL AMPLIFIERS

The design of the linear OA is similar to that of the semiconductor optical amplifier. The device has an active waveguide gain region; the input and output fibres are aligned to this waveguide. Unlike a semiconductor OA, the linear optical amplifier also features an integrated orthogonal laser that shares the gain region of the waveguide, as shown in Figure 8.11. This laser makes the amplifier gain linear. It also acts as an ultra-fast optical feedback circuit that responds to changes in the network. During the operation, the multiple wavelength signals to be amplified pass horizontally through the device, directly through the path of the laser, which is pumping photons of light vertically in the same device. The linear OAs are designed to be small and low cost. They are used in high data rate telecommunication systems.

8.4 OTHER TYPES OF OPTICAL AMPLIFIERS

There are many other types of OAs under research and development. The main objective is to increase the wavelength range and to increase the bit rates in communication systems along with improving the performance and decreasing the cost.

FURTHER READING

Agrawal, G. P., *Fiber-Optic Communication Systems*, 2nd ed., John Wiley & Sons, New York, 1997.

Bar-Lev, A., *Semiconductors and Electronic Devices*, 2nd ed., Prentice Hall International, New York, 1984.

Becker, P. M., Olsson, A. A., and Simpson, J. R., *Erbium-Doped Fiber Amplifiers: Fundamentals and Technology*, Elsevier, Burlington, MA, 1999.

Becker, P. C., Olsson, N. A., and Simpson, J. R., *Erbium-Doped Fiber Amplifiers: Fundamentals and Technology*, Academic Press, San Diego, CA, 1999, chap. 6, See also pp. 66–75, 184, 161–197.

Bise, R. T., Tunable photonic bandgap fiber, *Presented at Proceedings of the Optical Fiber Communication Conference and Exhibit*, Los Angeles, CA, March 17, pp. 466–468, 2002.

Cantore, F., Corte, D., and Francesca, G., 1.55-μm silicon-based reflection-type waveguide-integrated thermo-optic switch, SPIE: The International Society for Optical Engineering, *Opt. Eng.*, 42 (10), 2835–2840, 2003.

Chee, J. K. and Liu, J. M., Polarization-dependent parametric and Raman processes in a birefringent optical fiber, *IEEE J. Quantum Electron.*, 26, 541–549, 1990.

Chen, K. P., In-fiber light powers active fiber optical components, *Photon. Spectra*, pp. 78–90, April 2005.

Cole, M., *Telecommunications*, Prentice Hall, Upper Saddle River, NJ, 1999.

Connelly, M. J., *Semiconductor Optical Amplifiers*, 1st ed., Kluwer Academic Publishers, New York, 2002.

Derfler, F. J., Jr. and Freed, L., *How Networks Work*, Millennium Edition, Que Corporation, Indianapolis, IN, 2000.

Derickson, D., *Fiber Optic Test and Measurement*, Prentice Hall PTR, Upper Saddle River, NJ, 1998.

Digonnet, M. J. F., *Rare-Earth Fiber Lasers and Amplifiers*, Marcel Dekker, New York, 2001.

Dutton, H. J. R., *Understanding Optical Communications, IBM*, Prentice Hall, Upper Saddle River, NJ, 1998.

Dutta, N. K. and Wang, Q., *Semiconductor Optical Amplifiers*, 1st ed., World Scientific, Singapore, 2005.

Eggleton, B. J., Kerbage, C., Westbrook, P. S., Windeler, R. S., and Hale, A., Microstructured optical fiber devices, *Opt. Express*, 17, 698–713, 2001.

Fedder, G. K. and Howe, R. T., Multimode digital control of a suspended polysilicon microstructure, *IEEE J. MEMS*, 5 (4), 283–297, 1996.

Fedder, G. K., Iyer, S., and Mukherjee, T., Automated optimal synthesis of microresonators, *Presented at the IEEE International Conference on Solid-State Sensors and Actuators (Transducers '97)*, June 16–19, Chicago, IL, *Tech. Dig.*, 2, 1109–1112, 1997.

Fludger, C. R. S., Handerek, B., and Mears, R. J., Pump to signal RIN transfer in Raman fiber amplifiers, *Lightwave Technol.*, 19, 1140–1148, 2001.

Goff, D. R. and Hansen, K. S., *Fiber Optic Reference Guide: A Practical Guide to the Technology*, 2nd ed., Butterworth-Heinemann, Science and Technology Books Home, Boston, MA, 1999.

Green, P. E., *Fiber Optic Networks*, Prentice Hall, Englewood Cliffs, NJ, 1993.

Headley, C. and Agrawal, G., *Raman Amplification in Fiber Optical Communication Systems*, Academic Press, San Diego, CA, 2005.

Hecht, J., *Understanding Fiber Optics*, 3rd ed., Prentice Hall, Upper Saddle River, NJ, 1999.

Hioki, W., *Telecommunications*, 3rd ed., Prentice Hall, Englewood Cliffs, NJ, 1998.

Ho, M. C., Uesaka, K., Marhic, M., Akasaka, Y., and Kazovsky, L. G., 200-nm-bandwidth fiber optical amplifier combining parametric and Raman gain, *J. Lightwave Technol.*, 19, 977–981, 2001.

Hoss, R. J., *Fiber Optic Communications—Design Handbook*, Prentice Hall, Englewood Cliffs, NJ, 1990.

IGI Consulting, Inc., *Optical Amplifiers: Technology and Systems*, Global Information, Coquitlam, BC, Canada, 1999.

Jeong, Y. et al., Ytterbium-doped double-clad large-core fibers lasers with kW-level continuous-wave output power, *Conference on Lasers and Electro-Optics Proceedings of the SPIE*, San Francisco, CA, 5232, 140–150, 2004.

Kao, C. K., *Optical Fiber Systems: Technology, Design, and Applications*, McGraw-Hill, New York, 1982.

Kashyap, R., *Fiber Bragg Gratings*. Academic Press, New York, p. 458, 1999.

Keiser, G., *Optical Fiber Communications*, 3rd ed., McGraw-Hill, New York, 2000.

Keiser, G., *Optical Communications Essentials*, 1st ed., McGraw-Hill, New York, 2003.

Kolimbiris, H., *Fiber Optics Communications*, Prentice Hall, Englewood Cliffs, NJ, 2004.

Li, X., Chen, J., Wu, G., and Ye, A., Digitally tunable optical filter based on DWDM thin film filters and semiconductor optical amplifiers, *Opt. Express*, 13 (4), 1346–1350, 2005.

Litchinitser, N. M., Dunn, S. C., Steinvurzel, P. E., Eggleton, B. J., White, T. P., McPhedran, R. C., and Martijn de Sterke, C., Application of an ARROW model for designing tunable photonic devices, *Opt. Express*, 19, 1540–1550, 2004.

Nortel, Networks, Products, Services, and Solutions Change Notification, Nortel Equipment, Nortel Products Manual, Canada, 2005.

Palais, J. C., *Fiber Optic Communications*, 4th ed., Prentice Hall, Englewood Cliffs, NJ, 1998.

Ralston, J. M. and Chang, R. K., Spontaneous-Raman-scattering efficiency and stimulated scattering in silicon, *Phys. Rev. B.*, 2, 1858–1862, 1970.

Saini, S., Michel, J., and Kimerling, L. C., Index scaling for optical amplifiers, *IEEE J. Lightwave Tech.*, 21 (10), 2368–2376, 2003.

Salah, B. E. A. and Teich, M. C., *Fundamentals of Photonics*, Wiley, New York, 1991.

Senior, J. M., *Optical Fiber Communications*, 2nd ed., Prentice Hall, Englewood Cliffs, NJ, 1993.

Shamir, J., *Optical Systems and Processes*, SPIE-The International Society for Optical Engineering, SPIE Press, Bellingham, WA, 1999.

Shotwell, R. A., *An Introduction to Fiber Optics*, Prentice Hall, Englewood Cliffs, NJ, 1997.

Tran, A. V., Chae, C. J., and Tucker, R. S., A bidirectional optical add-drop multiplexer with gain using multiport circulators. Fiber Bragg gratings, and a single unidirectional optical amplifier, *IEEE Photon. Technol. Lett.*, 15, 975–977, 2003.

Uskov, A. V., Berg, T. W., and Mrk, J., Theory of pulse train amplification without patterning effects in quantum dot semiconductor optical amplifiers, *IEEE J. Quantum Electron.*, 40 (3), 306–320, 2004.

Uskov, A. V., Mork, J., Tromborg, B., Berg, T. W., Magnusdottir, I., and O'Reilly, E. P., On high-speed cross-gain modulation without pattern effects in quantum dot semiconductor optical amplifiers, *Opt. Commun.*, 227 (4–6), 363–369, 2003.

Uskov, A. V., O'Reilly, E. P., Manning, R. J., Webb, R. P., Cotter, D., Laemmlin, M., Ledentsov, N. N., and Bimberg, D., On ultra-fast optical switching based on quantum dot semiconductor optical amplifiers in nonlinear interferometers, *IEEE Photon. Technol. Lett.*, 16 (5), 1265–1267, 2004.

Vail, E., Wu, M. S., Li, G. S., Eng, L., and Chang-Hasnain, C. J., GaAs micromachined widely tunable fabryperot filters, *Electron. Lett.*, 31 (3), 228–229, 1995.

Yariv, A., *Optical Electronics*, Wiley, New York, 1997.

Yariv, A., Universal relations for coupling of optical power between microresonators and dielectric waveguides, *Electron. Lett.*, 36, 321–322, 2000.

Yeh, C., *Handbook of Fiber Optics: Theory and Applications*, Academic Press, San Diego, CA, 1990.

Yeh, C., *Applied Photonics*, Academic Press, San Diego, CA, 1994.

Zirngibl, M., Joyner, C. H., and Glance, B., Digitally tunable channel dropping filter/equalizer based on waveguide grating router and optical amplifier integration, *IEEE Photon. Technol. Lett.*, 6, 513–515, 1994.

9 Optical Receivers and Senders

9.1 INTRODUCTION

Optical receivers are an essential part of a communication system. An optical receiver converts an optical signal, transmitted through an optical fibre cable, into an electrical signal suitable for a receiving device installed at the other end of the communication system. The conversion process in the receiver is performed by two essential parts: a detector and an electronic signal processor. The detector converts the optical signal into an electrical signal. The electronic signal processor converts the raw detector signal into a form decipherable by the receiving device, such as a telephone, camera or scanner.

This chapter presents two experimental cases: measuring light power using two types of photo-detectors, and measuring the power output and calculating the efficiency of a solar cell with/without a filter and/or a lens.

9.2 FIBRE-OPTIC RECEIVERS

The function of the optical receiver is to pick up an optical signal and convert it into an electrical signal suitable for a receiving device at the end of the communication line. Optical signals can be data, video or audio. This conversion of an optical signal into an electrical signal is performed by an optical detector. Therefore, optical receivers are sometimes called optical detectors. Optical detectors perform the opposite function of optical transmitters, such as light emitting diodes and semiconductor lasers.

One application of an optical receiver is the conversion of an optical signal to digital form, as shown in Figure 9.1. The incoming optical signal is converted to an electronic signal using a photodetector, such as a p-i-n photodiode or an avalanche photodiode. The signal is then amplified by a pre-amplifier, and passed through a bandpass filter that removes unwanted wavelengths. Further amplification, with a gain feedback control circuit, provides stable signal levels for the rest of the process. This control circuit controls the bias current, and thus the sensitivity of the photodiode. A phase-locked loop (PLL)

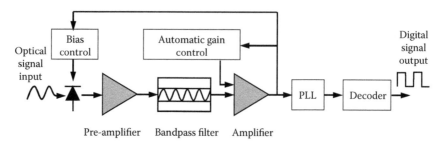

FIGURE 9.1 Optical signal conversion process.

recovers the data bit stream and the timing information. The stream of bits needs to be decoded, from the coding used on the line, into its data format coding. This decoding process varies, depending on the encoding, and is occasionally integrated with the PLL, depending on the code in use.

The most common type of optical detector is the semiconductor photodiode, which produces current in response to incident light. Detectors operate based on the principle of the semiconductor diode (*p-n* junction). An incident photon striking the diode gives an electron in the valence band of the atom. If the photon has sufficient energy to move the electron to the conduction band, this creates a free-moving electron and a hole. If the creation of these carriers occurs in a depleted region, the charge carriers (electrons and holes) will quickly separate and create a current. As they reach the edge of the depleted area, the electrical forces diminish and current ceases.

9.3 PRINCIPLES OF SEMICONDUCTORS

The following section describes the fundamental principles of semiconductors. In semiconductors, *doping* means adding a small amount of a material to the semiconductor. Doping changes the conductivity by creating an excess of electrons (*n*-doped) or holes (*p*-doped). This creates a *p*-region or *n*-region. The junction between the *n* and *p* forms a diode. Superscripts of + and − indicate light and heavy doping. The intrinsic semiconductor has no doping.

Semiconductors, conductors and insulators differ from each other based on their conductivity to electrical current. The electrical current in a material depends directly on the number of free electrons in the material. A comparison of the semiconductors, conductors and insulators is shown in Figure 9.2. This figure shows the differences amongst them based on energy gap thickness.

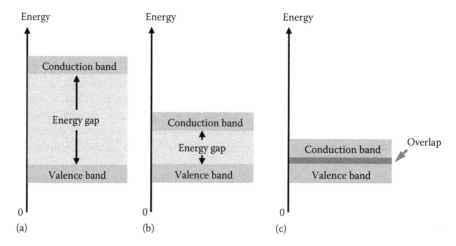

FIGURE 9.2 Energy diagram for the three types of materials: (a) insulator, (b) semiconductor and (c) conductor.

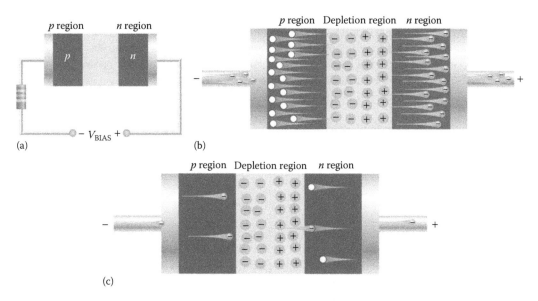

FIGURE 9.3 Junctions and depletion region with a reverse voltage bias: (a) reverse-bias connection, (b) there is transient current as depletion region widens and (c) majority current ceases when barrier potential equals bias voltage. There is an extremely small reverse current due to minority carriers.

9.3.1 *N*-TYPE AND *P*-TYPE SEMICONDUCTORS

Semiconductor materials are not as successful at conducting current. This is because of the limited number of free electrons in the conduction band and holes in the valence band. The number of free electrons and holes must be increased in the intrinsic silicon by using a process called doping. The doping process increases the number of electrons or holes by adding two categories of impurities *n*-type and *p*-type. There are many dopant and co-dopant (also called doping agent) are available for manufacturing semiconductors, such as erbium, boron, gallium, arsenic, phosphorus and gallium arsenide.

If you take a block of pure silicon and dope half of it with a trivalent impurity (*p*-type) and the other half with a pentavalent impurity (*n*-type), a boundary called the *p-n* junction is formed between the resulting *p*-type and *n*-type portions of a semiconductor diode.

A bias voltage *V* is applied to the *p-n* junction. Electrons diffuse away from the *n*-region into the *p*-region, leaving behind positively-charged ionised atoms (called donors), as shown in Figure 9.3. In the *p*-region, these electrons recombine with the abundant holes. Similarly, holes diffuse away from the *p*-region, leaving behind negatively-charged ionised atoms (called acceptors). In the *n*-region, the holes recombine with the abundant mobile electrons.

This diffusion process cannot continue indefinitely, however, because it causes a disruption of the charge balance in the two regions. As a result, a narrow region on both sides of the junction becomes almost totally depleted of mobile charge carriers (electrons and holes). This region is called the depletion region. This region has a built-in electric field E_o, due to the acceptors and donors beyond its edges. This occurs even without an applied voltage.

9.3.2 THE FUNDAMENTALS OF PHOTODETECTION

The fundamental principle behind the photodetection process is called optical absorption. If the energy ($h\nu$) of an incident photon exceeds the energy of the band gap (between the conduction and valence bands) and is absorbed in the depletion region, then an electron moves up to the conduction band leaving a hole in the valence band. Thus an electron-hole pair is generated each time a photon is absorbed by the semiconductor. Under the influence of the electric field, electrons and holes are swept across the semiconductor in opposite directions. This flow of carriers results in the flow of electric current called *generated photocurrent* when connected to an electric circuit. An applied voltage serves to speed up the carrier movement, increasing the current.

The fraction of light absorbed by the photodiode depends on

1. Wavelength (λ) of the light, determined by the photon energy ($\varepsilon = h\nu = hc/\lambda$).
2. The thickness of the absorption material (depletion region width or depletion layer thickness).

9.3.3 Leakage Current

The current resulting from absorbed incident light is called *generated photocurrent*. The current passing through the detector in the absence of light is called *dark current* or *leakage current*. Low leakage current is an important measure of device quality. If the dark current is high, the generated photocurrent needs to be larger, in order to provide a good signal. Otherwise, the leakage current will dominate the detector current. Therefore, it is important to control the dark current below a certain value.

9.3.4 Sources of Leakage Current

The three essential sources of leakage current are

1. *Generation-recombination (g-r) current*: Arises from the generation and recombination of electron-hole pairs (EHPs) in the diode depletion region. The *g-r* current dominates the leakage current at low temperature.
2. *Diffusion current*: Arises from the diffusion of the minority carriers, towards or away from the junction, in the diode neutral region. In the case of P^+-n junction with the intrinsic region, width larger than the hole diffusion length, the intrinsic region alone may be considered. Then diffusion current dominates the leakage current at high temperature.
3. *Tunnelling current*: Refers to the band-to-band tunnelling in the presence of high electric fields. A high electric field reduces the effective band gap barrier, allowing carriers to cross the band gap.

9.4 PROPERTIES OF SEMICONDUCTOR PHOTODETECTORS

The following sections describe the most common properties of semiconductor photodiodes.

9.4.1 Quantum Efficiency

The quantum efficiency (QE) of a photodetector is a measure of how effectively the detector converts light into an electrical current. QE denoted by η, is defined as the ratio of the flux of generated EHPs that contribute to the detector current, to the flux of the incident photons. The QE of a detector is the ratio of the number of photons actually detected, to the number of incident photons. The QE range is $0 \leq \eta \geq 1$. The QE of the photodiode is defined as

$$\eta = \frac{\text{Number of free EHP generated and collected}}{\text{Number of incident photons}} \tag{9.1}$$

As QE is a function of photon energy, the QE is calculated at a particular photon energy. The measured photocurrent (I_{ph}) in the external circuit is due to the flow of electrons to the terminals of the photodiode. The number of electrons collected at the terminals per second is I_{ph}/e, where e is the charge of an electron. For incident optical power P_o, the number of incident photons arriving per second is $P_o/h\nu$. Thus, the QE η can also be defined as

$$\eta = \frac{I_{ph}/e}{P_o/h\nu} \qquad (9.2)$$

9.4.2 RESPONSIVITY

The responsivity (R) of a photodetector is defined as the ratio of the photocurrent (I_{ph}) flowing in the device, to the incident optical power (P_o). Responsivity is measured in amps per watt. Thus:

$$R = \frac{\text{Photocurrent}}{\text{Incident optical power}} = \frac{I_{ph}}{P_o} \qquad (9.3)$$

As R is a function of photon wavelength, R is calculated at a particular wavelength λ. Substituting Equation 9.2 into Equation 9.3 gives

$$R = \eta \frac{e}{h\nu} = \eta \frac{e\lambda}{hc} \qquad (9.4)$$

As given in Equations 9.2 and 9.4, the efficiency and responsivity depend on wavelength. R is also called the spectral responsivity or radiant sensitivity. The R versus λ characteristics represents the spectral response of the photodiode. Spectral response curves are generally provided by the manufacturer. The spectral response characteristics for various quantum efficiencies are shown in Figure 9.4, and can be calculated using Equation 9.4. The outer area of the detector has a higher responsivity than the centre area, which can cause problems when aligning the fibre cable to the detector.

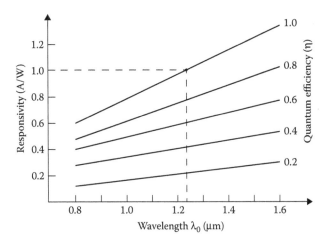

FIGURE 9.4 Responsivity versus wavelength for various quantum efficiencies.

9.4.3 Response Time

Response time is defined as the time needed for the photodiode to respond to an optical input by producing photocurrent. When a light incident on the photodiode generates an EHP in a photodetector material, an electrical charge is generated in an external circuit, as shown in Figure 9.5. This electrical charge, due to the electron and hole, equals $2e$ (e is the charge of an electron).

The charge delivered to the external circuit, by the movement of carriers in the photodetector material, is not provided instantaneously. The charge is delivered over an extended period. It is as if the motion of the charged carriers in the material draw charge slowly from the wire on one side of the device and pushes it slowly into the wire at the other side. In this way, each charge passing through the external circuit is spread out in time. This phenomenon is called transit-time spread. It is an important limiting factor for the speed of operation of all semiconductor photodetectors.

Consider an EHP generated (by photon absorption, for example) at position x in a semiconductor material. The semiconductor material has width W, to which a voltage V is applied, as shown in Figure 9.5. When an EHP is generated at position x, the hole moves to the left with velocity v_h, and the electron moves to the right with velocity v_e. This movement terminates when the carriers reach the edge of the material. The current (i) in the external circuit, generated by this movement, is given by

$$i(t) = -\frac{Q}{W}v(t) \tag{9.5}$$

where:

   t is the time
   Q is the total charge of the photo-generated electrons

If the voltage is increased the electron velocity increases, therefore increasing the current. This means that for an input light pulse, the output current pulse will have a faster response time for a higher applied voltage.

Response time can be affected by dark current, noise, responsivity linearity, back-reflection and detector-edge effect. Edge effect occurs because detectors only provide a fast response in their centre area.

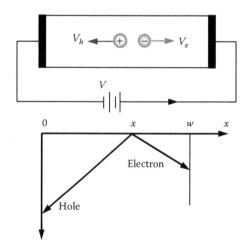

FIGURE 9.5   Generated electron-hole pair at position x.

9.4.4 SENSITIVITY

A photodetector is a device that converts photon energy into an electrical signal. A photodetector usually detects the energy of some photons better than others. Detection sensitivity is a function of the photon's energy being detected. The sensitivity is usually given as a function of the wavelength, and expressed as the QE. A high sensitivity allows a low level of light to be detected.

9.5 TYPES OF OPTICAL DETECTORS

There are many types of optical detectors including the phototransistor, photovoltaic, metal-semiconductor-metal (MSM), *pin* photodiode and avalanche photodiode (APD). These detectors are explained in the following sections.

9.5.1 PHOTOTRANSISTORS

Phototransistors are the simplest type of photodetector. The basic operating principle of a phototransistor is shown in the cross-section in Figure 9.6. The device consists of an *n-p-n* junction, in which *n* is the emitter, *p* is the base and the other *n* is the collector. The base terminal is normally open, and there is a voltage applied between the collector and emitter terminals (just as in the normal operation of common bipolar junction transistor [BJT]).

A large space charge layer (SCL) forms between the base and collector. The SCL region is called the absorption region. The operation of this device begins when an incident photon is absorbed in the SCL, and generates EHP. The electrical field E_o drifts the electron and hole in opposite directions. Phototransistors operate as a photodetector that amplifies the photocurrent. An applied voltage V will increase E to become E_o plus V. When the drifting electron reaches the collector, it gets collected (and thereby neutralised) by the power supply (applied voltage). On the other hand, when the hole enters the neutral base region, it can only be neutralised by injecting a large number of electrons into the base. It forces a large number of electrons to be injected from the emitter.

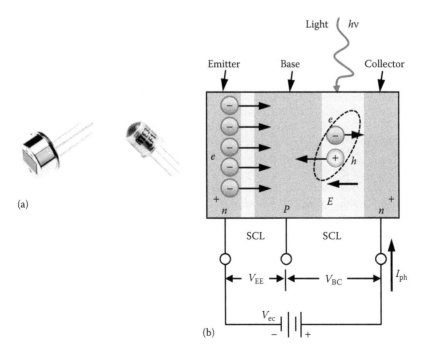

FIGURE 9.6   Phototransistor basic operation. (a) Phototransistors and (b) Phototransistor cross-section.

Normally, the electron recombination time in the base is very long, compared with the time it takes for electrons to diffuse across the base. This means that only a small fraction of electrons injected from the emitter can recombine with holes in the base. Thus, the emitter has to inject a large number of electrons to neutralise this extra hole in the base. These electrons diffuse across the base and reach the collector, and thereby create a photocurrent, which is amplified compared to the original electron. Thus, phototransistors have photocurrent gain.

9.5.2 Photovoltaics

Photovoltaic panels, or solar cells, convert the incident solar radiation, through the photovoltaic effect, into electrical current. The basic principle behind this effect relies on the small energy gap between the valence and conduction bands of the photovoltaic material. When light photons incident on a photovoltaic have enough energy to excite electrons from the valence to the conduction band, the resulting accumulation of charge leads to a flow of current.

Figure 9.7 shows a typical solar panel and its cross-section. Consider a *p-n* junction with a very narrow and more heavily doped *n*-region. Solar radiation is incident on the thin *n*-side. The electrodes attached to the *n*-side must allow illumination to enter the cell and at the same time have a small series resistance. The electrodes are deposited on the *n*-side in order to form an array of finger electrodes on the surface. A thin anti-reflection coating on the surface reduces reflections and allows more light to enter the cells.

The width (W) of the depletion region or the SCL extends primarily into the *p*-side. Most photons are absorbed within the *n*-region and depletion region. Thus, short and medium wavelengths are absorbed. The generated EHPs are swept away by the built-in field E_o in the depletion layer. This creates an open circuit voltage V_{oc} between the electrodes. If an external load is connected, a photocurrent results.

The efficiency of a solar cell is one of its most important characteristics; it allows the device to be assessed economically, in comparison to other energy conversion devices. The solar cell efficiency refers to the fraction of incident light energy converted to electrical energy. This conversion efficiency depends on the semiconductor material properties, the device structure and the incident light wavelength spectrum, which is mostly solar radiation. The efficiency of a solar cell decreases with increasing temperature. Therefore, the temperature of solar cells must be controlled for maximum efficiency.

Most solar cells are silicon based; because silicon-based semiconductor fabrication is a very developed technology, enabling cost-effective devices for energy production in remote applications.

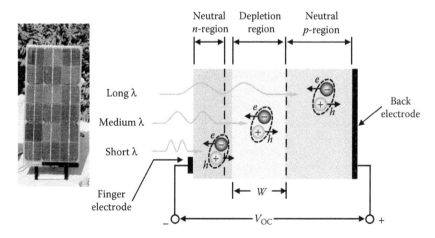

FIGURE 9.7 Photovoltaic panel.

A solar cell fabricated by making a *p-n* junction in the same crystal is called *homojunction*. A silicon homojunction solar cell is called a *single crystal passivated emitter rear locally diffused* (PERL) cell. It has higher efficiency than other types of semiconductor solar cells.

9.5.3   METAL-SEMICONDUCTOR-METAL DETECTORS

Metal-semiconductor-metal (MSM) detectors are probably the fastest and simplest optical detector to fabricate. To fabricate a MSM a Shottky barrier is first created, which forces the material at the surface to be depleted. This barrier is created by contacting a metal to the semiconductor surface. Figure 9.8 shows the cross-section of a metal area. The barriers are often in the form of inter-digitated metal fingers separated by a small distance, typically on the order of microns. The metal is usually opaque to the incoming light; the remainder of the surface area absorbs the light. All the depletion layers are connected together. Any absorbed light generates EHPs, which are quickly swept out to the contacts. As the response time is so quick, full-width at half maximum (FWHM) pulses are measured in picoseconds for such structures.

When using the MSM for the detection of 1300 nm and 1550 nm ranges, the MSM suffers from the following two serious drawbacks:

1. Shottky barriers on indium phosphate (InP) tend to have high dark current, and therefore low receiver sensitivity.
2. Low QE results, because the metal fingers prevent some of the incoming light from reaching the absorption layer.

9.5.4   THE *P-I-N* PHOTODIODES

In *p-i-n* photodiodes, the conversion of light into electrical current is achieved by the creation of free EHPs by the absorption of photons. This absorption process creates electrons in the conduction band and holes in the valence band. Figure 9.9 shows the simplified structure of a typical *p-i-n* junction photodiode. The structure of the photodiode is the p^+-intrinsic-n^+ junction. The intrinsic silicon (*i*-Si) layer has much less doping than both the p^+ and n^+ regions, and it is much wider than these regions. At long wavelengths, where penetration depth is large, the photons can be absorbed in the

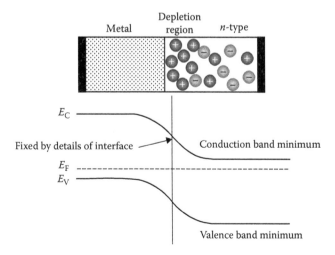

FIGURE 9.8   Shottky barrier and energy diagram.

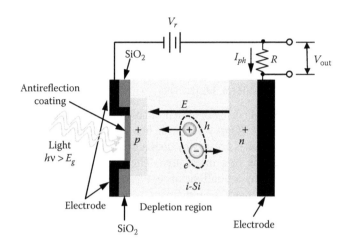

FIGURE 9.9   A *p-i-n* junction photodiode.

wide depletion region. Thus, the width depends on the particular wavelength used in the application. In contrast, a *p-n* junction has a narrow depletion region and fewer photons are absorbed.

Holes diffuse from the p^+ side, and electrons from the n^+ side, into the *i*-Si layer, also called the depletion region. In this region, they recombine (with other holes and electrons) and disappear. This leaves behind a thin layer of exposed negatively charged acceptor ions in the p^+ side, and a thin layer of exposed positively-charged donor ions in the n^+ side. The two charges are separated by the built-in electric field E in the *i*-Si layer. An exterior voltage increases E, which increases the response speed. Although the photogenerated carriers are drifting through the *i*-Si layer, they create an external photocurrent when a voltage is applied. This photocurrent can be detected as the voltage across a small external resistor R, as shown in Figure 9.9. A larger thickness of the *i*-Si layer increases QE, but slows the response time because carriers have further to travel.

In some photodiodes, such as pyroelectric detectors, the energy conversion generates heat, which increases the temperature of the detector. The temperature increase changes the polarisation and relative permittivity of the photodiode.

The *p-i-n* photodetectors offer high bandwidth, high QE and low dark current. High bandwidth and low dark current are important characteristics for good receiver sensitivity. However, the device has no gain, which places a lower limit on the sensitivity achievable, before dark current becomes significant. The *p-i-n* photodetectors are small devices with small capacitance, thus can detect high-speed signals with high sensitivity. The most important applications of the photodiodes are in optical communications.

9.5.5   AVALANCHE PHOTODIODES

Avalanche photodiodes (APDs) are high performance devices, and are widely used in many applications, such as optical communication, due to their high speed and internal gain. Although not as fast as *p-i-n* photodiodes, the devices offer superior receiver sensitivity in their own bandwidth range. The device bandwidth at high gain is limited by the gain-bandwidth product (GBW), for InGaAs-InP APDs. However, there are difficulties in fabricating the device; this process requires stringent process control. For this reason, commercial high performance APDs cost more than similar photodetectors.

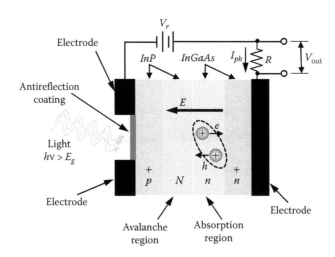

FIGURE 9.10 Avalanche photodiode.

Figure 9.10 shows the cross-section of the structure of an InGaAs-InP avalanche photodiode with separate absorption and multiplication (SAM) regions. The InP multiplication (avalanche) layer has a wider band gap than the InGaAs absorption layer. The p-type and n-type doping of InP is indicated by capital letters, P and N.

The main depletion layer forms between the P^+-InP and the N-InP layers within the N-InP. The electric field is greatest in this N-InP layer; this is therefore where avalanche multiplication takes place. With sufficient reverse voltage bias, the depletion layer in the n-InGaAs extends into the N-InP layer.

The electric field in the n-InGaAs depletion layer is not as great as that in the N-InP. Although long wavelength photons are incident on the InP side, they are not absorbed by InP, because the photon energy is less than the band gap energy of InP ($E_g = 1.35$ eV). Long wavelength photons pass through the InP layer, and are absorbed in the n-InGaAs layers.

The electric field E in the n-InGaAs layer drifts the holes to the multiplication region, where impact ionisation multiplies the carriers. The impact ionisation, from the physics point of view, is the mechanism that creates the internal current gain. Primary electrons and/or holes (carriers) are generated through the absorption of photons. Carriers can acquire large amounts of energy from a high E field, when the device has a strong reverse-voltage bias. This can be translated into high-speed motion. When a collision between a carrier and the lattice occurs, the energy from the carrier can be transferred to the lattice. Sufficient energy can be absorbed by the lattice for an electron to be promoted from the valence to the conduction band, creating an EHP. This process is called impact ionisation. These new carriers are swept out by E and can acquire high energy causing further EHPs to be created. The entire process in which many carriers are created from one initial carrier is called avalanche multiplication.

9.6 COMPARISON OF PHOTODETECTORS

The most common candidates for photodetectors are p-i-n photodiodes and APDs. They both have the same basic operating principle, where light is absorbed and converted to photocurrent. Table 9.1 highlights the major differences between p-i-n and APD detectors.

TABLE 9.1
Comparison of Photodetectors

p-i-n Detectors	APD Detectors
Fast	Not as fast
High bandwidth, up to 40 GHz at quantum efficiency >80%	Significantly less bandwidth
Low dark current	High dark current
No gain, which leads to lower sensitivity	Built-in gain, which extends the sensitivity to lower levels of received light
Conversion efficiency (responsivity) from 0.5 to 1.0 Amps/Watt	Conversion efficiency (responsivity) from 0.5 to 100.0 Amps/Watt
Less expensive	More expensive

9.7 EXPERIMENTAL WORK

The experiments on optical detectors include the following two experiments:

9.7.1 MEASURING LIGHT POWER USING TWO PHOTODETECTOR TYPES

The power of a light source can be measured using two types of photodetectors, and compare amongst them with data sheets provided by the manufacturers. Figure 9.11 illustrates the experimental set-up.

9.7.2 PHOTOVOLTAIC PANEL TESTS

The purpose of this experiment is to measure the intensity of the solar radiation using a solar radiation meter (black and white pyranometer). Students will determine the converted power and efficiency of a photovoltaic panel subjected to solar radiation. The students will also determine the efficiency of the photovoltaic panel using different set-ups described by the following cases:

1. A photovoltaic panel producing electricity without using a filter or lens. Figure 9.12 (without a filter or lens) shows the experiment set-up for a photovoltaic panel producing electricity.
2. A photovoltaic panel producing electricity using a lens. Figure 9.12 (without a filter) shows the experiment set-up for a photovoltaic panel producing electricity using a lens to concentrate solar radiation.

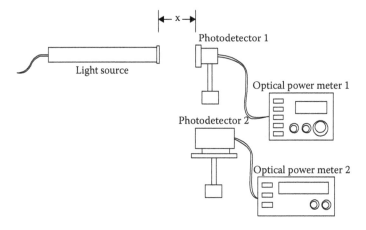

FIGURE 9.11   Measuring light power using a photodetector.

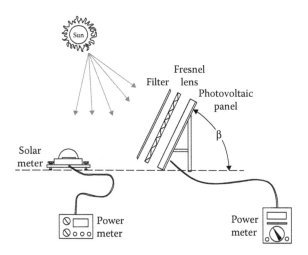

FIGURE 9.12   Photovoltaic panel with a lens and filter combination.

3. A photovoltaic panel producing electricity using four filters (one at a time). Figure 9.12 (without a lens) shows the experiment set-up for a photovoltaic panel producing electricity using four types of filters. These filters are used to filter different wavelengths of the solar radiation.

4. A photovoltaic panel producing electricity using a combination between a lens and four filters (one at a time). Figure 9.12 shows the experiment set-up for a photovoltaic panel producing electricity using a combination of a lens and one filter. These filters are used to filter different wavelengths of the solar radiation.

9.7.3   Technique and Apparatus

Appendix A presents the details of the devices, components, tools and parts.

1. Fresnel lens and lens holder, as shown in Figure 9.13.
2. Four colour filters and filter holder, as shown in Figure 9.14.
3. Light source, as shown in Figure 9.15.
4. Photodetector, as shown in Figure 9.15.
5. Power light meter, as shown in Figure 9.16.
6. Photovoltaic panel and stand, as shown in Figure 9.16.
7. Solar radiation sensor (black and white pyranometer), as shown in Figure 9.16.
8. Solar radiation meter, as shown in Figure 9.16.
9. Hardware assembly (clamps, screws, etc.).
10. Ruler.

9.7.4   Procedure

Follow the laboratory procedures and instructions given by the professor and/or instructor.

9.7.5   Safety Procedure

Follow all safety procedures and regulations regarding the use of pyranometer, photovoltaic panel, lens, filters and power meters.

FIGURE 9.13   A Fresnel lens.

FIGURE 9.14   Four colour filters.

9.7.6   Apparatus Set-Up

9.7.6.1   Measuring Light Power Using Two Photodetector Types

1. Figure 9.15 shows the experimental apparatus set-up.
2. Prepare a light source on the table and connect it to the power supply.
3. Mount a photodetector, connected to an optical power meter, at a distance of 2 cm facing the light source.

FIGURE 9.15 Measuring light power using a photodetector.

FIGURE 9.16 Photovoltaic panel set-up.

4. Align the photodetector and light source to face each other.
5. Turn on the power supply, to provide electrical power to light source.
6. Turn off the lights of the lab.
7. Measure light power emitted by the light source, using the photodetector. Fill out Table 9.2.
8. Repeat steps 3–7 for the second type of the photodetector.
9. Turn on the lights of the lab.
10. Turn off the power supply, to provide electrical power to light source.
11. Illustrate the location of the light sources and the photodetector head in a diagram.

TABLE 9.2

Measuring Light Power Using Two Photodetector Types

| Light Source Power (unit) | Optical Power-Measurements by | |
	Photodetector #1 (unit)	Photodetector #2 (unit)

9.7.6.2 Photovoltaic Panel Tests

1. *Photovoltaic panel*
   a. Figure 9.16 (without a filter or lens) shows the experimental apparatus set-up.
   b. Mount a photovoltaic panel on the stand. Place the photovoltaic panel and stand on a flat surface. Place the photovoltaic panel facing the sun at the south orientation. Normally, the photovoltaic panel is positioned inclined at an angle equal to the geographic location of the site, to collect maximum solar energy without tracking the sun movement during the day.
   c. Measure the electrical power produced by the photovoltaic panel. Fill out Table 9.3.
   d. Place a solar radiation meter on the flat surface near to the photovoltaic panel.
   e. Measure the solar radiation intensity, using the solar radiation meter. Fill out Table 9.3.
   f. Illustrate the locations of the photovoltaic panel and solar radiation meter in a diagram.
2. *Photovoltaic panel with a lens*
   a. Repeat the procedure explained in Case (a) of this experiment, but use a Fresnel lens to concentrate solar radiation.
   b. Mount the Fresnel lens on the photovoltaic panel. Keep a distance between the lens and the photovoltaic panel. The distance should be less than the focal length of the lens.
   c. Measure the electrical power produced by the photovoltaic panel. Fill out Table 9.4.
   d. Measure the solar radiation intensity, using the solar radiation meter. Fill out Table 9.4.
   e. Illustrate the locations of the photovoltaic panel, Fresnel lens, and solar radiation meter in a diagram.

TABLE 9.3

Photovoltaic Panel Set-Up

Solar Radiation Intensity (unit)	Electrical Power (unit)

TABLE 9.4

Photovoltaic Panel with a Lens Set-Up

Solar Radiation Intensity (unit)	Electrical Power (unit)

TABLE 9.5
Photovoltaic Panel with a Filter Set-Up

Filter Types	Solar Radiation Intensity (unit)	Electrical Power (unit)
Red		
Yellow		
Green		
Blue		

TABLE 9.6
Photovoltaic Panel with a Lens and Filter Combination Set-Up

Filter Types	Solar Radiation Intensity (unit)	Electrical Power (unit)
Red		
Yellow		
Green		
Blue		

3. *Photovoltaic panel with a filter*
   a. Repeat the procedure explained in Case (a) of this experiment, but use a filter without a lens. Using one colour filter, filter out some wavelengths of the solar radiation spectrum.
   b. Measure the electrical power produced by the photovoltaic panel. Fill out Table 9.5.
   c. Measure the solar radiation intensity, using the solar radiation meter. Fill out Table 9.5.
   d. Repeat steps 2–3 using the three other colour filters.
   e. Illustrate the locations of the photovoltaic panel, filter, and solar radiation meter in a diagram.
4. *Photovoltaic panel with a lens and filter combination*
   a. Repeat the procedure explained in Case (a) of this experiment, but use a Fresnel lens and one filter.
   b. Mount the Fresnel lens and one colour filter on the photovoltaic panel.
   c. Measure the electrical power produced by the photovoltaic panel. Fill out Table 9.6.
   d. Measure the solar radiation intensity, using the solar radiation meter. Fill out Table 9.6.
   e. Repeat steps 3–4 using three other colour filters.
   f. Illustrate the locations of the photovoltaic panel, Fresnel lens, filter and solar radiation meter in a diagram.

9.7.7 DATA COLLECTION

9.7.7.1 Measuring Light Power Using Two Photodetector Types

Measure the light power emitted by the light source, using two types of photodetectors. Fill out Table 9.2.

9.7.7.2 Photovoltaic Panel Tests

1. *Photovoltaic panel*
   a. Measure the solar radiation intensity, using the solar radiation meter.
   b. Measure the electrical power converted by the photovoltaic panel.
   c. Fill out Table 9.3.

2. *Photovoltaic panel with a lens*
   a. Measure the solar radiation intensity, using the solar radiation meter.
   b. Measure the electrical power converted by the photovoltaic panel.
   c. Fill out Table 9.4.

3. *Photovoltaic panel with a filter*
   a. Measure the solar radiation intensity, using the solar radiation meter.
   b. Measure the electrical power converted by the photovoltaic panel.
   c. Fill out Table 9.5.

4. *Photovoltaic panel with a lens and filter combination*
   a. Measure the solar radiation intensity, using the solar radiation meter.
   b. Measure the electrical power converted by the photovoltaic panel.
   c. Fill out Table 9.6.

9.7.8 CALCULATIONS AND ANALYSIS

9.7.8.1 Measuring Light Power Using Two Photodetector Types

No calculations or analysis are required for this experiment.

9.7.8.2 Photovoltaic Panel Tests

No calculations or analysis are required for Cases (a), (b), (c) and (d).

9.7.9 RESULTS AND DISCUSSIONS

9.7.9.1 Measuring Light Power Using Two Photodetector Types

1. Study the specifications of the light source and the photodetectors.
2. Discuss the measurement and accuracy of the two types of photodetectors.
3. Compare the two types of photodetectors.
4. Verify the optical-power measurements with the data provided by the manufacturer.

9.7.9.2 Photovoltaic Panel Tests

1. *Photovoltaic panel*
   a. Discuss the measurements and weather conditions.
   b. Illustrate the locations of the photovoltaic panel and solar radiation meter in a diagram.

2. *Photovoltaic panel with a lens*
   a. Discuss the measurements and how the power output is affected by the weather conditions. Verify the measurements obtained in this case with Case (a).
   b. Illustrate the locations of the photovoltaic panel, lens and solar radiation meter in a diagram.

3. *Photovoltaic panel with a filter*
   a. Discuss the measurements and how the power output is affected by the weather conditions. Verify the measurements obtained in this case with Cases (a) and (b).
   b. Illustrate the locations of the photovoltaic panel, filter and solar radiation meter in a diagram.

4. *Photovoltaic panel with a lens and filter combination*
   a. Discuss the measurements and how the power output is affected by the weather conditions. Verify the measurements obtained in this case with Cases (a), (b) and (c).
   b. Illustrate the locations of the photovoltaic panel, lens, filter and solar radiation meter in a diagram.

9.7.10 Conclusion

Summarise the important observations and findings obtained in this lab experiment.

9.7.11 Suggestions for Future Lab Work

List any suggestions for improvements using different experimental equipment, procedures and techniques for any future lab work. These suggestions should be theoretically justified and technically feasible.

9.8 LIST OF REFERENCES

List any references that were used in the report. Use one format in writing the references. Never mix reference formats in a report.

9.9 APPENDIX

List all of the materials and information that are too detailed to be included in the body of the report.

FURTHER READING

Agrawal, G. P., *Fiber-Optic Communication Systems*, 2nd ed., John Wiley & Sons, New York, 1997.
Agrawal, G. P., and Dutta, N. K., *Semiconductor Lasers*, 2nd ed., Van Nostrand, New York, 1993.
Bar-Lev, A., *Semiconductors and Electronic Devices*, 2nd ed., Prentice Hall International, Upper Saddle River, NJ, 1984.
Bean, J., Optical wireless: Secure high-capacity bridging, *Fiber Opt. Technol.*, 10–13, 2005.
Derfler, F. J., Jr. and Freed, L., *How Networks Work*, Millenium Edition, Que Corporation, Indianapolis, IN, 2000.
Donati, S., *Photodetectors*, Prentice Hall PTR, Upper Saddle River, NJ, 2000.
Dutton, H. J. R., *Understanding Optical Communications*, IBM, Prentice Hall, Upper Saddle River, NJ, 1998.
Fedder, G. K., and Howe, R. T., Multimode digital control of a suspended polysilicon microstructure, *IEEE J. MEMS*, 5 (4), 283–297, 1996.
Fedder, G. K., Iyer, S., and Mukherjee, T., Automated optimal synthesis of microresonators, *Technical Digest of the IEEE International Conference on Solid-State Sensors and Actuators (Transducers' 97)*, Chicago, IL, June 16-19, Vol. 2, pp. 1109–1112, 1997.
Fedder, G. K., Santhanam, S., Rud, M. L., Eagle, S., Guillon, D. F., Lu, M., and Carliy, L. R., Laminated high aspect-ratio microstructures in a conventional CMOS process, *Sens. Actuators A*, 57 (2), 103–110, 1996.
Floyd, T. L., *Electronic Devices: Electron Flow Version*, 9th ed., Prentice Hall, Upper Saddle River, NJ, 2012.
He, S., and Mrad, R. B., A vertical bi-directional electrostatic comb driver for optical MEMS devices, *Paper presented at Opto-Canada, SPIE Regional Meeting on Optoelectronics, Photonics and Imaging. SPIE*, Vol. TD01, 2002.
Heidrich, H., Albruht, P., Hamachr, H. P., Notting, H., Schroiter-Janssery, and Uneinert, C.M., Passive mode converter with a periodically tilted InP/GaInAsP rib waveguide, *IEEE Photon. Technol. Lett.*, 4 (34), 36, 1992.
Him, F. C., Effects of passivation of GaN transistors, *Paper presented to Epitaxy Group of the IMS at NRC*, Ottawa, Canada, 2005.
Hioki, W., *Telecommunications*, 3rd ed., Prentice Hall, Upper Saddle River, NJ, 1998.
Horng, H. H., Designing optical-fiber modulators by using magnetic fluids, *Opt. Lett.*, 30 (5), 543–545, 2005.
Hoss, R. J., *Fiber Optic Communications-Design Handbook*, Prentice Hall, Englewood Cliffs, NJ, 1990.

Kaiser, R., Trommer, D., Heidrich, H., Fidorra, F., and Hamachr, M., Heterodyne receiver PICs as the first monolithically integrated tunable receivers, *Opt. Quantum Electron.*, 28, 565–573, 1996.

Kasap, S. O., *Optoelectronics and Photonics Principles and Practices*, Prentice Hall PTR, IN, 2001.

Kolimbiris, H., *Fiber Optics Communications*, Prentice Hall, Englewood Cliffs, NJ, 2004.

Litchinitser, N. M., Dunn, S. C., Steinvurzel, P. E., Eggletan, B. S., White, T. P., Mophedran, R. C., and de Sterky, C. M., Application of an ARROW model for designing tunable photonic devices, *Opt. Express*, 12, 1540–1550, 2004.

Mouthaan, T., *Semiconductor Devices Explained Using Active Simulation*, Wiley & Sons, New York, 1999.

Ralston, J. M., and Chang, R. K., Spontaneous-Raman-scattering efficiency and stimulated scattering in silicon, *Phys. Rev. B*, 2, 1858–1862, 1970.

Razavi, B., *Design of Integrated Circuits for Optical Communications*, McGraw-Hill Higher Eucation, OH, 2003.

Salah, B. E. A., and Teich, M. C., *Fundamentals of Photonics*, Wiley, New York, 1991.

SCIENCETECH, Modular Optical Spectroscopy, *Designers and Manufacturers of Scientific of Scientific Instruments Catalog*, SCIENCETECH, London, Ontario, Canada, 2005.

Senior, J. M., *Optical Fiber Communications: Principle and Practice*, 2nd ed., Prentice Hall, Englewood Cliffs, NJ, 1986.

Setian, L., *Applications in Electro-Optics*, Prentice Hall PTR, Upper Saddle River, NJ, 2002.

Simin, G., *High Speed Semiconductor Devices, Basics of heterostructures*, University of South Carolina, Columbia, SC, Online Course 02/ELCT882, 2005.

Singh, J., *Semiconductor Devices: Basic Principles*, Wiley & Sons, New York, 2001.

Sze, S. M., *Physics of Semiconductor Devices*, 2nd ed., Wiley, New York, 1981.

Thompson, G. H. B., *Physics of Semiconductor Laser Device*, Wiley & Sons, Chichester, 1980.

Tooley, M., *Electronic Circuits: Fundamentals and Applications*, 2nd ed., Newnes, Boston, MA, 2002.

Watanabe, T., Inoue, Y., Kaneko, A., Ooba, N., and Kurinara, T., Polymeric arrayed-waveguide grating multiplexer with wide tuning range, *Electron. Lett.*, 33 (18), 1547–1548, 1997.

Yariv, A., *Quantum Electronics*, 3rd ed., John Wiley & Sons, New York, 1989.

Yariv, A., *Optical Electronics*, John Wiley & Sons, New York, 1997.

10 Optical Switches

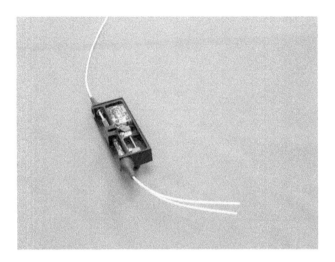

10.1 INTRODUCTION

Many optical networks integrate optical switches into their design. Opto-mechanical switches redirect optical signals from one port to another by moving a fibre tube assembly or an optical component, such as a mirror or prism. There are many different types of optical switches incorporated into networks. In practice, most optical switches are still operated mechanically and controlled by an electronic control circuit. Speed is a crucial parameter in network applications, because a high-speed data transmission of tenths of milliseconds is required. In the near future, dynamic optical routing will require much faster switching speeds. More technology exists for optical switches than any other functional component within the optical network. Researchers are developing optical switches to increase the number of outputs, and to reduce size, cost and switching time. Presently, optical switches include many types, for example: opto-mechanical switches, thermo-optic switches, electro-optic switches, micro-electro-mechanical switches (MEMS) and micro-opto-mechanical switches (MOMS). New types of optical switches are in the research and development stages.

This chapter illustrates a few switch designs, which are manufactured for use in communication systems, and other applications. Opto-mechanical and electro-mechanical switches are the oldest type of optical switch and the most widely deployed at this time. These devices achieve switching by moving fibre or other optical components by means of stepper motors or relay arms. This causes them to have relatively slow switching time; however, their reliability is excellent, and they offer low insertion and crosstalk losses.

This chapter presents four cases in building opto-mechanical switches using a movable mirror or prism to switch between the input and output ports.

10.2 OPTO-MECHANICAL SWITCHES

Figure 10.1 illustrates common switch configurations. The input signal comes through the input fibre cable on the left side of the switch. A mechanical slider moves that fibre up and down, latching into one of the two output fibre cables on the right side of the switch. In OFF/ON positions,

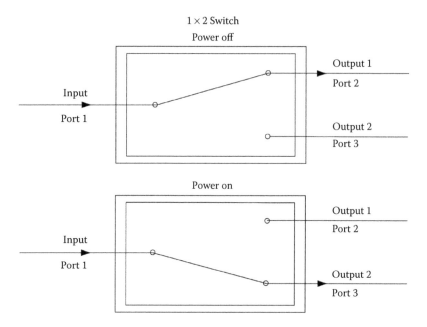

FIGURE 10.1   A typical 1 × 2 switch configuration.

the switch directs light from the input fibre into one of the two outputs. This arrangement is called *1 × 2 switch configuration*. As input at Port 1, the signal can be switched to either Port 2 or Port 3.

For the following definitions, assume the switch is configured to couple to Port 2. The insertion loss L_{IL} (in decibels) is defined by Equation 10.1. Insertion loss depends on fibre-cable alignment at the input and output ports. Low insertion loss value can be obtained on switches with good mechanical alignment. A good switch provides similar values of insertion loss for all switch positions.

$$L_{IL} = -10 \times \log_{10} \frac{P_2}{P_1} \qquad (10.1)$$

where P_1 is the power going into Port 1 and P_2 is the power exiting from Port 2.

Crosstalk loss L_{CT} is one of the important losses, which should be considered in opto-mechanical switches. Crosstalk loss is a measure of how well the uncoupled port is isolated. The crosstalk loss L_{CT} (in decibels) is defined by Equation 10.2. Crosstalk loss values depend on the particular design of the switch.

$$L_{CT} = -10 \times \log_{10} \frac{P_3}{P_1} \qquad (10.2)$$

where P_1 is the power going into Port 1 and P_3 is the power exiting from Port 3.

There are other important optical parameters that need to be specified for each switch type. These parameters include: polarisation-dependent loss (PDL), return loss (RL) and the Etalon effect. The PDL is defined as the maximum difference in insertion loss between any two polarisation states. It is caused by mechanical stress and temperature variation on optical components or fibre cables. This causes changes in the birefringence and a gradient of index of refraction (*n*) of the optical material. The RL is defined as the light reflected back into the input path. It is caused by scattering and reflection from optical surfaces like mirrors, lenses and connectors or from defects, such as cracks and scratches. The back reflection is equal to the RL with a negative quantity. Elaton effect is defined as light resonance (ripple) at a certain wavelength. It is caused by reflection of light from parallel optical surfaces and interference between the signals. All the above losses are measured in decibels (dB). Special optical parameters can be specified by the customers.

Another important parameter of the optical switches is the repeatability—achieving the same insertion loss each time the switch is returned to the same position. Switching speed is also another important specification of a switch. The switching speed is defined as how fast the switch can change the signal from one port to the other. It is an important factor in some switch applications in communication systems.

Figure 10.2 shows a schematic diagram of a mechanical switch configuration with two inputs and two outputs. The inputs are located on the one side and the outputs on the other side of the switch. This configuration is called a 2 × 2 switch. The signal enters Port 1 and Port 4, and exits from Port 2 and Port 3, respectively. This case is called the bypass state, in the OFF position. When the latching mechanism changes position between Port 2 and Port 3, signal enters Port 1 and exits Port 3 and from Port 4 to Port 2. This case is called the operate state, in the ON position.

Opto-mechanical switches collimate the optical beam from each input and output fibre and moves these collimated beams around inside of the switch. This creates low optical loss, and allows distance between the input and output fibre. These switches have more bulky components compared to newer alternatives, such as the micro-opto-mechanical switches.

Figure 10.3 shows a schematic diagram of a two-position switch. The switch consists of a sliding prism and quarter pitch graded-index (GRIN) lenses at the input and output ports. The components

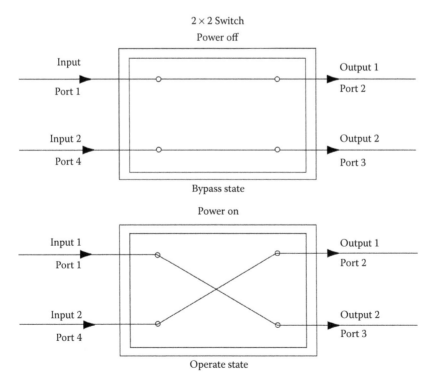

FIGURE 10.2   A typical 2 × 2 switch configuration.

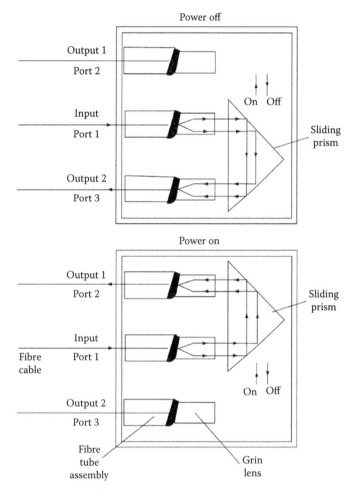

FIGURE 10.3 A 1 × 2 opto-mechanical switch.

are assembled in a packaging base and sealed with a lid. Each GRIN lens is connected to the fibre tube assembly using an epoxy. Figure 10.3 illustrates the OFF/ON positions of a 1 × 2 switch. As explained above, the GRIN lens collimates the divergence beam exiting from the input fibre. The right angle prism deflects the light by total internal reflection (TIR) at its two slanting surfaces. The GRIN lens refocuses the collimated beam onto a fibre cable at one of the output ports. To direct the signal from Port 1 to Port 3, the prism slides to a new position, as shown in Figure 10.3 in the OFF position. Figure 10.3 also shows the signal directed between Port 1 and Port 2, in the ON position, when the prism changes position.

Opto-mechanical switches drive optical fibre networks mechanically. They can switch between light paths at high speed and with low insertion loss. They are widely used in rapidly developing areas of the fibre-optic field, such as optical cross-connection and wavelength multiplexing. Figure 10.4 shows the design of a 2 × 4 opto-mechanical switch. The switch uses an electromagnetic actuator with a latching function to drive a movable block to change the light path between the ports. Figure 10.4 shows the switch in the OFF position. The light passes through from Port 1 and Port 2 to Port 3 and Port 4, respectively. When the power is turned ON, an electromagnetic actuator

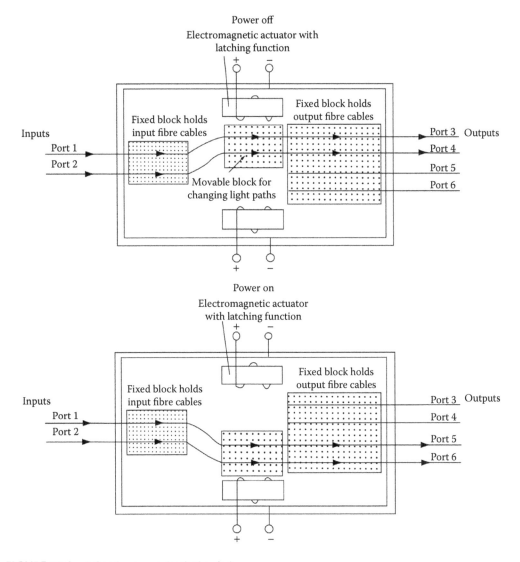

FIGURE 10.4   A 2 × 4 opto-mechanical switch.

(with latching function) drives a movable block to change light path from Port 1 and Port 2 to Port 5 and Port 6, respectively, as shown in Figure 10.4. The optical and mechanical components of a switch are assembled in a packaging box with minimal alignment work. There are three configurations of this switch: 1 × 2, 2 × 2 and 2 × 4.

A practical electromagnetic bypass switch is illustrated in Figure 10.5. The switch contains a quarter pitch GRIN lens connected to fibre tube assembly at the input and output ports, a relay and an iron bar with mirror end faces. The components are assembled in a packaging base, which is sealed with a lid. When the power is turned OFF, a spring pulls the iron bar out of the signal path, returning the switch to the bypass condition. This is called the bypass state. In the bypass state, the signal passes directly from Port 1 and Port 4. When the power is ON, the electromagnet is activated

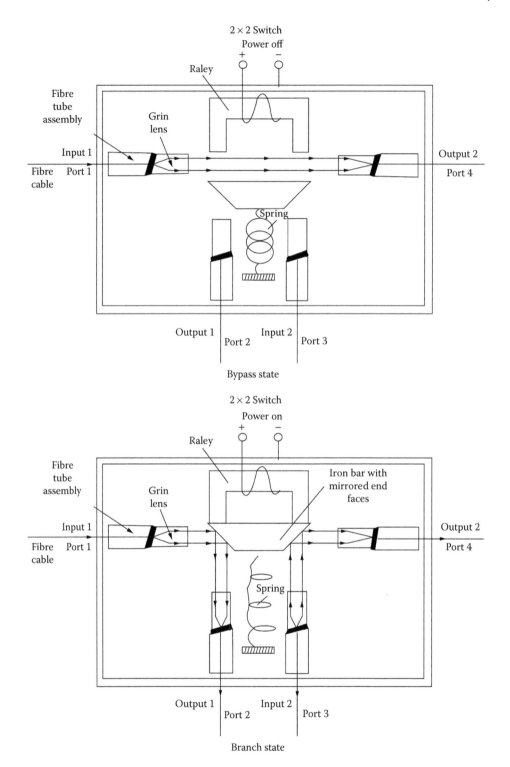

FIGURE 10.5 An electromagnetic bypass switch.

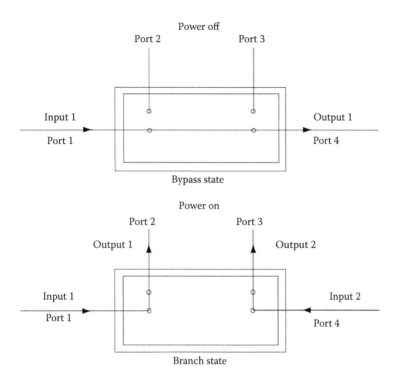

FIGURE 10.6 A bypass switch.

and the iron bar is raised. This is called the branch state. In the branch state, mirrors direct the signal between Port 1 and Port 2, and between Port 3 and Port 4.

Another type of bypass switch is also used in communication networks. Figure 10.6 illustrates the function of this type of bypass switch. When the power is in the OFF position, the input signal comes through the input fibre cable on Port 1 on the left side and leaves through the output Port 4 on the right side of the switch. This is called the bypass state. When the power is ON, a mechanical slider moves two fibre connections to the up position, latching into two output fibre cables at Port 2 and Port 3 on the side of the switch. In this position, the input signals from Port 1 and Port 4 are launched into Port 2 and Port 3, respectively. This position is called branch state. As input at Port 1, the signal can be directed to either Port 4 or Port 2. Also, an input signal at Port 4 can be directed to output Port 3.

This section presents new switch designs of a 1 × 8 latching switch configuration using prisms. These switches are commercially available in the market. There are two types of models: the linear and triangular models. The linear model directs the signal from the input to the outputs by arranging the prisms linearly, as shown in Figures 10.7 and 10.8. The triangular model directs the input to the outputs by arranging the prisms triangularly, as shown in Figures 10.9 and 10.10. These models have come into wide use because they are simple, offer 8 outputs, and are cost effective. They are also used in backup systems to re-route signals around broken fibre-optic cable and in fibre-optical instruments.

Figure 10.7 illustrates a schematic diagram for one configuration of a linear model of a 1 × 8 latching switch. The common element of this type of opto-mechanical switch is that their operation involves

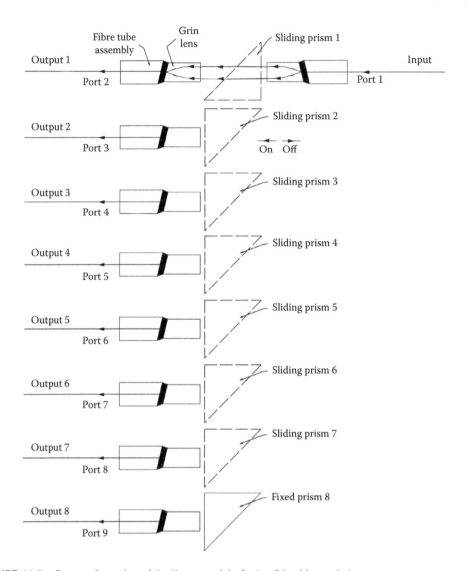

FIGURE 10.7 One configuration of the linear model of a 1 × 8 latching switch.

mechanical sliding motion of prisms in OFF/ON positions to direct the signal from one port to another. Figure 10.7 shows the input located on one side and the outputs on the other side of the switch. This configuration is called a 1 × 8 switch in the linear model. Light enters Port 1 and exits from Port 2 when sliding prism 1 is in the OFF position. When the latching mechanism places the sliding prisms 1 and 2 in position, the light enters Port 1 and exits from Port 3, as shown in Figure 10.8. Similarly, light enters Port 1 and exits from Port 4, when the sliding prism 2 is in the OFF position and sliding prism 3 is in the ON position. This switch configuration is more complicated than the second switch configuration because the input is located on one side and the outputs on the other side of the switch. This configuration includes a complex mechanism, controls, seven sliding prisms and one fixed prism.

Figure 10.9 illustrates a schematic diagram of the second configuration of the linear model of a 1 × 8 latching switch. This configuration is different because the input and the outputs are located on the same side of the switch, as shown in Figure 10.9. Prisms are also used in the operation of this

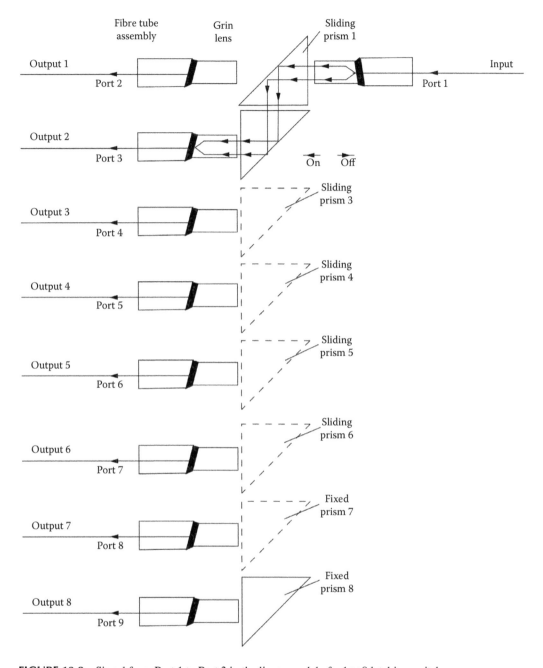

FIGURE 10.8   Signal from Port 1 to Port 3 in the linear model of a 1 × 8 latching switch.

type of switch configuration. Light enters Port 1 and exits from Port 2 when fixed prism 1 and sliding prism 2 are in position. When the latching mechanism places the sliding prism 2 in the OFF position and sliding prism 3 in the ON position, the light enters Port 1 and exits from Port 3. Similarly, light enters Port 1 and exits from Port 4 when the sliding prism 3 is in the OFF position and sliding prism 4 is in the ON position. The same procedure is used for the signal exiting from other ports. This switch configuration is simpler than the first configuration because the input and outputs are located

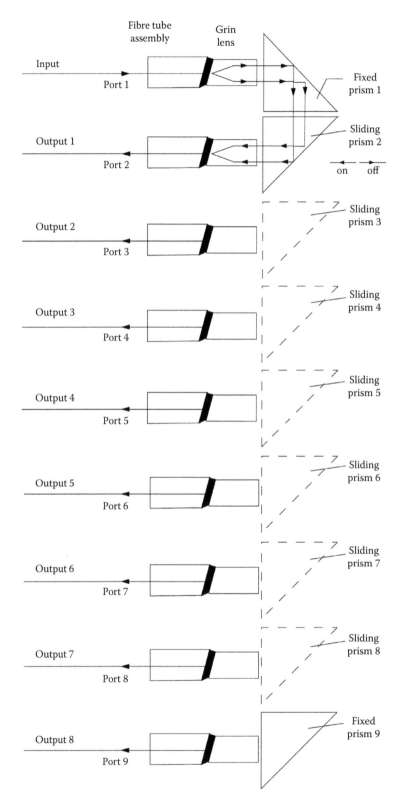

FIGURE 10.9   Second configuration of the linear model of a 1 × 8 latching switch.

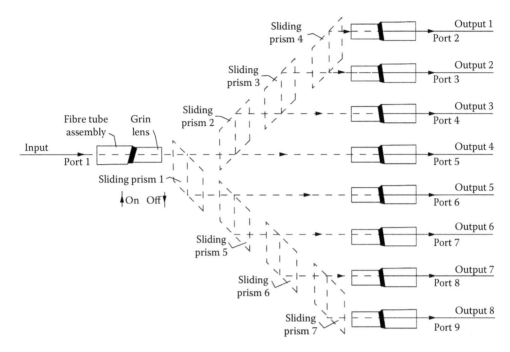

FIGURE 10.10   The triangular model of a 1 × 8 latching switch.

on the same side of the switch. This configuration includes less complex mechanism, controls, seven sliding prisms and two fixed prisms.

Figure 10.10 illustrates a schematic diagram of a configuration of the triangular model of a 1 × 8 latching switch. The common element of this type of opto-mechanical switch is that the operation involves mechanical sliding motion of parallelogram prisms in OFF/ON positions to direct the signal from one port to another. Figure 10.10 shows the input located on one side and the outputs on the other side of the switch. This configuration is called a 1 × 8 switch in the triangular model. Light enters Port 1 and exits from one of the outputs. When the latching mechanism places the sliding parallelogram prism 1 in the OFF position and sliding prism 2 into position, the light enters Port 1 and exits from Port 4, as shown in Figure 10.10. Similarly, light enters Port 1 and exits from Port 3 when the sliding prisms 2 and 3 are in the ON position. Figure 10.11 shows light entering Port 1 and exiting from Port 5, when sliding prism 1 is in the OFF position. This switch configuration is more complicated than the linear model because the input is located on one side, and the outputs on the other side of the switch. Seven sliding parallelogram prisms with additional mechanisms and controls form this configuration. Both models have difficulty achieving precise alignment and low losses during the manufacturing processes.

Many other modern opto-mechanical switches are used in telecommunication networks management, monitoring, restoration and protection. They have excellent optical performance and the high reliability necessary for network applications. They feature low insertion loss, high RL and channel isolation, excellent repeatability and fast switching speeds. The switches are available in single-mode and multi-mode, and cover wide wavelength ranges. They are available in 1 × 1, 1 × 2 and 2 × 2 configurations. The switching mechanism is latching and remains in its selected state following a loss of power. The switch consists of a quarter pitch GRIN lens glued to a double bore fibre-tube assembly, relay and mirror mounted on a shaft. The components are assembled inside a packaging base box and covered by a lid. Figure 10.12 illustrates a schematic diagram

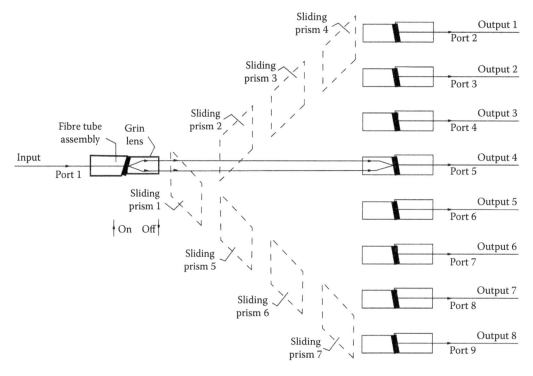

FIGURE 10.11   Signal from Port 1 to Port 5 in a 1 × 8 latching switch.

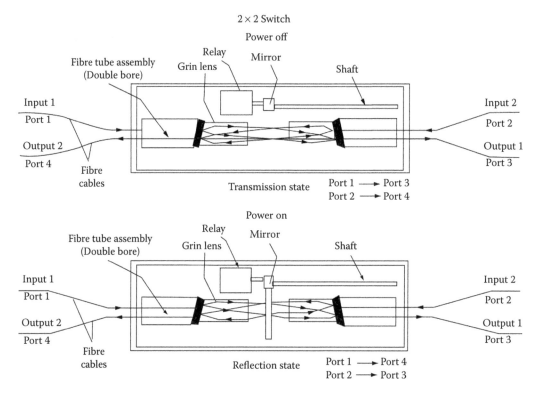

FIGURE 10.12   A 2 × 2 opto-mechanical switch.

of a 2×2 switch. This figure illustrates the switch in the OFF/ON positions. When the power is off, light transmits from Port 1 to Port 2 and Port 4 to Port 3. This configuration is called the transmission state. When the power is ON, the mirror is in position, the light is reflected by the mirror, light exits Port 1 and reflects to Port 4 and similarly, Port 2 reflects to Port 3. This state is called the reflection state.

10.3   ELECTRO-OPTIC SWITCHES

Switches with no moving parts can be built by using some of the passive devices, such as Mach–Zehnder Interferometers (MZIs) and couplers. Some optical materials, such as lithium niobate crystal (LiNiO$_3$), avalanche photo diode (APD) (NH$_4$H$_2$PO$_4$) and KDP (KH$_2$PO$_4$) exhibit an electro-optic effect. The index of refraction (RI) of the optical material changes in the presence of an electric field. These optical materials are used in building devices, such as the MZI, APD and KDP. An electric field applied across the lithium niobate crystal causes a variation in the RI. This changes the transit time, creating a phase shift of the optical signal passing through the lithium niobate crystal.

Mach–Zehnder interferometers are used in building optical devices, which are used in a wide variety of applications in optical communication systems. The basic requirement of the Mach–Zehnder interferometer is to have a balanced configuration of a splitter and a combiner connected by a pair of optically matched waveguides, as shown in Figure 10.13. The optical signal entering the Mach–Zehnder interferometer input port is split through a Y splitter section into two equal components. Each component goes to one of the two arms of the Y splitter. When there is no phase change in signal components after passing through both arms of the interferometer, the signal component is recombined at the Y coupler immediately before the optical signal exits the Mach–Zehnder interferometer. The re-combination of the two signal components takes place as constructive interference between two components and regenerates the original optical signal. In this case the Mach–Zehnder interferometer acts as a passive device.

When an electric field is applied to one arm of the Mach–Zehnder interferometer, the RI changes and causes 180 degree shift in the phase of the signal component, due to the change in optical path length of this arm. As shown in Figure 10.14, when there is a difference in phase at the destination Y coupler, the signal components will be out of phase with one another. The signal components re-combination will be lost because the components will cancel each other in destructive interference. If the phase difference is a full 180 degree, then the output will be zero. In other words, applying the electric field to one of the arms of the Mach–Zehnder interferometer will make the phase shift one of the signal components. The Mach–Zehnder interferometer acts as an active device, when an external electric voltage is applied causing the switching.

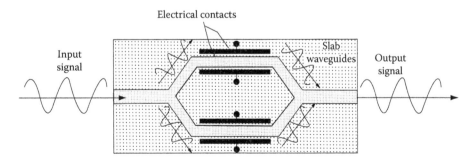

FIGURE 10.13   Mach–Zehnder interferometer acts as a passive device.

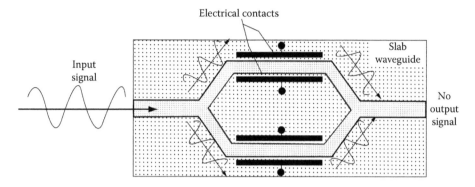

FIGURE 10.14 An electro-optic switch using a Mach–Zehnder interferometer.

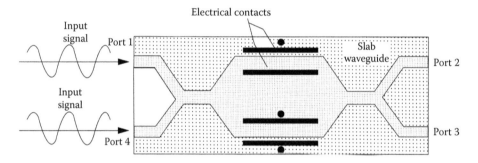

FIGURE 10.15 A 2 × 2 electro-optic switch.

TABLE 10.1
Input and Output Signals Connections

Voltage	Connections
V_1	Port 1 to Port 2 and Port 4 to Port 3
V_2	Port 1 to Port 3 and Port 4 to Port 2

Using the same principle as discussed above, one can build an electro-optic switch using two branching waveguides arranged like a 3 dB coupler to switch one input between two outputs. You can replace the one input with two parallel outputs coupled to the pair of switching waveguides by a combining coupler, as shown in Figure 10.15. An electric field applied to one arm of the waveguide causes a 180degree shift in the phase of the signal component. The electrical voltage is raised or lowered to shift the delay between waveguides by 180 degree. This directs the output from one waveguide on the right side to the other output. As signal interference depends on phase shift, it is possible to further increase the voltage to switch the signal back to the other output. Table 10.1 presents the possible outcomes achieved by applying different voltages across the waveguide arms.

10.4 THERMO-OPTIC SWITCHES

A novel rib waveguide-integrated thermo-optic switch has appeared recently. The device is based on the TIR phenomenon and the thermo-optic effect (TOE) in hydrogenated amorphous silicon (a-Si:H) and crystalline silicon (c-Si). It takes advantage of a bandgap-engineered a-Si:H layer to explore the properties of an optical interface between materials showing similar refractive indexes (RIs) but different thermo-optic coefficients. In particular, the modern plasma-enhanced chemical vapour deposition techniques, the refractive index of the amorphous film can be properly tailored to match that of c-Si at a given temperature. TIR may be achieved at the interface by acting on the temperature, because the two materials have different thermo-optic coefficients. The switch is integrated in a 4-pm-wide and 3-μm-thick single-mode rib waveguide, as shown in Figure 10.16. The substrate is a silicon-on-insulator wafer with an oxide thickness of 500 nm. The active middle region has an optimal length of 282 μm. The device performance is analysed at a wavelength of 1.55 μm. As shown in Figure 10.12, the optical waveguide-integrated switch consists of a 2 × 2 waveguide structure with an input Y branch and an output Y branch. They are joined by a middle active region, although in this work, only the 1 × 2 switch operation will be considered. The device is composed of rib channel waveguides, which guarantee both an effective optical confinement and low propagation losses. When properly designed, single-mode operation can be achieved in the input and output of the Y branches.

As shown in the top-left inset of Figure 10.16, the device structure is symmetric with respect to the $(y-z)$ plane. It consists of a core layer of c-Si in the upper half, and a core layer of a-Si:H in the lower half, both laying on a SiO_2 layer grown on a highly doped crystalline silicon substrate. The thickness of the two guiding layers together is 3 μm. Due to the RI of Sia_2 ($n_{SiO2} = 1.48$), a 500-nm-thick under-cladding layer ensures the optical confinement for both waveguides, as suggested by electromagnetic field propagation simulations. In the top-right inset of Figure 10.16, a detail of the interface between the a-Si:H waveguide and the crystalline Si waveguide in the active region is also shown. The irregular profile at the TIR interface takes into account surface roughness that may result from the fabrication process.

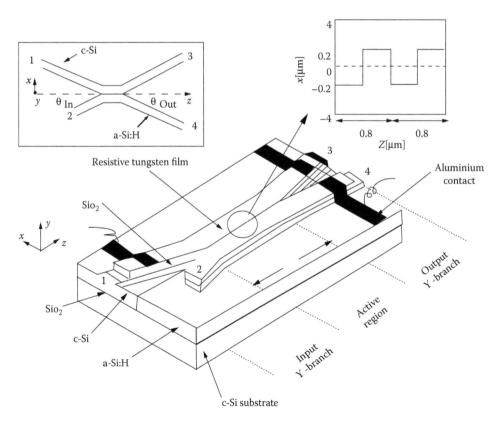

FIGURE 10.16 A waveguide-integrated thermo-optic switch.

We can exploit the TOE in a-Si:H and c-Si by changing the RI of the core layers and thereby switching the light beam at the output of the structure. A 300-nm-thick tungsten heating film is introduced on top of the stacked structures. It is separated from the active region by a 100-nm-thick SiO$_2$ film. This reduces the optical absorption by the electrodes due to the evanescent field of the optical mode. Finally, the heating structure is completed by aluminium bonding pads.

The operating principle of the device is the TIR, which can be activated or dropped by exploiting the different thermo-optic coefficients in a-Si:H and c-Si. In particular, by choosing a proper gas phase composition during the deposition process of a-Si:H, the two materials develop the same RI at a given temperature. By changing the device operating temperature, a RI discontinuity is created at the (y–z) interface, producing the desired optical switching. At room temperature, an incident channel-guided light beam coming from Port 1 will encounter a RI discontinuity between c-Si and a-Si:H, and the reflection (straight state) exists. Under these conditions TIR will occur and the incident light beam at Port 1 will be reflected to Port 3.

Another type of thermo-optical switch uses the Mach–Zehnder interferometer for the switching process. This type of thermo-optical switch is used in communications systems. Figure 10.17 illustrates the logic of an 8 × 8 thermo-optical switch. The 8 × 8 optical matrix switch employs Mach–Zehnder interferometer with a thermo-optic phase shifter as the switching mechanism. The small switch offers low loss, low crosswalk, low RL, excellent stability, high reliability and low power consumption. Applications of such switches include: space-division switching systems (with

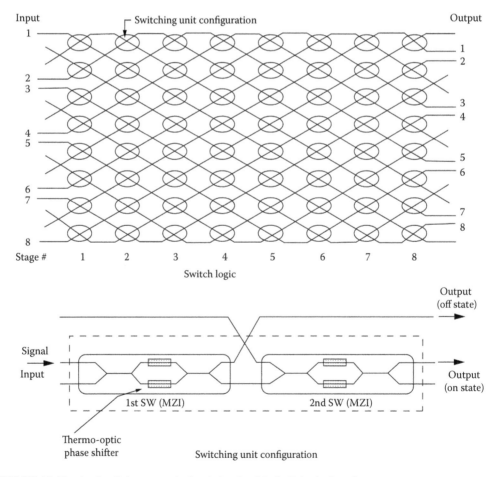

FIGURE 10.17 An 8 × 8 thermo-optical switch using Mach–Zehnder interferometer.

analogue and/or digital signals), wavelength routing (such as cross-connect and add-drop), protective switching, video switching and inter-module connection.

10.4.1 Switch Logic

10.4.1.1 Switching Unit Configuration

A high-speed all-optical switch using a fibre-optic coupler and a light-sensitive variable-index material is illustrated in Figure 10.18. RI variation with light is the principle of the switch operation. Evanescent-wave coupling between two mono-mode fibres is extremely sensitive to the RI of the material surrounding the coupling region. Two ground and polished fibres, producing an evanescent field, can be brought in close proximity so that light in one fibre will couple into the other fibre in any desired ratio. Such polished couplers have been constructed to produce very low losses.

A similar coupler may be made by timed etching of the fibres. Hydrofluoric acid may be used to remove as much cladding as desired; this exposes the core and produces an evanescent field. Within the etched regions, fibre-to-fibre optical coupling will occur for fibres placed in close proximity. The coupling efficiency will depend on the RI of the surrounding medium, the core-to-core separation, the length of the interaction and the amount of etching.

Another evanescent-field coupler is based on a non-etched, fused and drawn coupler. This type of coupler is drawn to such an extent that the core is essentially lost and the cladding reaches a diameter near that of the undrawn core. The claddings become the core, and the evanescent field is forced outside the new core (waist region) into the air. Twisting two fibres together and fusing by using heat makes fused bi-conical tapered couplers.

The contribution of charge carriers (electrons) to the RI provides a simple way to modulate the index by the introduction or withdrawal of such carriers in the material. This can easily be done by the creation of electron-hole pairs (EHP) if the material is also photoconductive (optical modulation).

One of the important parameters of the couplers is the coupling length. The coupling length is wavelength dependent. Thus the shifting of power between the two parallel waveguides will take place at different places along the coupler for different wavelengths. Figure 10.19 shows two wavelengths entering at Port 1 and Port 2. When coupler length is made exactly to match the wavelength of the signal, the coupler works to combine wavelengths. Combined signals exit from Port 2.

Figure 10.20 shows the reverse process where two different wavelengths arrive on the same input fibre at Port 1. At a particular location along the coupler, the wavelengths will be in different waveguides. Then the wavelengths separate exactly and each wavelength exits from a different

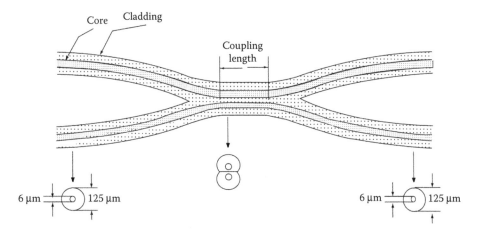

FIGURE 10.18   A high-speed optical switch using a fibre-optic coupler.

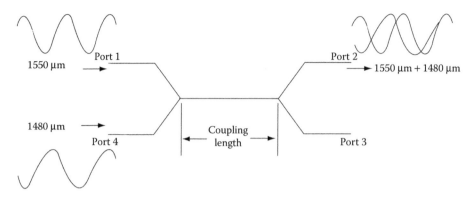

FIGURE 10.19   Two wavelengths entering at Port 1 and Port 2.

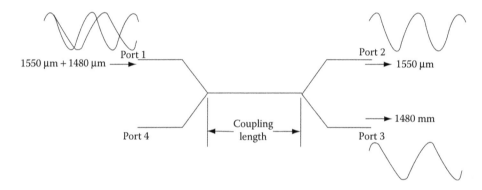

FIGURE 10.20   Two wavelengths entering at Port 1.

port. In this case, one wavelength exits from Port 2 and the other from Port 3. The processes described in Figures 10.19 and 10.20 are performed in the same coupler. This process is bi-directional. The coupler in Figure 10.19 works as a splitter; the same coupler in Figure 10.20 also works as a combiner.

There are other types of switches that use the same elements of either the micro-opto-mechanical switch or micro-electro-optical switch. They employ couplers for switching between the inputs and outputs. These types of optical switches are used in communications systems. Figure 10.21 shows the configuration of a 4 × 4 optical space-division switch. The switch is designed to connect any input port to any output port as desired by the user. Any input may be switched to any output; however, two inputs may not go to the same output at the same time. The device is bi-directional such that once a connection has been established between an input port and an output port, that particular connection may be used in either one or both directions. The switches have no moving parts, are very stable, and reliable while exhibiting very low loss; thereby reducing the need for expensive amplifiers.

Figure 10.22 shows the cross-connect switch, which selects outputs by optical cross connecting. This results in a significant reduction of overall complexity and number of required elements. These switches are used in protection/backup switching, optical cross connecting, network testing and monitoring, optical routing and optical burst switching. A 4 × 4 switch configuration can be cascaded to build 16 × 16 − 256 × 256 switch configurations.

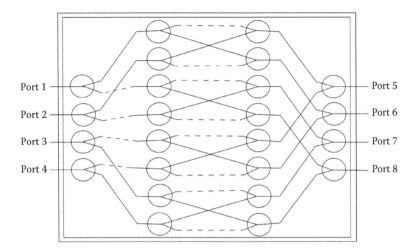

FIGURE 10.21 A 4 × 4 optical space-division switch.

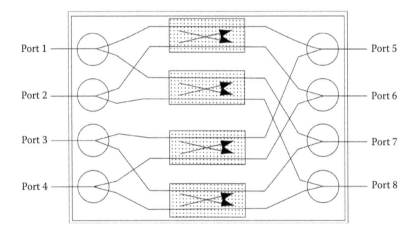

FIGURE 10.22 A cross-connect switch.

10.5 ACOUSTO-OPTIC SWITCHES

Sound waves are generated when a material is in mechanical vibration mode. They can also be generated by acoustic transducers. Like any light wave, a sound wave is a moving wave, which has a frequency. Light waves travel at the speed of light; sound waves travel at the speed of sound, which is slower than light waves. Sound waves are used to control light transmission in acousto-optic switches and modulators. The RI of some optical material is altered by the presence of sound waves. The sound wave causes regular zones of compression and tension within the optical material. This creates a regular pattern of changes in IR n of the optical material; this is called a Bragg diffraction grating. Within the optical material, there is interference between sound and light waves. The power of the deflected light is controlled by the intensity of the sound wave. The angle of deflection is controlled by the frequency of the sound wave. Figure 10.23 illustrates a design of an acoustic-optic switch. The figure shows that an incident light can be controlled by the frequency of the sound wave. The incident light exits the switch from one or more selected output ports depending on the sound wave intensity.

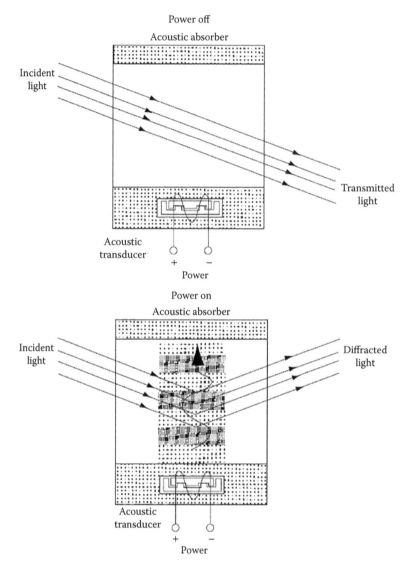

FIGURE 10.23   A design of an acoustic-optic switch.

10.6   MICRO-ELECTRO-MECHANICAL SYSTEMS

Micro-electro-mechanical systems (MEMS), is a rapidly growing technology for the fabrication of miniature devices using processes similar to those used in the integrated circuit industry. MEMS are widely used in optical switching in telecom networks. The appeal of MEMS goes beyond just switching applications in defence, aerospace and medical industries. MEMS technology provides a way to integrate electrical, electronic, mechanical, optical, material, chemical and fluids engineering on very small devices ranging in size from a few microns to 1 mm. MEMS devices have many important advantages over conventional opto-mechanical switches. First, like integrated circuits, they can be fabricated in large numbers, so that cost of production can be reduced substantially.

Second, they can be directly incorporated into integrated circuits, so that far more complicated systems can be made than with other technologies. Third, MEMS have small size, low cost and high reliability and stability. Fourth, MEMS have the important capability of high-density digital transmission communication with different bandwidths.

There are two categories of MEMS switches: MEMS 2D and 3D. They are typically fabricated onto a substrate that may also contain electronics needed to drive the MEMS switching element. MEMS-based optical switches route light from one fibre to another to enable equipment to switch traffic completely in the optical domain without requiring any optical-to-electrical conversion. At the core of MEMS, 2D matrix switches is an array of micro-mirrors capable of re-directing light either in free space or within a waveguide framework. The 2D switch architecture shown in Figure 10.24 employs one mirror for every possible switched node in a matrix switch, and thus requires N^2 mirrors for an $N \times N$ array. 2D mirror arrays are characterised by two-state mirror positioning. One state is inactive and requires only that the mirror can be parked out of the optical path. During the switching state, the mirror re-directs the light path. Mirror positioning accuracy, repeatability and stability are critical in determining switch performance. Unlike 3D switch architectures that require servo positioning of individual mirrors, a 2D switch can rely on passive positioning control of the switched mirror, simplifying the control scheme. But a successful MEMS 2D approach must provide means for actuating the mirror into a highly predictable, stable state and hold it there indefinitely.

MEMS use an array of pop-up MEMS mirrors fabricated on the surface of a silicon wafer. The mirror is hinged to allow its rotation off the plane of the substrate to an angle of 90 degree where it redirects a light channel from the through to cross-state, as shown in Figure 10.25. An addressing scheme is required to select individual mirrors for actuation into the popped-up state and also for positioning them with sufficient accuracy for efficient coupling into the switched channel. The 2D MEMS array described here, called MagO × C, which stands for *magnetically optical actuated cross-connect*, uses a combination of magnetic and electrostatic actuation to rotate the mirrors and to select and de-select individual mirrors for clamping into the up or down state.

To rotate unclamped mirrors into the up state, magnetic actuation is implemented globally by applying an external field generated with a small electromagnet. The magnetic signal only needs

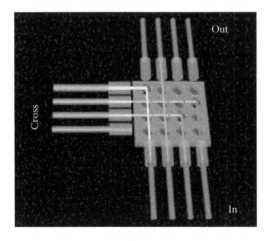

FIGURE 10.24 MEMS 2D switch architecture.

FIGURE 10.25   Pop-up mirror array with passive mirror alignment.

to be applied momentarily. Using a global field avoids the need to fabricate individual magnetic actuators for each chip. Mirrors are fabricated with a layer of attached nickel to produce torque on the mirror hinge in response to the applied field. The nickel plate aligns with the magnetic field lines and generates a magnetostatic torque on the mirror. This lifts the mirror off the substrate and orients it near the desired vertical position, where electrostatic force can take over in setting and holding the final desired mirror position. An electrostatic field is applied mirror-by-mirror, either to hold the mirror down against the torque produced by the magnet, or to hold the mirror in the up position against the restoring force of the elastic hinge. As the magnetic field is applied globally, all mirrors will attempt to rotate when the field is turned on. Only the mirrors to be rotated into the up position are unclamped; all others are held down electrostatically. Similarly, mirrors clamped in the up state remain so until the electrostatic signal is removed; the magnet is no longer needed to hold the mirror up. The combination of magnetic and electrostatic actuation provides an effective means for configuring a mirror array without resorting to complex individual actuators for each mirror. As electrostatic clamping of the mirrors virtually require no current flow, the switch array consumes very little steady state power. Power is consumed only during transitions when the magnet is activated. The components of the switch are packed in a packaging base and lid.

10.7   3D MEMS-BASED OPTICAL SWITCHES

3D MEMS-based optical switch routes light from any of 80 input fibres to any of 80 output fibres. Designed for fibre-based test and measurement, 10-Gbit/s Ethernet, high-definition video and telecom applications, this all-optical micro-photonic sub-system fits in the palm of a hand. The switch design is based on 3D MEMS mirror arrays, which is called reflexion. It can switch signals within 10 ms, which is well within the telecommunications requirements for communications applications. 3D designs have switching elements accommodating hundreds, even thousands, of ports. Figure 10.26 shows a 3D MEMS optical switch. The design of the 3D MEMS is simple, solves

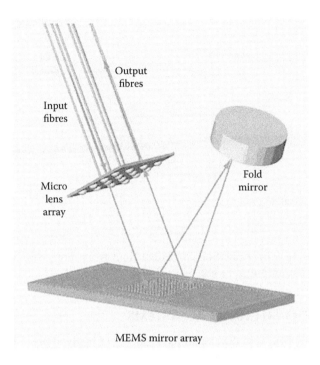

FIGURE 10.26   A 3D MEMS optical switch architecture.

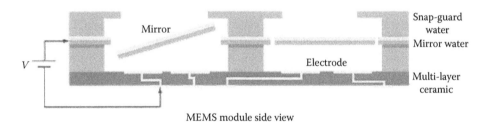

FIGURE 10.27   A 3D MEMS optical switch design.

mechanical and optical issues, is easy to fabricate and achieves manufacturing tolerances that are accepted by the telecommunications industry.

A 3D MEMS design is shown in Figure 10.27. A micro-mirror array rests atop a single piece of silicon on a ceramic substrate. No bonding pads or other integrated electronics exist on the chip. All routing to the rest of the actuation electronics is done on the back end of the ceramic substrate. Additional electronics is located on a photodetector card, along with constant-delay two-pole Bessel filters, and an analogue-to-digital converter with a conversion-phasing time. Parallel-plate electrostatic actuation of the mirrors, with potentials of about 200 V, is provided by high-voltage linear amplifiers.

Figure 10.28 shows a torsional micro-mirror, which is driven by a vertical bi-directional comb driver (micro-electrostatic actuator). Underneath the mirror plate is the substrate electrode. During operation, a voltage is applied to the electrode in order to generate an electrostatic attractive force on the mirror plate. The mirror plate rotates around the supporting axis in two

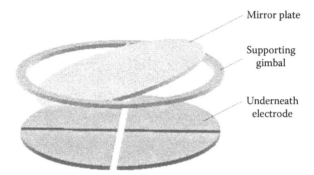

FIGURE 10.28 A torsional micro-mirror.

dimensions. Such tilting mirror position can direct light from many distinct input ports to any of the many distinct output ports.

10.8   MICRO-OPTO-MECHANICAL SYSTEMS

On-chip integrated MOMS were developed for a variety of applications for optical telecommunications. It is a new technology that allows for the integration of multiple passive and active components at the chip level. The technology is an extension of integrated optics and is based on suspended waveguides fabricated on chips, which are integrated with other optical components. It is used for a variety of optical solutions, including optical switching and cross-connect, signal dispersion correction, configurable optical add/drop multiplexers and signal intensity equalisers. The technology bases its main technological concept on integrating optics at the chip level. The technology explores low-cost, high-performance, planar optical waveguide switches. Waveguides are used to channel light, rather than allowing it to propagate in free space. The use of waveguides allows a degree of freedom in reducing size, while at the same time operating in a controlled physical environment. Switching time and losses are very low. By using waveguides, photons can be channelled in a controlled fashion, making it possible to lay out a photonic network within the chip with low losses. As silicon is used as the propagating material, wavelength transparency for the 1300–1600 nm is utilised. The use of silicon for the waveguide material creates a tight confinement of light that allows the use of very small curvature radii, enabling a significant reduction of footprint chip size.

10.9   EXPERIMENTAL WORK

The purpose of this lab is to build and test an opto-mechanical switch. The following cases illustrate the experimental set-up for the different types of switch arrangements that can be performed in an experimental laboratory. Different optical components (mirror, double side silvered mirror and prism) may be used with one of the experimental set-ups to create different opto-mechanical switch configurations as follows:

1. A 1 × 2 switch with one laser source.
2. Two 1 × 2 switches with two laser sources.
3. A 2 × 2 switch using a movable mirror.
4. A 1 × 2 switch using a prism.

10.9.1 A 1 × 2 SWITCH WITH ONE LASER SOURCE

Figure 10.29 illustrates the design of a 1 × 2 switch using fixed and movable mirrors to redirect the reflected light onto one of the output ports. The input light comes through the input fibre cable on one side of the switch and leaves on one of the two output ports on the same side. A movable mirror assembly activated by a relay moves the mirror up and down in the OFF and ON positions. When the movable mirror is in the OFF position, the incident light on the fixed mirror reflects onto the output Port 1. When the switch is in the ON position, the movable mirror reflects the incident light onto the second output Port 2, as shown in Figure 10.29. This arrangement is called a 1 × 2 switch configuration.

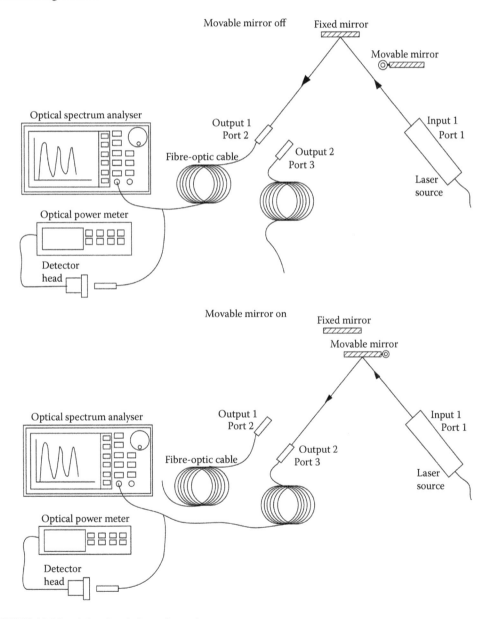

FIGURE 10.29 A 1 × 2 switch configuration.

10.9.2 TWO 1 × 2 SWITCHES WITH TWO LASER SOURCES

We can use the same arrangement as explained in this section to create a two-1 × 2 switch, as shown in Figure 10.30. This set-up has two laser light sources at Ports 1 and 2. Each laser source has two independent outputs. The outputs are at Ports 3 and 4. The switch operates as a two-1 × 2 switch. Each laser light input gives two outputs by the movable mirror in OFF/ON positions.

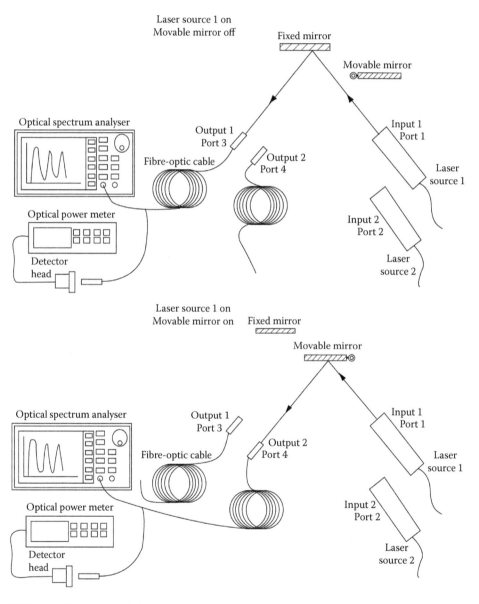

FIGURE 10.30 Two 1 × 2 switches configuration.   (*Continued*)

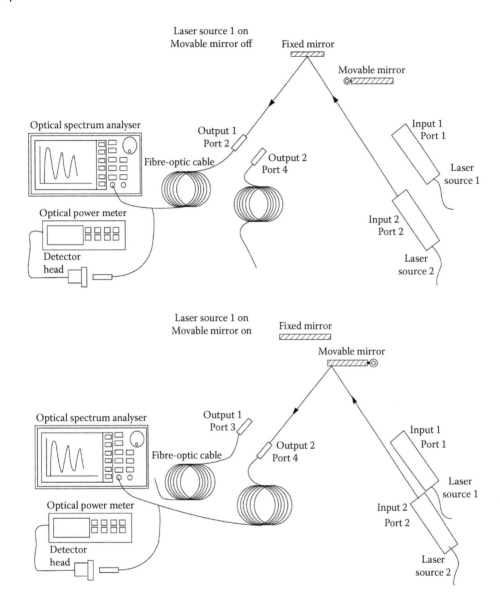

FIGURE 10.30 (Continued) Two 1 × 2 switches configuration.

10.9.3 A 2 × 2 Switch Using a Movable Mirror

As explained in Section 10.9.1 for 1 × 2 switch, we can use a movable mirror to create a 2 × 2 switch, as shown in Figure 10.31. We need two laser sources, two outputs and a mirror mounted on a movable arm assembly arrangement to build a 2 × 2 switch. Figure 10.31 shows a schematic diagram of a two-position switch that can be built in the laboratory. The following experimental set-up has

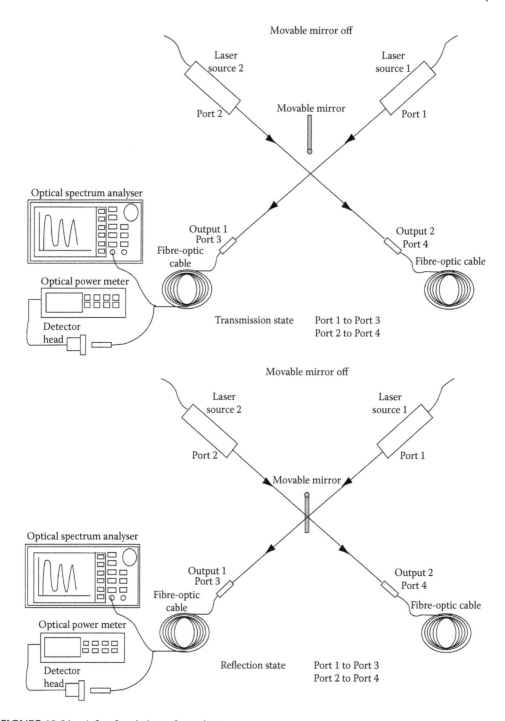

FIGURE 10.31 A 2 × 2 switch configuration.

two laser light sources at Ports 1 and 2, and two outputs at Ports 3 and 4. Each laser source has two independent outputs. The figure shows the switch in the OFF/ON states. When the power is OFF, light is transmitted from Port 1 to Port 3 and from Port 2 to Port 4. This case is called the transmission state. When the power is ON and the mirror moves into position, the light is reflected by the mirror, light from laser source 1 is reflected to Port 4 and light from laser source 2 is reflected to Port 3. This case is called the reflection state.

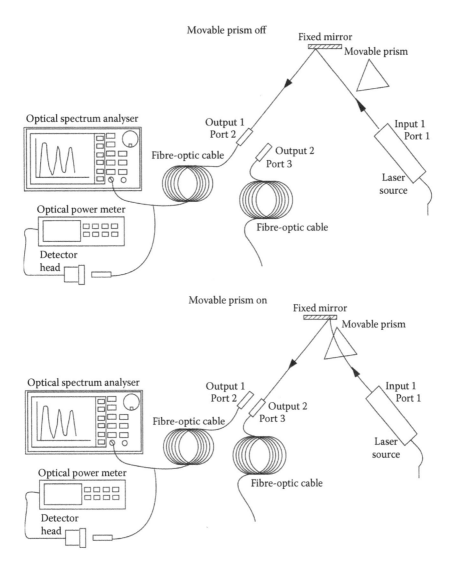

FIGURE 10.32 A 1 × 2 switch using a prism.

10.9.4 A 1 × 2 Switch Using a Prism

As explained in Section 10.9.1, instead of a movable mirror, we can use a movable prism to create a 1 × 2 switch, as shown in Figure 10.32. We need one laser source, one fixed mirror, two outputs and a prism mounted on a movable arm assembly to build a 1 × 2 switch. Figure 10.32 shows a schematic diagram of a two-position switch that can be built in the laboratory. This experimental set-up has one laser light source as a Port 1 and two outputs at Ports 2 and 3. The figure shows the switch in the OFF/ON positions. When the power is OFF, light from Port 1 is reflected to Port 2. When the power is ON, the prism moves into position. The light refracts through the prism and is reflected by the mirror onto output Port 3.

10.9.5 Technique and Apparatus

Appendix A presents the details of the devices, components, tools and parts:

1. 2 × 2 ft Optical breadboard.
2. Two laser light sources.

3. Two laser power supplies.
4. Two laser mount assemblies.
5. Laser light detector.
6. Laser light power meter.
7. Optical spectrum analyser.
8. Four multi-axis translation stages.
9. Hardware assembly (clamps, posts, screw kits, screwdriver kits, fibre-cable holder/positioner, positioners, brass fibre-cable holders, laser holder/clamp, etc.).
10. Two mirrors and mirror holder assemblies (large size mirrors can be used because of the size of the multi-axis translation stage and the space between the components).
11. Double side-silvered mirror.
12. Prism and prism holder assembly.
13. Fibre-optic cable end preparation kit.
14. Fibre cable end holder.
15. Two fibre-optic cables, 250 micrometres diameter, 500 metres long.
16. Fibre-cable holder/positioner assembly.
17. Movable arm assembly for mirror and prism.
18. Black/white card and cardholder.
19. Ruler.

10.9.6 Procedure

Follow the laboratory procedures and instructions given by the professor and/or instructor.

10.9.7 Safety Procedure

Follow all safety procedures and regulations regarding the use of fibre-optic cables, fibre-optic devices and instruments, optical components and instruments, light source devices and optical cleaning chemicals.

10.9.8 Apparatus Set-Up

10.9.8.1 A 1 × 2 Switch with One Laser Source

1. Figure 10.33 shows the experimental apparatus set-up. (Note: You could arrange this experimental set-up to fit with the measurement instruments that you have in the lab).
2. Mount three multi-axis translation stages on the breadboard. Multi-axis translation stages are for Port 1 (laser source), Port 2 and Port 3.
3. Place a laser source into the laser mount clamp and tighten the screw.
4. Bolt the laser source to the multi-axis translation stage.
5. Place a fibre-cable holder/positioner assembly on the multi-axis translation stages to hold Ports 2 and 3.
6. Mount a fixed mirror and mirror holder facing the laser source at an angle with the normal line of the mirror.
7. Turn the laser on. Follow the operation and safety procedures of the laser device in use.
8. Turn off the laboratory lights before taking measurements.
9. Align the laser beam reflected off the mirror to fibre cable at Port 2.
10. Use the optical spectrum analyser to measure the laser power and wavelength of the laser source (Port 1). Figure 10.34 shows a sample of collected data by the optical spectrum analyser printed on a paper chart. Ask your supervisor/instructor for the optical spectrum analyser calibration procedure and referencing the source. Otherwise use the optical power

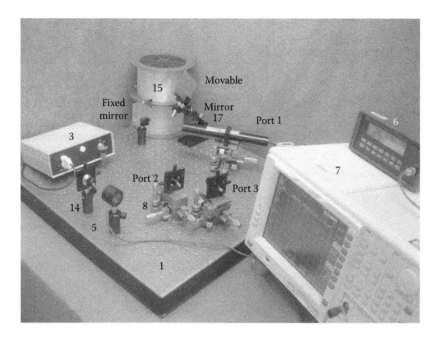

FIGURE 10.33 A 1 × 2 opto-mechanical switch.

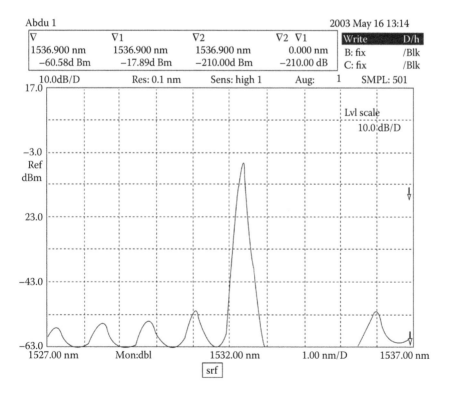

FIGURE 10.34 A sample of collected data printed on a chart.

TABLE 10.2

A 1 × 2 Switch Data Collection

Port 1		Port 2			Port 3		
Power (unit)	Wavelength (unit)	Power (unit)	Wavelength (unit)	Loss (dB)	Power (unit)	Wavelength (unit)	Loss (dB)

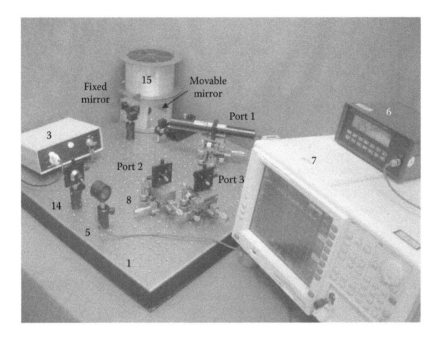

FIGURE 10.35 Movable mirror is in the ON position.

meter for quick power out results. When using the optical power meter, you need to calibrate it; ask your supervisor/instructor for the calibration procedure. Record your results for Port 1 in Table 10.2.

11. Measure the laser power and wavelength of the laser light at Port 2. Record your results for Port 2 in Table 10.2.

12. Mount the movable mirror assembly between the laser source and the fixed mirror. Locate the movable mirror parallel to the fixed mirror.

13. Move down the movable mirror, as shown in Figure 10.35.

14. Align the fibre cable at Port 3 to receive the reflected laser beam from the movable mirror.

15. Measure the laser power and wavelength of the laser light at Port 3. Record your results for Port 3 in Table 10.2.

10.9.8.2   Two 1 × 2 Switches with Two Laser Sources

Continue the procedure as explained in Section 10.9.1 of this experiment. Figure 10.36 shows the experimental apparatus set-up for Section 10.9.2. Add the following steps:

1. Mount the forth multi-axis translation stage on the breadboard.

2. Place a second laser source into the laser mount clamp and tighten the screw.

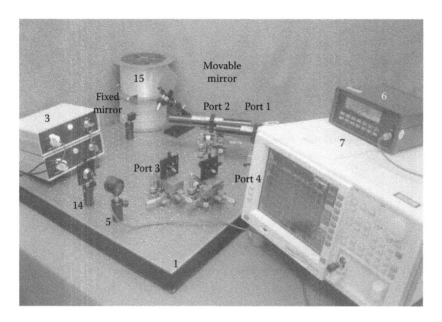

FIGURE 10.36 Two 1 × 2 switches.

3. Bolt the laser source to the multi-axis translation stage.
4. Align each laser beam to point at a fibre cable.
5. As explained in Section 10.9.1, measure the first laser power and wavelength of the laser light at Port 1. Record your results in Table 10.3.
6. Measure the laser power and wavelength of the laser light at Port 3. Record your results for Port 3 in Table 10.3.
7. Put movable mirror in position, as shown in Figure 10.37.
8. Measure the laser power and wavelength of the laser light at Port 4. Record your results in Table 10.3.
9. Turn off the first laser and the second laser on. Swing the movable mirror upwards.
10. Align laser source 2 to reflect onto Port 3.
11. Measure the second laser power and wavelength of the laser light at Port 2. Record your results in Table 10.3.

TABLE 10.3

Two 1 × 2 Switches Data Collection

Port 1		Port 3			Port 4		
Power (unit)	**Wavelength (unit)**	**Power (unit)**	**Wavelength (unit)**	**Loss (dB)**	**Power (unit)**	**Wavelength (unit)**	**Loss (dB)**
Port 2		**Port 3**			**Port 4**		
Power (unit)	**Wavelength (unit)**	**Power (unit)**	**Wavelength (unit)**	**Loss (dB)**	**Power (unit)**	**Wavelength (unit)**	**Loss (dB)**

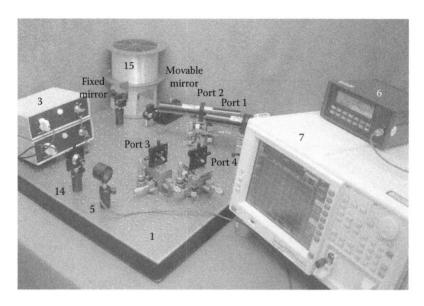

FIGURE 10.37   A movable mirror is in the ON position.

12. Measure the laser power and wavelength of the laser light at Port 3. Record your results for Port 3 in Table 10.3
13. Put movable mirror in position, as shown in Figure 10.37.
14. Measure the laser power and wavelength of the laser light at Port 4. Record your results in Table 10.3.

10.9.8.3   A 2 × 2 Switch Using a Movable Mirror

You can build a 2 × 2 opto-mechanical switch using a double-sided mirror mounted on a movable arm mechanism, to direct the light from one light source onto two outputs in two modes. Figure 10.38 shows the experimental apparatus set-up for Section 10.9.3.

1. Mount four multi-axis translation stages on diagonals of the breadboard close to the corners.
2. Place two laser sources into the laser mount clamps and tighten the screw. Mount each laser source on one multi-axis translation stage. Bolt each multi-axis translation stage on the same side of the breadboard pointing to the centre.
3. Mount a movable mirror and mirror holder assembly in a position where the two laser beams intersect.
4. Place a fibre-cable holder/positioner assembly on the remaining two multi-axis translation stages.
5. Bolt each multi-axis translation stage on the same side of the breadboard on the opposite side of the laser sources (i.e., each laser source is facing one fibre-cable holder assembly). Arrange so that laser source 1 (Port 1) faces one fibre-cable holder assembly (Port 3) and laser source 2 (Port 2) faces the second fibre-cable holder assembly (Port 4).
6. Turn on the two laser sources. Follow the operation and safety procedures of the laser device in use.
7. Turn off the laboratory lights before taking measurements.
8. As explained in Section 10.9.1, measure the laser power and wavelength of the laser light at Ports 1 and 2. Record your results in Table 10.4.

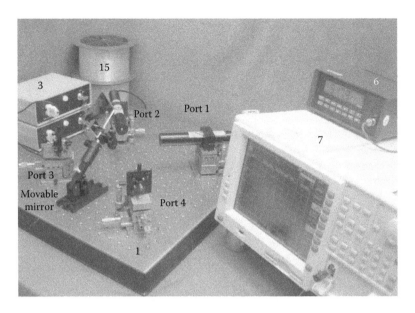

FIGURE 10.38   A 2 × 2 switch using a movable mirror.

9. Align the laser beam 1 to point at fibre cable on Port 3; align the laser beam 2 to point at fibre cable on Port 4.
10. Measure the laser power and wavelength of the laser sources at Port 3 and Port 4. Record your results in Table 10.4.
11. Put the movable mirror in position, as shown in Figure 10.39.
12. Carefully align the position of the mirror assembly so the laser beam from laser source 1 reflects by one side of the movable mirror to Port 4 and the laser beam from laser source 2 reflects by other side of the movable mirror to Port 3, as shown in Figure 10.39.
13. Measure the laser power and wavelength of the laser beam at Port 3 and Port 4. Record your results in Table 10.4.

TABLE 10.4
2 × 2 Switch Using a Movable Mirror Data Collection

Port 1		Port 3			Port 4		
Power (unit)	Wavelength (unit)	Power (unit)	Wavelength (unit)	Loss (dB)	Power (unit)	Wavelength (unit)	Loss (dB)

Port 2		Port 3			Port 4		
Power (unit)	Wavelength (unit)	Power (unit)	Wavelength (unit)	Loss (dB)	Power (unit)	Wavelength (unit)	Loss (dB)

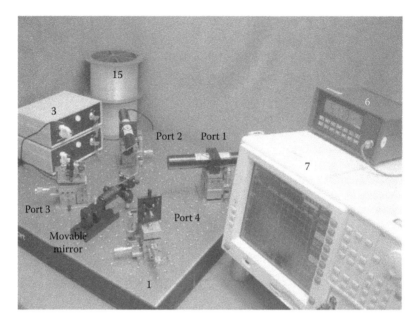

FIGURE 10.39   A 2 × 2 switch when the movable mirror is in the ON position.

10.9.8.4   A 1 × 2 Switch Using a Prism

You can build a 1 × 2 opto-mechanical switch using a prism mounted on a movable arm mechanism to direct the light from the light source onto the second output. Figure 10.40 shows the experimental apparatus set-up for the experiment described in Section 10.9.4.

1. Mount three multi-axis translation stages on the breadboard.
2. Place a laser source into the laser mount clamps and tighten the screw. Mount the laser source on one multi-axis translation stage. Bolt the multi-axis translation stage to the breadboard.

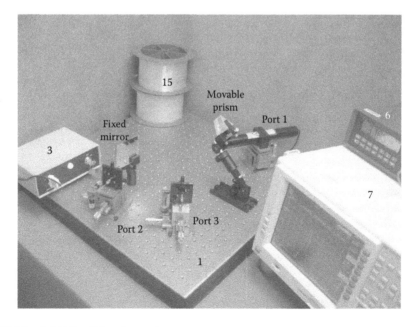

FIGURE 10.40   A 1 × 2 switch using a prism.

3. Turn the laser on. Follow the operation and safety procedures of the laser device in use.
4. Turn off the laboratory lights before taking measurements.
5. As explained in Section 10.9.1, measure laser power and wavelength of the laser light at Port 1. Record your results in Table 10.5.
6. Place a fibre-cable holder/positioner assembly on each multi-axis translation stage for Ports 2 and 3.
7. Mount a fixed mirror and mirror holder facing the laser source at an angle with the normal line of the mirror.
8. Bolt each multi-axis translation stage to the breadboard, as shown in Figure 10.40 (i.e., the laser source is facing one fibre-cable holder assembly). Arrange the laser source (Port 1) to face the fixed mirror by an angle with the normal. The light from Port 1 is reflected by the fixed mirror onto Port 2.
9. Measure the laser power and wavelength of the laser source at Port 2. Record your results in Table 10.5.
10. Mount a prism and prism-holder mechanism into a position such that the laser beam refracts from the prism, is incident on the fixed mirror, and reflects onto the second output at Port 3.
11. Put the prism in position, as shown in Figure 10.41.

TABLE 10.5

1 × 2 Switch Using a Prism Data Collection

Port 1		Port 2			Port 3		
Power (unit)	Wavelength (unit)	Power (unit)	Wavelength (unit)	Loss (dB)	Power (unit)	Wavelength (unit)	Loss (dB)

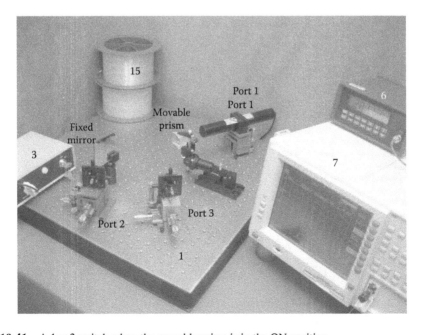

FIGURE 10.41 A 1 × 2 switch when the movable prism is in the ON position.

12. Carefully align the position of the prism assembly, so the laser beam from the laser source refracts through the prism onto the fixed mirror and reflects by the fixed mirror onto Port 3, as shown in Figure 10.41.
13. Measure the laser power and wavelength of the laser source at Port 3. Record your results in Table 10.5.
14. Turn on the laboratory lights.

10.9.9   DATA COLLECTION

10.9.9.1   A 1 × 2 Switch with One Laser Source
1. Measure the power and wavelength of the laser source beam at Port 1.
2. Measure the power and wavelength of the laser beam at Ports 2 and 3.
3. Fill out the Table 10.2 with the collected data.

10.9.9.2   Two 1 × 2 Switches with Two Laser Sources
1. Measure the power and wavelength of the laser source beams at Ports 1 and 2.
2. Measure the laser power and wavelength of the laser beams at Ports 3 and 4.
3. Fill out the Table 10.3 with the collected data.

10.9.9.3   A 2 × 2 Switch Using a Movable Mirror
1. Measure the power and wavelength of the laser source beams at Ports 1 and 2.
2. Measure the power and wavelength of the laser beams at Ports 3 and 4.
3. Fill out the Table 10.4 with the collected data.

10.9.9.4   A 1 × 2 Switch Using a Prism
1. Measure the power and wavelength of the laser source beam at Port 1.
2. Measure the power and wavelength of the laser beams at Ports 2 and 3.
3. Fill out the Table 10.5 with the collected data.

10.9.10   CALCULATIONS AND ANALYSIS

To calculate the power output loss at each port, use the light power loss formula given by Equation 10.3.

$$\text{Loss(dB)} = -10 \times \log_{10} \frac{P_{\text{out}}}{P_{\text{in}}} \tag{10.3}$$

10.9.10.1   A 1 × 2 Switch with One Laser Source
1. Calculate the loss at each port using the light power loss formula Equation 10.3.
2. Fill out the Table 10.2 with the calculated data.

10.9.10.2   Two 1 × 2 Switches with Two Laser Sources
1. Calculate the loss at each port using the light power loss formula Equation 10.3.
2. Fill out the Table 10.3 with the calculated data.

10.9.10.3   A 2 × 2 Switch Using a Movable Mirror
1. Calculate the loss at each port using the light power loss formula Equation 10.3.
2. Fill out the Table 10.4 with the calculated data.

10.9.10.4   A 1 × 2 Switch Using a Prism
1. Calculate the loss at each port using the light power loss formula Equation 10.3.
2. Fill out the Table 10.5 with the calculated data.

10.9.11   Results and Discussions

10.9.11.1   A 1 × 2 Switch with One Laser Source

1. Determine the power and wavelength of the laser source beam at each port.
2. Compare the power and wavelength of the laser source beam at the input and output ports.

10.9.11.2   Two 1 × 2 Switches with Two Laser Sources

1. Determine the power and wavelength of the laser source beams at each port.
2. Compare the power and wavelength of the laser source beams at the input and output ports.

10.9.11.3   A 2 × 2 Switch Using a Movable Mirror

1. Determine the power and wavelength of the laser source beam at each port.
2. Compare the power and wavelength of the laser source beam at the input and output ports.

10.9.11.4   A 1 × 2 Switch Using a Prism

1. Determine the power and wavelength of the laser source beam at each port.
2. Compare the power and wavelength of the laser source beam at the input and output ports.

10.9.12   Conclusion

Summarise the important observations and findings obtained in this lab experiment.

10.9.13   Suggestions for Future Lab Work

List any suggestions for improvements using different experimental equipment, procedures and techniques for any future lab work. These suggestions should be theoretically justified and technically feasible.

10.10   LIST OF REFERENCES

List any references that were used in the report. Use one format in writing the references. Never mix reference formats in a report.

10.11   APPENDIX

List all of the materials and information that are too detailed to be included in the body of the report.

FURTHER READING

Allan, R., *3D MEMS-Based Optical Switch Handles 80-by-80 Fibers*, Electronic Design, Infineon Technologies, 2002.

Bar-Lev, A., *Semiconductors and Electronic Devices*, 2nd ed., Prentice Hall International, Englewood Cliffs, NJ, 1984.

Berlin, A. and Gabriel, K. J., Distributed MEMS: New challenges for computation, *IEEE Comput. Sci. Eng. J.*, 4 (1), 12–16, 1997.

Boyd, W. T., *Fiber Optics Communications, Experiments and Projects*, 1st ed., Howard W. Sams & Co., Indianpolis, IN, 1982.

Brower, D. L., Ding, W. X., and Deng, B. H., *Fizeau Interferometer for Measurement of Plasma Electron Current*, Electrical Engineering Department, University of California, Los Angeles, CA, Vol. 75, issue 10, pp. 3399–3401, 2004.

Camperi-Ginestet, C., Kim, Y. W., Wilkinson, S., Allen, M., and Jokerst, N. M., Micro-opto-mechanical devices and systems using epitaxial lift off, *JPL, Proceedings of the Workshop on Microtechnologies and Applications to Space Systems*, California, pp. 305–316, June 1993.

Derickson, D., *Fiber Optic Test and Measurement*, Prentice Hall PTR, Upper Saddle River, NJ, 1998.

Dutton, H. J. R., *Understanding Optical Communications*, Prentice Hall, Englewood Cliffs, NJ, 1998.

Eggleton, B. J., Kerbage, C., Westbrook, P. S., Windeler, R. S., and Hale, A., December microstructured optical fiber devices, *Opt. Express*, 17, 698–713, 2001.

Fedder, G. K., Santhanam, S., Reed, M. L., Eagle, S. C., Guillou, D. F., Lu, M. S. C., and Carley, L. R., Laminated high-aspect-ratio microstructures in a conventional CMOS process, *Sen. Actuator, A*, 57 (2), 103–110, 1996.

Francon, M., *Optical Interferometry*, Academic Press, New York, pp. 97–99, 1966.

Gabriel, K. J., Tabata, O., and Sugiyama, S., Surface-normal electrostatic pneumatic actuators, *Technical Proceedings of MEMS'92*, Travemunde, Germany, pp. 110–114, February 1992.

Gebhard, B. and Knowles, C. P., Design and adjustment of a 20cm Mach-Zehnder interferometer, *Rev. Sci. Instrum.*, 37 (1), 12–15, 1966.

Goff, D. R. and Hansen, K. S., *Fiber Optic Reference Guide: A Practical Guide to the Technology*, 2nd ed., Butterworth-Heinemann, London, UK, 1999.

Hariharan, P., Modified Mach-Zehnder interferometer, *Appl. Optics*, 8 (9), 1925–1926, 1969.

He, S. and Mrad, R. B., A vertical bi-directional electrostatic comb driver for optical MEMS devices, *Paper presented at Opto-Canada, SPIE Regional Meeting on Optoelectronics, Photonics and Imaging, SPIE TD01*, Ottawa, Canada, May 9–10, 2002.

Hecht, J., *Understanding Fiber Optics*, 3rd ed., Prentice Hall, Englewood Cliffs, NJ, 1999.

Hibino, K., Error-compensating phase measuring algorithms in a Fizeau interferometer, *Opt. Rev.*, 6 (6), 529–538, 1999.

Hoss, R. J., *Fiber Optic Communications-Design Handbook*, Prentice Hall, Englewood Cliffs, NJ, 1990.

Jackel, J., Goodman, M. S., Baran, J. E., Tomlinson, W. J., Chang, G.-K., Iqbal, M. Z., and Song, G. H., Acousto-optic tunable filters (AOTF's) for multiwavelength optical cross-connects: Crosstalk considerations, *J. Lightwave Technol.*, 14, 1056–1066, 1996.

Jackson, R. A., The laser as a light source for the Mach-Zehnder interferometer, *J. Sci. Instrum.*, 42, 282–283, 1965.

Jacobs-Cook, A. J., MEMS versus MOMS from a systems point of view, *J. Micromech. Microeng.*, 6, 148–156, 1996.

JDS Uniphase Corporation, Switch, SN Series, 2003.

Johnstone, R. D. M. and Smith, W., A design for a 6 in. Mach-Zehnder interferometer, *J. Sci. Instrum.*, 42, 231–235, 1965.

Kao, C. K., *Optical Fiber Systems: Technology, Design, and Applications*, McGraw-Hill, New York, 1982.

Kolimbiris, H., *Fiber Optics Communications*, Prentice Hall, Englewood Cliffs, NJ, 2004.

Kranz, M. S. and Fedder, G. K., Micromechanical vibratory rate gyroscopes fabricated in conventional CMOS, *Paper presented at the Symposium Gyro Technology*, September 16–17, Stuttgart, Germany, 1997.

Li, J., Kahrizi, M., and Landsberger, L. M., In-plane electro-thermal actuated optical switches, *Paper presented at Opto-Canada, SPIE Regional Meeting on Optoelectronics, Photonics and Imaging, SPIE TD01*, Ottawa, Canada, May 9–10, 2002.

Loreggia, D., Gardiol, D., and Gai, M., Fizeau interferometer for global astrometry in space, *Appl. Optics*, 43 (4), 721–728, 2004.

Marrakchi, A., Ed., *Photonic Switching and Interconnects*, Marcel Dekker, New York, 1994.

Mehregany, M. and Roy, S., *Introduction to MEMS*, Aerospace Press, AIAA, Inc., El Segundo, CA, 1999.

NEL NTT Electronics Corporation, *Thermo-Optic Switches Catalog*, Tokyo, Japan, 2004.

Page, D. and Routledge, I., November using interferometer of quality monitoring, *Photon. Spectra*, pp. 147–153, November 2001.

Palais, J. C., *Fiber Optic Communications*, 4th ed., Prentice Hall, Englewood Cliffs, NJ, 1998.

Panepucci, R. R., *Integrated Micro Opto-Mechanical Systems (MOMS)*, Abstracts 2002/Mechanical and MEMS Devices, Emerald Group Publishing Limited, Vol. 22, Issue 1, p. 74, 2002.

Salah, B. E. A. and Teich, M. C., *Fundamentals of Photonics*, Wiley, New York, 1991.

Senior, J. M., *Optical Fiber Communications: Principle and Practice*, 2nd ed., Prentice Hall, Englewood Cliffs, NJ, 1992.

Setian, L., *Applications in Electro-Optics*, Prentice Hall PTR, Upper Saddle River, NJ, 2002.

Shamir, J., *Optical Systems and Processes*, SPIE - The International Society for Optical Engineering, SPIE Press, Optical Engineering Press, Bellingham, WA, 1999.

Uskov, A. V., O'Reilly, E. P., Manning, R. J., Webb, R. P., Cotter, D., Laemmlin, M., Ledentsov, N. N., and Bimberg, D., On ultra-fast optical switching based on quantum dot semiconductor optical amplifiers in nonlinear interferometers, *IEEE Photon. Technol. Lett.*, 16 (5), 1265–1267, 2004.

Vandemeer, J. E., Kranz, M. S., and Fedder, G. K., Nodal simulation of suspended MEMS with multiple degrees of freedom, *Paper presented at The Winter Annual Meeting of ASME in the 8th Symposium on MEMS*, Dallas, TX, pp. 16–21, November 1997.

Wood, R. L., Mahadevan, R., and Hill, E., *MEMS 2-D Matrix Switch*, MEMS Business Unit, JDS Uniphase Corporation, Research Triangle Prak, NC, 2000.

Yariv, A., Universal relations for coupling of optical power between microresonators and dielectric waveguides, *Electron. Lett.*, 36, 321–322, 2000.

11 Laser Technology

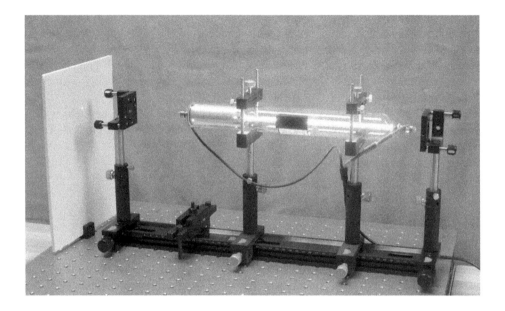

11.1 INTRODUCTION

The term *laser* is an acronym for *light amplification by stimulated emission of radiation*. Laser technology is one of the most rapidly developing areas in modern technology. When the first laser was developed by Maiman in the United States in 1960, it was called *a solution in search of many problems* because there were no applications for the laser at that time. Since then, the laser has become the solution for many diverse applications from a simple laser pointer to a very sophisticated satellite tracking laser system and military laser missile launchers.

To explain how the laser can be applied in such diverse areas, we need to understand the basic physical principles of the operation of a laser. Basically, the laser is a device that transforms various forms of energy into electromagnetic radiation. The energy put into the laser can be in many forms, such as electromagnetic radiation, electrical energy or chemical energy. Energy is always emitted from the laser as electromagnetic radiation. Laser devices produce a narrow, intense beam of monochromatic coherent light. Laser beams are used to produce signals for fibre-optic telecommunication systems and also industrial, medical, military, micro-machining and so on.

This chapter has two experimental sections. One section involves learning how to direct a laser beam in space using different techniques. In the second section, presents practical applications of laser-beam expansion by building a Galilean telescope and a Keplerian telescope. These sections would present the theory applications, alignment process and hands-on experience.

11.2 LIGHT EMISSION

Light can be produced by rapidly changing the state of an electron from a state of relatively high energy to a lower energy or ground state. The energy of the electron has to leave the atom, usually in the form of a photon of light. A series of rapid energy state transitions will produce a stream of photons, which can provide illumination. This principle of light emission is used in the operation of light bulbs, light emitting diodes and lasers.

Artificial light can be produced by either spontaneous emission or a stimulated emission process. Absorption of energy occurs before either of the emission process.

When an atom at a low energy state E_1 absorbs energy, it will be elevated to a higher energy state E_2, as shown in Figure 11.1. The absorbed energy can come from many energy sources, such as electrical, thermal, chemical, optical or nuclear energy sources.

The atom will eventually fall to a lower energy state; photons will be emitted, with photon energy being the difference between the two atomic energy states. This is the process of spontaneous emission. The energy of the photon is given by

$$E_{\text{photon}} = E_2 - E_1 = hf \tag{11.1}$$

where:
  h is Planck's constant (6.6261×10^{-34} Js)
  f is the frequency of light (Hz)

The spontaneous emission of light serves as the basis of most lighting systems. For example, by passing an electric current through a metal wire (filament), the filament will begin to glow. The type and temperature of the filament will determine the wavelength range of the light being generated. This process occurs when an atom is initially in the higher energy level; it may drop spontaneously to the lower energy level and release its energy in the form of photon, as shown in Figure 11.2. As such, light that is generated by this method is composed of many different wavelengths and distributed in all directions.

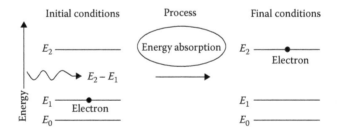

FIGURE 11.1 Absorption process.

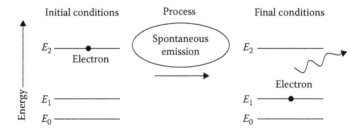

FIGURE 11.2 Spontaneous light emission.

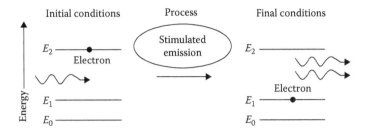

FIGURE 11.3 Stimulated light emission.

The stimulated emission of light occurs when an incoming photon of energy hf stimulates the emission process by suddenly inducing the electron at E_2 to transit down to E_1. The emitted photon is in phase with, is in the same direction as, and has the same energy as the incoming photon, as shown in Figure 11.3. These two stimulated photons will stimulate more transitions. If conditions are appropriate, this process continues and results in monochromatic light being amplified to produce laser light. Again, laser is an acronym for *light amplification by stimulated emission of radiation*. Stimulated emission can be achieved by using either light, electricity or both. Lasers are the most common example of stimulated radiation emission.

11.3 PRINCIPLE OF THE LASER

The following is a brief introduction to the development of the laser. The laser was conceived after the maser, which stands for microwave amplification by stimulated emission of radiation. The first maser, using vibrations of ammonia molecules (NH_3), was developed by James Gordon, Zeiger and Charles Townes in 1954. The ammonia molecules, when pumped to an excited state, radiate electromagnetic waves (microwaves) at a precise frequency (the same frequency used in the first atomic clock and the basis of extremely sensitive microwave amplifiers used by astronomers). Development of the maser was connected with radar development for military research. The first high power laser developed was a pink ruby rod with its ends silvered and placed inside a helical shaped flash lamp. This laser was only capable of pulsed operation. The laser is still widely used in many applications. Some of these applications will be presented later in this chapter.

The emitted laser beam is a nearly perfect plane wave. In contrast, an ordinary light wave emits light in all directions, and the emitted light is incoherent. The underlying aspects of laser production are based on quantum theory.

Electrons exist at specific energy levels, or states, characteristic of the particular atom or molecule. The energy levels can be imagined as rings or orbits around a nucleus. Electrons in the outer rings are at higher energy levels than those in the inner rings. When external optical or electrical energy is applied to a material, the atoms of the material get excited and are bumped up to higher energy levels. An atom that is already in the excited state may fall spontaneously to the lower state, resulting in the emission of energy in the form of a photon. However, if a photon with this same energy encounters the excited atom, the incident photon can stimulate the electron to make the transition to the lower state sooner.

Figure 11.4 shows dots that represent the energy state of one atom. In a normal situation, most of the electrons are found in a lower, or ground state, energy level of the atoms, as shown in Figure 11.4a. If an external energy source is applied to excite many electrons into a higher energy level, a population inversion can be created in which more electrons are in the higher level than in the lower level, as shown in Figure 11.4b. With the normal population in thermal equilibrium, as shown in Figure 11.4a, some stimulated emission still occurs. But only in the non-equilibrium case, as shown in Figure 11.4b, light amplification will occur due to stimulated emission. In stimulated

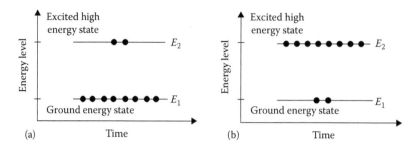

FIGURE 11.4 Two-energy levels for a collection of atoms: (a) normal population and (b) inverted population.

emission, the original photon and a second one of the same frequency exist. These two photons are exactly in phase, and are moving in the same direction, producing coherent light as a laser beam. These two photons are exactly in phase and move in the same direction, producing coherent light as a laser beam.

For further explanation, assume that the atoms have been excited to a higher state, as shown in Figure 11.5. Some of the excited atoms drop down spontaneously soon after being excited. If an emitted photon strikes another atom in the excited state, it stimulates this atom to emit a photon of the same frequency, moving in the same direction, and phase. These two photons then move on to strike other atoms, causing more stimulated emissions. As the process continues, the number of photons multiplies. A laser device typically has two mirrors at each end. When the photons strike the end mirrors, most are reflected back. As they move in the opposite direction, the photons continue to stimulate more atoms to emit photons. As the photons move back and forth between the mirrors, a small percentage passes through the partially transparent mirror at one end, creating a narrow, coherent laser beam.

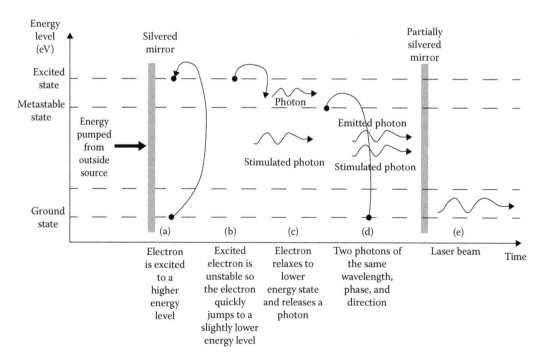

FIGURE 11.5 Energy level diagram showing excited atoms stimulated to emit laser light.

Depending on the particular lasing material, specific wavelengths of light are absorbed and specific wavelengths are emitted. As a result, laser beams are very different from normal light. Laser beams have the following properties:

1. *Laser light is monochromatic*: This means it contains one specific wavelength of light. The wavelength of laser light is determined by the amount of energy released when the electron drops to the lower orbit.
2. *Laser light is coherent*: This means it is organised with each photon moving in step with the others. All of the photons' wave fronts launch in unison.
3. Laser light is very directional and has a tight concentrated beam. An ordinary light bulb, on the other hand, releases light in many directions, and the light is relatively weak and diffuse.

11.4   THERMAL EQUILIBRIUM AND POPULATION INVERSION

It is only possible to achieve gain in a laser, as described above, if the population of electrons in the laser's upper energy state is greater than the population in the lower state. This condition is called a population inversion. In thermal equilibrium, which is the typical state of normal matter, such as the air around us, the population ratio between two states is governed by the Maxwell–Boltzmann equation, which shows that higher energy levels always have lower populations than lower energy levels:

$$\frac{N_2}{N_1} = \exp\left(\frac{-(E_2 - E_1)}{k_B T}\right) = \exp\left[-\frac{h\nu}{k_B T}\right] \qquad (11.2)$$

where:
N_2 is the population in the upper state
N_1 is the population in the lower state
k_B is Boltzmann's constant = 8.61738×10^{-5} (eV/K)
T is the temperature

For example, consider two energy states: $E_2 = 1.30$ eV and $E_1 = 1.00$ eV. Assume that there are 2×10^{15} electrons/cm^3 in E_1. At a temperature of 600°K, the number of electrons in state E_2 is calculated using Equation 11.2 as follows:

$$N_2 = 2.0 \times 10^{15} \times \exp\left(\frac{-(1.30 - 1.00)}{8.61738 \times 10^{-5} \times 600}\right) = 6.067 \times 10^{12} \left(\text{electron/cm}^3\right)$$

The negative sign in the exponent suggests that a population inversion is only permitted under the conditions of negative temperature. This result was very disturbing to early laser researchers, as negative temperatures are not physically possible.

However, the Boltzmann equation only describes conditions of thermal equilibrium. Lasers are not operated in thermal equilibrium. Instead, the upper state is populated by pumping it via some non-equilibrium process. A pulse of light, an electrical spark, light or a chemical reaction can all be used to populate the upper laser state.

In order to determine when lasing will occur, the rate of change in population between the two energy levels in an atom needs to be estimated. The approach for this was initially done by Einstein for his new vision of the thermal radiation law. A few of his analysis steps are outlined below, using an approach very similar to his original paper.

Essentially Einstein was rethinking the blackbody thermal radiation law of Planck when he produced his seminal paper. Start with the concept of a flux of radiation u_ν impinging on a medium. The medium is absorbing radiation to a higher energy state and re-emitting radiation when moving to a lower energy state. For simplicity, only consider two states: a higher energy state E_2 and a lower state E_1. The rate of change, called the transition rate, of the number of atoms is proportional to the number of atoms in that state (N_1). The more atoms there are in a state, the more atoms that can leave this state per second.

When the transition occurs due to absorption, the process is driven by the photon flux. This is called stimulated absorption. The transition rate of absorption (from state 1 to 2) is given by

$$\left(\frac{dN_1}{dt}\right) = -B_{12}N_1 u_v \tag{11.3}$$

where B is a constant of proportionality between the two states. The minus sign is required because N_1 is decreasing.

Similarly, the transition rate of stimulated emission (from state 2 to 1) is given as

$$\left(\frac{dN_2}{dt}\right) = -B_{21}N_2 u_v \tag{11.4}$$

In the case of spontaneous emission, the rate is independent of the radiation field and is given as:

$$\left(\frac{dN_2}{dt}\right) = -A_{21}N_2 \tag{11.5}$$

The A and B constants are called the Einstein coefficients. Einstein assumed that there is thermodynamic equilibrium between the radiation field and the atoms, the energy density has blackbody characteristics and the electron densities of the two states are governed by the Maxwell–Boltzmann distribution.

In equilibrium, the down transitions (emission) must equal the up transitions (absorption). In other words, the rate of absorption equals the rate of stimulated emission plus the rate of spontaneous emission. Therefore

$$B_{12}N_1 u_v = B_{21}N_2 u_v + A_{21}N_2 \tag{11.6}$$

Dividing both sides by N_1 and rearranging gives

$$\frac{N_2}{N_1} = \frac{B_{12}u_v}{A_{21} + B_{21}u_v} \tag{11.7}$$

Using the Maxwell–Boltzmann Equation 11.2, the following equation can be obtained:

$$e^{-hv/k_B T} = \frac{B_{12}u_v}{A_{21} + B_{21}u_v} \tag{11.8}$$

Solving for u_v and simplifying, (assuming that B_{12} equals B_{21}, without proof) gives

$$u_v = \frac{A}{B}\left[\frac{1}{e^{hv/k_B T} - 1}\right] \tag{11.9}$$

Planck's law of blackbody radiation describes radiant energy as a function of frequency. This law produces the following equations to generate the radiation curve for a blackbody as a function of frequency:

$$U_v \, dv = \frac{8hv^3}{c^3}\left[\frac{dv}{\exp\left(\frac{hv}{kT}\right) - 1}\right] \tag{11.10}$$

where:

ν is the frequency

U_ν is the energy density of radiation between the frequencies ν and $\nu + \delta\nu$

h is Planck's constant, 6.63×10^{-34} Js

c is the speed of light, 3.0×10^{8} m/s

k is Boltzmann's constant, 1.38×10^{-23} J

Using Planck's law and comparing Equations 11.9 and 11.10, the following equation can be derived:

$$\frac{A}{B} = \frac{8h\nu^3}{c^3}$$

(11.11)

Therefore, there is a way of finding A in terms of B. If A or B can be measured, the rates for the system of interest can be found. Then, the pump rate (the rate of energy pumped into a laser) needed to make a system laze can be estimated. However, to obtain conditions for population inversion (in a radiation field), a similar, but more complex, analysis of three and four systems is required.

Also note that the spontaneous emission rate increases as ν^3. This is why it is much more difficult to build short wavelength lasers (e.g. UV and x-ray) than long wavelength (e.g. infrared) lasers. At short wavelengths, electrons jump to a lower energy state spontaneously at a higher rate; therefore, fewer are available to emit lasing light by stimulation.

11.5 TRANSVERSE AND LONGITUDINAL MODES

The output spot of the laser beam, observed on a screen, is termed the transverse electromagnetic mode (TEM). This intensity curve is a round mode with a Gaussian (bell-shaped) profile in cross-section. However, it is possible to operate a laser having a wide variety, or combination, of other transverse modes. In these cases, the output beam may have a strange shape. Figure 11.6a shows that an off-axis transverse mode is able to self-replicate (resonate) after one round trip along the optical cavity. Figure 11.6b shows wave fronts in a self-replicating wave. Figure 11.6c shows four possible low-order transverse cavity modes. Figure 11.6d shows the intensity patterns of these modes. Most commercial lasers produce the transverse electromagnetic TEM_{00} mode. Figure 11.6e shows a photograph of the TEM_{11} mode.

A laser can only laze, however, at those wavelengths for which an integral number of half wavelengths fit into the laser (Fabry–Perot) cavity. The set of possible integral multiples of the cavity length is called the longitudinal electromagnetic modes of the cavity. The frequencies of these modes are given by

$$\nu = \frac{m.c}{2nL}$$

(11.12)

where:

m is the mode number of an integer as 1, 2, 3 and so on

n is the reflective index in the cavity

L is the length of the cavity (resonator)

Notice that m will be a very large number in a typical laser system.

The spacing between longitudinal modes is frequently of interest, and is given as

$$\delta\nu = \frac{c}{2nL}$$

(11.13)

The spacing between modes is independent of wavelength. The longitudinal mode spacing can be expressed in wavelength, denoted by $\delta\lambda_m$ in Figure 11.7b and c.

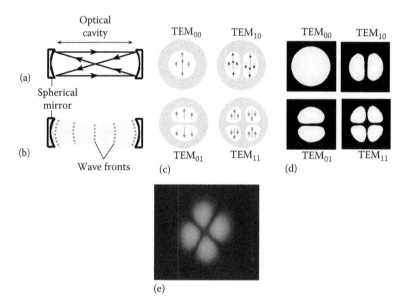

FIGURE 11.6 Laser cavity and transverse modes. (a) Optical cavity between two spherical mirrors, (b) Wave fronts travelling between two spherical mirrors, (c) and (d) Transvers Electro-Mechanical Modes in a laser beam, and (e) Transvers Electro-Mechanical Mode (TEM$_{11}$).

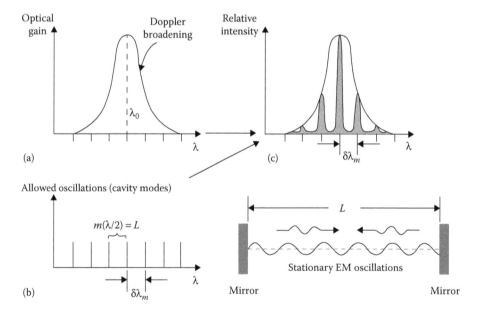

FIGURE 11.7 (a) Optical gain curve, (b) optical cavity modes and (c) output spectrum.

11.6 GAIN

A given laser does not have gain at all frequencies. The function that describes the frequency dependence of the gain, g, is called the gain profile $g(\nu)$, or in terms of wavelength λ, $g(\lambda)$. Figure 11.7a shows an optical gain curve of the gain versus wavelength; this curve is characteristic of the lasing medium. Figure 11.7b shows the allowed modes and their wavelengths due to stationary (resonating) electromagnetic waves within the optical cavity. There are an infinite number of integral multiples (modes) of the cavity length. However, only a finite number will fit into the gain profile of the laser

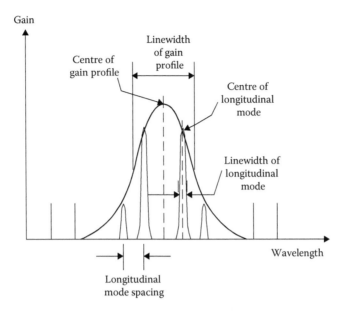

FIGURE 11.8 Optical gain curve of the lasing medium.

gain material. The actual output of the laser is the intersection of the set of possible longitudinal modes with the gain profile, as shown in Figures 11.7c and 11.8. These figures show that the output spectrum (relative intensity versus wavelength) is determined by satisfying (a) and (b) simultaneously, assuming no cavity losses. Figure 11.8 shows that the centre of longitudinal mode is not necessarily the same as the centre of the gain curve.

11.7 THRESHOLD CONDITION

As noted earlier, a laser consists of a pair of mirrors, between which some active lasing medium is made into an amplifier in a limited frequency range.

One of the reflectors is usually made slightly transparent at the wavelength of interest, and it has a reflectivity that can be denoted by r. At each reflection, a fraction $(1 - r)$ of the intensity is not returned to the resonator. The reflection coefficients are r_1 and r_2 for the pair of mirrors. Then the wave energy diminishes at each set of reflections by $r_1 r_2$.

In most lasers there are losses. But for the sake of simplicity consider only the reflective losses in this case. The fraction of intensity remaining after a full round trip passage through the laser cavity is

$$e^{-2\gamma} = r_1 r_2 \tag{11.14}$$

The value of γ is positive and represents the amount of loss in a single passage. Solving for this loss gives

$$\gamma = -\frac{1}{2}\log r_1 r_2 \tag{11.15}$$

If the medium is able to amplify the signal and compensate for the losses in the cavity, then the intensity is increased by $e^{\alpha L}$ by virtue of the amplification in the gain material. Therefore, by taking the losses into consideration, the intensity changes from 1 to $F = e^{(\alpha L - \gamma)}$, in which F is the intensity factor. If F is greater than 1, the oscillations will grow; if less than 1, they will die out. At the threshold, F is equal to 1, thus threshold is clearly defined as $\alpha L = \gamma$. Lasing occurs when $\alpha L \geq \gamma$.

The amplification α in the laser material, as a function of frequency and relative population inversion n, can be shown to be

$$\alpha = k(\nu)_o \, n \tag{11.16}$$

where $k(\nu)_o$ is the absorption in the unexcited laser material. The relative population inversion is given as

$$n = \frac{1}{N_o}\left(\frac{g_1}{g_2}N_2 - N_1\right) \tag{11.17}$$

where g is the degeneracy or multiplicity of the laser level.

A laser of a given length and mirror reflectivity will operate only if the population inversion is large enough to ensure that

$$\alpha_m = nk_0 \geq \frac{\gamma}{L} \tag{11.18}$$

When the absorption coefficient k_0 has been determined experimentally, this equation can be used with a substitution of the relative population inversion to determine the threshold population inversion:

$$\left(\frac{g_1}{g_2}N_2 - N_1\right) = \frac{N_0 \gamma}{k_0 L} \tag{11.19}$$

Note here, without proof, that the minimum inversion necessary for laser oscillation can also be written (in terms of frequency) as

$$\left(\frac{g_1}{g_2}N_2 - N_1\right) = \frac{4\pi^2}{\lambda^3}\frac{\Delta\nu}{\nu}\frac{g_1}{g_2}\frac{t_2}{t_p} \tag{11.20}$$

where $\Delta\nu$ is the linewidth of the longitudinal mode, t_2 is related to the probability of spontaneous emission $= 1/A21$, and t_p is the photon lifetime, and the term $(g_1/g_2 N_2 - N_1)$ is called the population inversion factor.

The probability of spontaneous emission t_2 depends on the active material. The photon lifetime t_p depends on the laser decay/loss in the cavity and also depends on the construction of the laser (in terms of length and reflectivity). There is also a relationship to the linewidth. The main point to be made is that one can manipulate certain aspects of the laser design to assist the population inversion process; specifically decreasing the photon lifetime which increases the population inversion.

11.8   POWER AND ENERGY

Lasers can operate either in a continuous wave (CW) or pulsed mode. Pulsed operation is occasionally used to reduce the heating of the laser (common for semiconductor diode lasers). However, in most cases, pulsed operation is combined with techniques such as Q-switching (that concentrates the laser energy into the pulse) and mode-locking (that shortens the width of the pulse in time). Q-switched and mode-locked lasers can concentrate very high peak power densities, due to the relatively short length of the pulse.

The excitation of the atoms in a laser can be done continuously or in pulses. In a pulsed laser, the atoms are excited by periodic inputs of energy. The multiplication of photons continues until all the atoms have been stimulated to jump down to the lower energy state; this process is repeated with each input pulse. In a continuous laser, the energy input is continuous so that atoms are stimulated to jump down to the lower level.

Some care must be taken in discussing the properties of pulsed versus CW wave lasers because the use of watts describing peak power can be easily confused with the use of watts describing average power. The following definitions will be used:

1. In a CW laser, the average power is the power (watts).
2. For pulsed lasers, the average power equals the energy per pulse (joules) times the pulse repetition rate (Hz):

$$P_{avg} = E_{pulse} \cdot R_{reprate} \tag{11.21}$$

3. The peak power is the energy per pulse (joules) divided by the temporal length of the pulse (seconds), given as

$$P_{peak} = \frac{E_{pulse}}{t_{pulse}} \tag{11.22}$$

where P_{avg} is the average power density, in watts per unit area (cm^2 or m^2), and P_{peak} is the peak power density, in watts per unit area (cm^2 or m^2).

11.9   THREE- AND FOUR-ENERGY LEVEL LASER SYSTEMS

Lasers are classified, by the number of energy levels involved in the actual lasing process. Lasers classified as three- or four-energy level lasers are shown in Figure 11.9. As noted previously, it is not possible to achieve population inversion with a two-level system. The three- and four-level lasers are the most typical classes found in commercial systems.

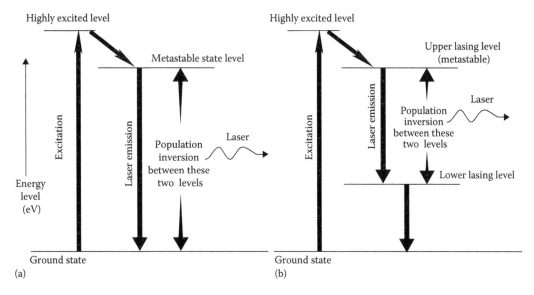

FIGURE 11.9   (a) Three-level laser systems and (b) four-level laser systems.

To produce the required population inversion for laser activity, atoms or molecules must be excited to specific energy levels. Either light or electricity can provide the external (pumped) energy necessary to excite atoms to higher energy levels.

As previously discussed, the amount of time spent by an atom in an excited state is critical in determining whether it will be stimulated to emission or lose of its energy through spontaneous emission. Excited states commonly have lifetimes of only nanoseconds before they release their energy by spontaneous emission, a period that is not long enough to likely undergo stimulation by another photon. A critical requirement for laser action, therefore, is an upper energy state that is long-lived. Such states do exist for certain materials and are referred to as metastable states. The average lifetime before spontaneous emission occurs for a metastable state is on the order of a microsecond to a millisecond, quite a lengthy period of time on the atomic timescale. With long lifetimes, excited atoms can produce significant amounts of stimulated emission. Laser action is only possible if population builds up faster than it decays in the upper energy level, thus maintaining a population larger than that of the lower level. The longer the spontaneous emission lifetime, the more suitable the atom is for laser applications.

Figure 11.9a illustrates the simplest functional energy-level structure for laser operation, the three-level system. In this system, the ground state is the lower-lasing level, and a population inversion is created between the ground level and a higher-energy metastable state. Most of the atoms are initially excited to a short-lived high-energy state that is higher than the metastable level. From this state they quickly decay to the intermediate metastable level, which has a much longer lifetime than the higher state (often on the order of 1000 times longer). As each atom resides in the metastable state for a relatively long time, the population tends to increase that leads to a population inversion between the metastable state and the lower ground state (that is being de-populated to the highest level). Stimulated emission occurs because more atoms are available in the upper excited (metastable) state, than in the lower state where absorption of light would most likely occur. Then, laser emission occurs between the metastable level and the ground state.

Although the three-level laser system works for all practical purposes, as exemplified by the first laser (Ruby), a number of problems limit the effectiveness of this approach. The central problem is that the laser has difficulties operating efficiently. This occurs because the lower-lasing level is the ground level, which is the normal state for most atoms or molecules. In order to produce the population inversion, a majority of ground state electrons must be promoted to the excited energy level, requiring a significant input of external energy. In addition, the population inversion is difficult to sustain for an appreciable time, and, therefore, three-level lasers generally must be operated in pulsed mode rather than continuously.

Figure 11.9b illustrates a four-level laser structure. The energy level structure is similar to that of the three-level system. The difference occurs after the atoms drop from the highest level to the metastable upper state; they do not drop all the way to the ground state in a single step. As the population inversion is not created between the ground state and the upper level, the number of atoms that must be elevated is dramatically reduced. In a typical four-level laser system, if only one or two percent of the atoms or molecules reside in the lower laser level (that is above the ground state), then exciting only two to four percent of the total to the higher level will achieve the required population inversion. Another advantage of separating the lower laser level from the ground level is that the lower level atoms will naturally fall to the ground state. If the lower laser level has a lifetime that is much shorter than the upper level, atoms will decay to the ground level at a rate sufficient to avoid accumulation in the lower laser level. Then, laser emission occurs between the metastable level and the lower level. Many of the lasers designed under these constraints can be operated in a continuous mode to produce an uninterrupted laser beam.

11.10 TYPES OF LASER

Most lasers are constructed of three elements: an active medium, a pumping source and a resonant cavity. The active medium is a collection of atoms or molecules, which can be excited into a population-inversion situation, and can release electromagnetic radiation by the stimulated emission process. The active medium can be any of the four states of matter: solid, liquid, gas or plasma. The wavelengths of the emitted light are determined by the specific transitions between the laser energy levels in the material. The basic physics of the laser is similar for all types of lasers, and the term active medium will be used. It is assumed that the active medium is composed of atoms. In reality, the active medium can be atoms, molecules, ions or semiconductors, depending on the laser type.

The pumping source provides the energy required to pump atoms to higher energy levels, so that stimulated emission can occur. Lasers can be optically pumped, electrically pumped (also by using an electric discharge) or pumped by a chemical reaction. There are three sources for optical pumping:

1. Flash lamps, which are built from a quartz tube filled with gas (such as mercury vapour) at low pressure.
2. Noble gas discharge tubes, which usually use xenon gas (but sometimes when higher energy is required, other noble gases with lower atomic weights, such as krypton or helium, are used).
3. Another laser or any other light source.

The resonant cavity provides a regenerative path for the photons. In essence, the functions of the resonant cavity are to shorten the laser and to construct the electromagnetic mode. Although the resonant cavity is a key part of most lasers, there are lasers for which the resonant cavity is not essential. It is certainly possible to make a laser long enough so that a reasonable intensity emerges without a resonant cavity. Nitrogen lasers and most x-ray lasers are made this way. However, such lasers tend to have poor output-beam quality.

11.10.1 GAS LASERS

The gas laser is probably the most frequently encountered type of laser. The red-orange, green and blue beam of the He–Ne, Ar^+ and He–Cd gas lasers, respectively, are common for many applications.

The Kr^+ laser readily produces hundreds of milliwatts of optical power at wavelengths ranging from 350 nm in the ultraviolet to 647 nm in the red. It can be operated simultaneously on a number of lines to produce white laser light. Gas lasers are classified into two categories: pulsed gas lasers, such as N_2, Excimer and TEA CO_2 and CW gas lasers, such as copper vapour, CO_2, Argon Ion and Helium–Neon.

11.10.1.1 Helium–Neon (He–Ne) Laser

Ali Javan, William Bennet and Donald Herriot made the first gas laser using helium and neon. The helium–neon (He–Ne) laser is used for classroom demonstrations, laboratory experiments and was initially used in optical scanning systems at supermarket checkouts. In the He–Ne laser, the lasing material is a gas, a mixture of about 15% He and 85% Ne. The low pressure gas mixture is placed in a glass tube that has two parallel mirrors one at each end. Figure 11.10 shows a He–Ne laser beam.

Essentially, helium is used for energising and neon for amplification. Atoms in the gas mixture are excited, by applying a high voltage to the tube so that an electric discharge takes place within the gas mixture. In the process, some of the He atoms are raised to the metastable state E_3, as shown in Figure 11.11, which corresponds to a jump of 20.61 eV, almost exactly equal to an excited state in

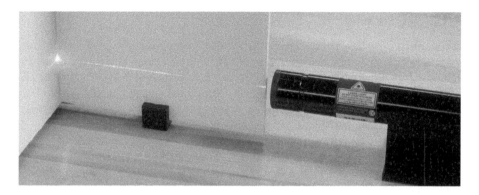

FIGURE 11.10   Helium–Neon (He–Ne) laser beam.

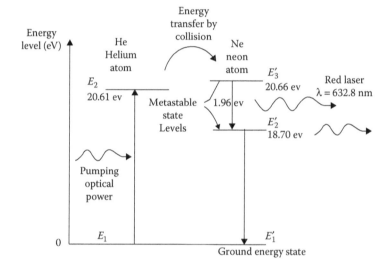

FIGURE 11.11   Energy levels for the He–Ne laser.

neon, 20.66 eV. The He atoms do not quickly return to the ground state by spontaneous emission, but instead often give their excess energy to the Ne atoms when they collide. In such a collision, the He drops to the ground state and the Ne atom is excited to the state E_3' (the prime refers to neon states). The slight difference in energy (20.66 – 20.61 = 0.05 eV) is supplied by the kinetic energy of the moving molecules. In this manner, the E_3' state in Ne, which is metastable, becomes more populated than the E_2' level. This inverted population between E_3' and E_2' is needed to produce the laser beam. The He–Ne laser produces a characteristic reddish-pink sidelight. The most common He–Ne lasing wavelength is red, at 632.8 nm. The He–Ne laser will also laze at other wavelengths (notably green light).

11.10.1.2   Argon Ion (Ar⁺) Laser

The argon laser was invented in 1964 by William Bridges in the United States. The argon ion (Ar⁺) laser provides powerful, CW-visible coherent light of several watts. The laser operation starts when the argon atoms are ionised by electron collisions in a high current electrical discharge. Further, multiple collisions with electrons excite the argon ion to an energy level, about 35 eV above the atomic ground state, as shown in Figure 11.12. The atoms cannot stay long at the high energy level. Thus a population inversion forms between the high energy and the next energy level, about 33.5 eV above the argon ground level. Consequently, the stimulated radiation between the

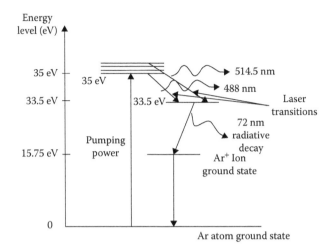

FIGURE 11.12   Energy levels for the Ar-ion laser.

two energy levels contains a series of wavelengths ranging from 351.1 to 528.7 nm. Maximum light power is concentrated in the 488 and 514.5 nm emissions. The Ar^+ ion at 33.5 eV, the lower laser level, returns to its neutral atomic ground state via a radiative decay to the Ar^+ ion ground state. The Ar^+ ion then recombines with an electron to form the neutral atom, Ar. The Ar atom is then ready for pumping again. The operation cycle continues, as long as pumping power is available.

11.10.1.3   Carbon Dioxide Gas Laser

Carbon dioxide (CO_2) gas lasers are capable of high power output in the infrared region, tunable between 9 and 11 microns. Although, the high and medium power CO_2 laser has many industrial applications, such as cutting and welding, there are also many medical uses. CO_2 laser therapy removes tattoo pigment after vapourisation of the epidermis and superficial dermis and is also used for skin resurfacing. Although effective, this laser has a high incidence of scarring and pigmentary disturbances. CO_2 lasers can operate in continuous or pulsed mode.

11.10.2   Solid State Lasers

There are several types of solid state lasers, such as ruby, Ti sapphire, alexandrite and rare-earth lasers, such as neodymium–YAG (Nd^{3+}–YAG), neodymium–glass and erbium–silica (Er^{3+}–Si) fibre. When placed in an optical resonator that provides feedback, all of these materials behave as laser oscillators. Each has their own set of advantages and disadvantages. The following explains the principles and operation of the ruby laser.

11.10.2.1   Ruby Laser

The first successful optical maser or laser was developed in 1960 by Theodore Maiman, using a ruby crystal. Ruby is an aluminium oxide crystal (Al_2O_3) in which some of the aluminium atoms has been replaced with chromium atoms. Chromium gives ruby its characteristic red color and is responsible for the lasing behaviour of the crystal. Chromium atoms absorb green and blue light and emit red light. Figure 11.13 shows the components of the first ruby laser device. A cylindrical crystal of ruby is used. A fully reflective mirror is placed on one end, and a partially reflective mirror on the other. A high-intensity lamp is spiralled around the ruby cylinder, to provide a flash of white light that triggers the laser action. The green and blue wavelengths in the flash light excite electrons in the chromium atoms to a higher energy level. On returning to their normal state, the electrons emit their characteristic ruby red light. The mirror reflect some of this light back and forth inside the ruby

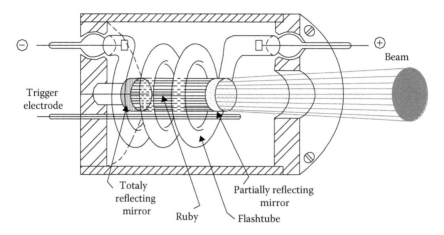

FIGURE 11.13   The first ruby laser produces pulses of light.

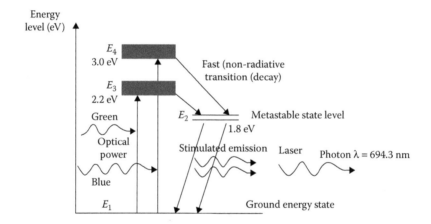

FIGURE 11.14   Energy level diagram for the ruby laser.

crystal, stimulating other excited chromium atoms to produce more red light, until the light pulse builds up to high power and drains the energy stored in the crystal.

Figure 11.14 shows energy level diagram for ruby laser. The atoms are excited from state E_1 to state E_3. This process is called optical pumping. The atoms quickly decay either back to E_1 or to the intermediate state E_2, which is metastable with a lifetime of about 3×10^{-3} second (compared to 10^{-8} second for ordinary levels). With strong pumping, more atoms can be forced into the E_2 state than are in the E_1 state. Thus we have the inverted population needed for lasing. As soon as a few atoms in the E_2 state jump down to E_1, they emit photons that produce stimulated emission of the other atoms and the lasing action begins. A ruby laser thus emits a beam whose photons have 1.8 eV of energy and a wavelength of 694.3 nm (ruby-red light).

The lasing operation process of the ruby laser is typical for most laser types, and is illustrated below:

1. Figure 11.15 shows the ruby laser in its non-lasing state.
2. The flash tube fires and injects external light into ruby crystal. The chromium ions (Cr^{3+}) in a ruby crystal are excited by light from the flash tube, when high-voltage electricity is applied to this external light source, as shown in Figure 11.16. This light source must have photons of the right frequency to raise to the atoms to excited states.

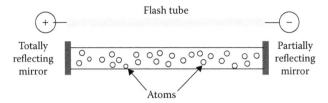

FIGURE 11.15   The ruby laser in its non-lasing state.

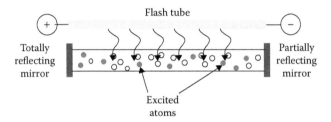

FIGURE 11.16   The flash tube injects external light into the Ruby rod.

3. The Cr^{3+} ions are raised in the pumping process to energy level E_3, from which they decay to the metastable level E_2 by losing energy to other atoms in the crystal. Some of these atoms emit photons, as shown in Figure 11.17.

4. A few Cr^{3+} ions in the E_2 level then spontaneously fall to the E_1 level. Some of these photons propagate parallel to the axis of the ruby crystal, so they bounce back and forth between the reflecting end mirrors of the ruby rod, as shown in Figure 11.18. This initiates the continuing process of stimulated emission and amplification.

5. The presence of light of the right frequency now stimulates the other Cr^{3+} ions in the E_2 level to radiate; the result is an avalanche of photons that produces a large pulse of red light. Some of the photons leave through the partially silvered mirror at one end. The red light is the laser light, as shown in Figure 11.19.

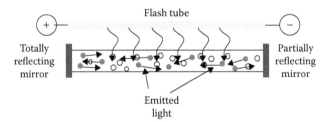

FIGURE 11.17   Some atoms emit photons spontaneously.

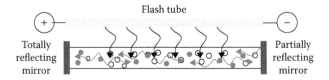

FIGURE 11.18   Some photons traverse the ruby's axis and reflect between the mirrors.

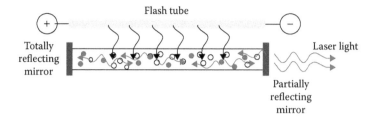

FIGURE 11.19 Laser light leaves the ruby through the partially silvered mirror.

11.10.2.2 Neodymium–YAG Laser

Another example of a solid-state laser, the neodymium–YAG, is shown in Figure 11.20. The neodymium ion (Nd^{3+}) is doped into the yttrium–aluminium–garnet (YAG) host crystal. Neodymium–YAG lasers have become very important because they can be used to produce high power laser light. The energy diagram of this laser is shown in Figure 11.21. Such lasers have been constructed to produce over a kilowatt of continuous laser power at 1064 nm, and can achieve extremely high power in a pulsed mode. Neodymium–YAG lasers are used in pulse mode in laser oscillators to produce short pulses of several nanoseconds. Furthermore, neodymium laser light can be passed through a second harmonic generating crystal that doubles its frequency, thereby providing a strong source of green light at 532 nm. It can also be quadrupled, providing ultraviolet radiation at 266 nm.

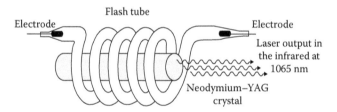

FIGURE 11.20 Neodymium–YAG laser.

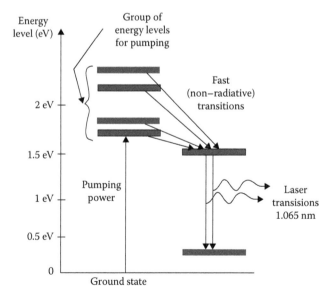

FIGURE 11.21 Energy level diagram for the Neodymium–YAG laser.

Aside from ruby, Nd^{3+} and Er^{3+}, other commonly encountered optically pumped solid-state laser amplifiers and oscillators include the following:

1. Alexandrite (Cr^{3+}: Al_2BeO_4), which has a tunable output in the wavelength range between 700nm and 800 nm.
2. Ti^{3+}: Al_2O_3 (Ti: sapphire), which is tunable over an even broader range, from 660 to 1180 nm.
3. Er^{3+}: YAG, which is often operated at 1660 nm. Ti: sapphire is often chosen for ultra-short pulse generation because of its broad bandwidth, because pulse length is inversely proportional to bandwidth.

11.10.2.3 Alexandrite Laser

The alexandrite laser is a solid-state laser, in which the crystal ($BeAl_2O_4$) is doped with 0.01%–0.40% of chromium ions (Cr^{+3}). The alexandrite laser has an energy level structure similar to that of the ruby laser. The alexandrite laser operates as a four level laser, which can be tuned over a range of wavelengths: 720–800 nm. It was the first tunable solid-state laser to reach the market.

11.10.3 Dye Lasers

The dye laser was first demonstrated in 1965 at IBM laboratories in the United States a, by Peter Sorokin and Lankard. They discovered the dye laser action during research into fluorescence of organic dye molecules that were excited by a ruby laser. In 1967 scientists discovered the possibility of tuning the emitted wavelength using a grating at the end of the optical cavity. A dye laser can be considered as a special device to convert electromagnetic radiation from one wavelength into another tuned wavelength. The output of a dye laser is always coherent radiation tunable over a specific spectral region, determined by the dye material.

The wavelength of the dye laser can be tuned for specific applications. For example, the wavelength range of one tunable medical dye laser is from 577 to 585 nm. This wavelength range increases the penetration depth from 0.5 to 1.2 mm in skin, when scanning skin tissue with light. By using microsecond pulses from a pulsed tunable dye laser, it is possible to confine thermal injury to the target blood vessels, without heat diffusion to surrounding tissue.

11.10.4 Semiconductor Lasers

All semiconductor lasers, sometimes called laser diodes, are built from semiconductor materials. The first semiconductor laser was created by Robert Hall in 1961 in the United States. In 1975 the first semiconductor laser capable of operating continuously at room temperature was introduced. This development has led to the use of semiconductor lasers in the CD-ROM, DVD-ROM, laser pointer and many other useful devices. Figure 11.22 shows a semiconductor laser device.

Researchers succeeded in getting coherent electromagnetic radiation from a forward-biased diode (*p-n* junction) made from semiconductor materials. The characteristics of the laser beam depend on the semiconductor materials used in building the device. The majority of semiconductor materials are based on a combination of elements in the third group of the periodic table (such as Al, Ga, In) and the fifth group (such as N, P, As, Sb), hence referred to as the III–V compounds. Examples include GaAs, AlGaAs, InGaAs and InGaAsP alloys. The CW laser emission wavelengths are normally within 630 to 1600 nm, but InGaN semiconductor lasers were found to generate CW at 410 nm (blue light) at room temperature. Semiconductor lasers can generate blue-green light using materials, which are the combination of elements of the second group (such as Cd and Zn) and the sixth group (S and Se), referred to as the II–VI compounds.

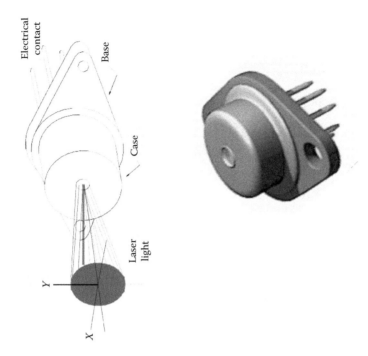

FIGURE 11.22 A packaged semiconductor laser.

The conductivity of a semiconductor increases with temperature, up to a certain temperature level. The efficiency of a semiconductor laser decreases when the temperature increases.

The Fabry–Perot laser structure is the simplest example of semiconductor lasers. Fabry–Perot single-spatial-mode laser diodes are available that operate from 730 to 770 nm, with tens of mega-watts of optical power. Applications of the Fabry–Perot laser include medicine, spectroscopy, gas detection, sensors, measurement instruments and high-power laser seeding.

11.10.4.1 Energy Bands

As explained in the principle of light-emitting diodes, when a diode is forward biased, holes from the p-region are injected into the n-region, and electrons from the n-region are injected into the p-region, as shown in Figure 11.23. If electrons and holes are present in the same region, they may radiatively recombine; that is, the electron falls into the hole and emits a photon with the energy of the band gap. This is called spontaneous emission and is the main source of light in a light-emitting diode. There is a spread in emitted wavelength.

Under suitable conditions, the electron and the hole may co-exist in the same area for quite some time (on the order of microseconds) before they recombine. If a photon of exactly the right frequency happens along within this time period, a recombination (of hole and electron) may be stimulated by the photon. This causes another photon of the same frequency to be emitted, with exactly the same direction, polarisation and phase as the first photon.

In a laser diode, the semiconductor crystal is fashioned into three layers very thin in thickness and rectangular in the other two dimensions. An optical waveguide is made in the middle layer such that the light is confined to a relatively narrow line. This is called the active layer. The top of the crystal is p-doped, and the bottom is n-doped, resulting in a p-n junction, which forms the diode. The two ends of the crystal are cleaved so as to form perfectly smooth, parallel edges like mirrors; these two reflective parallel edges form a resonator called a Fabry–Perot cavity, as shown in Figure 11.24.

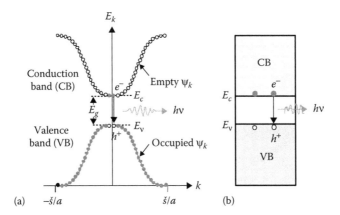

FIGURE 11.23   (a) *E-K* diagram and (b) energy band diagram for a diode.

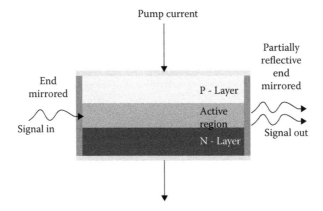

FIGURE 11.24   Laser diode.

Photons emitted in precisely the right direction will travel along the cavity and be reflected several times from each end before they exit the cavity. Each time photons pass through the cavity, the light is amplified due to stimulated emission. Hence, if there is more amplification than loss, the diode begins to laze. The wavelength emitted is a function of the band gap between the energy levels of the *p* and *n* regions. No photons with energy higher than the band gap will be emitted.

11.10.4.2   Types of Semiconductor Lasers

There are essentially two ways that laser diodes can direct laser light that can be suitable for optical communications, as shown in Figure 11.25. If the emitted light emerges from an area on the edge of the device, as shown in Figure 11.25a, that is, from an area on a crystal face perpendicular to the active layer, then the device is called an edge-emitting laser diode. If the emitted light emerges from an area on the plane of the active layer, as shown in Figure 11.25b, then the device is called a surface-emitting laser diode.

There are many types of laser diodes on the market, varying in device structure, wavelength of emitted light and application. The basic types of laser diodes are listed in the following sections:

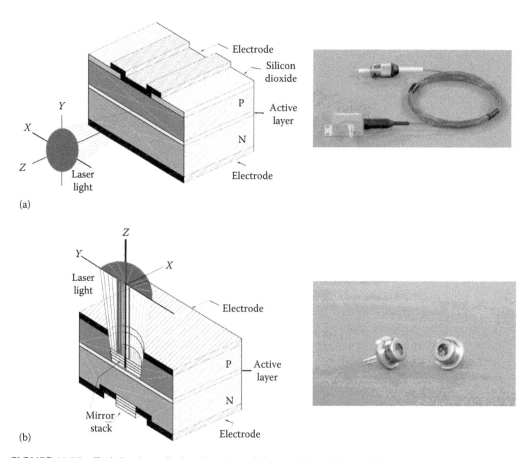

FIGURE 11.25 Emitting laser diodes: (a) edge emitting and (b) surface emitting.

11.10.4.3 Heterojunction Laser Diodes

The type of laser diode described above is called a heterojunction laser diode, as shown in Figure 11.26. The heterojunction is a junction between two different semiconductors with different band gap energies. Unfortunately, heterojunction laser diodes are extremely inefficient. They require so much power that they can only be operated in short pulses.

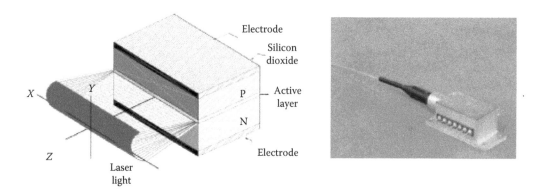

FIGURE 11.26 A heterostructure laser.

11.10.4.4 Double Heterostructure Laser Diodes

In double heterostructure devices, a layer of low band gap material is sandwiched between two high band gap layers. One commonly used pair of materials is GaAs with Al GaAs. Each of the junctions between different band gap materials is called a heterostructure. Hence the name double heterostructure laser describes this type of laser.

The advantage of a double heterostructure laser is that the region where free electrons and holes exist simultaneously, that is, the active region, is confined to the thin middle layer. As a result, many more of the electron-hole pairs (EHPs) can contribute to amplification, and fewer are left out in the poorly amplifying periphery. In addition, light is reflected from the heterojunctions. Hence, the light is confined to the active region where the amplification takes place.

11.10.4.5 Quantum Well Lasers

If the middle layer of a laser diode is made thin enough, it starts acting like a quantum well. This means that in the vertical direction, electron energy is quantised.

The quantum layers are usually stacked one on top of the other. The energy difference between quantum well levels can be used to produce the laser action instead of the band gap. This is very useful because the wavelength of light emitted can be tuned simply by altering the thickness of the layer. The efficiency of a quantum well laser is greater than that of a bulk laser, due to a tailoring of the distribution of electrons and holes that are involved in the stimulated emission (light producing) process.

11.10.4.6 Separate Confinement Heterostructure Lasers

The problem with heterostructure laser devices is that the thin layer is simply too small to effectively confine the light. To compensate, another two layers are added on outside of the first three. These layers have a lower refractive index than the centre layers and, hence, confine the light effectively. Such a design is called a separate confinement heterostructure (SCH) laser diode. Almost all commercial laser diodes since the 1990s have been SCH quantum well diodes.

11.10.4.7 Distributed Feedback Lasers

Distributed feedback lasers (DFB) are the most common laser type used to transmit signals in dense wave division multiplexer systems (DWDM). To create a precise lasing wavelength, a diffraction grating is etched close to the *p-n* junction of the diode. This grating acts like an optical filter, causing only a single wavelength to be fed back to the active region where the lasing is amplified. At least one face of a DFB is anti-reflection coated. The DFB laser has a stable wavelength that is set during manufacturing by the pitch of the grating, and can only be tuned slightly with temperature.

11.10.4.8 Tunable Laser Diodes

As the number of wavelengths required in networks increases, tunable lasers play an increasingly important role in the development of dynamic network systems. With the assistance of electronic controls, a tunable laser can be tuned to emit different wavelengths. This reduces the need for spare wavelength-specific line cards and thus lowers a network's overall cost. There are several different technologies for tunable lasers. Figure 11.27 shows the basic principle of a tunable laser.

The tuning can be described by the following equations:

The wavelength of the mode is given by

$$\frac{m\lambda}{2} = \bar{n}L$$

(11.23)

where:
    m is the mode number
    λ is the wavelength
    $\vec{n}$ is the effective refractive index
    L is the effective cavity length

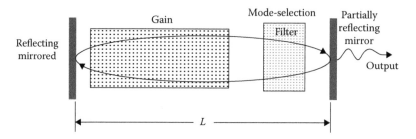

FIGURE 11.27 Common principle of a tunable laser.

From the Equation 11.23, the relative change in wavelength is given as

$$\frac{\Delta\lambda}{\lambda} = \frac{\Delta n}{n} + \frac{\Delta L}{L} - \frac{\Delta m}{m} \tag{11.24}$$

where:
  $\Delta n/n$ is the tuning portion due to the change in the net cavity index
  $\Delta L/L$ is the tuning portion due to the change in the physical length of the crystal
  $\Delta m/m$ is the tuning portion due to the mode-selection filtering (via index or grating angle)

11.10.5 OTHER TYPES OF LASERS

Other types of lasers include the following:

1. Chemical lasers, in which the energy input comes from the chemical reaction of highly reactive gases.
2. Ultrafast, gas dynamic, x-Ray and Bragg reflector (DBR) lasers, such as distributed, tunable and external cavity.
3. Vertical cavity surface smitting lasers (VCSELs).
4. In-fibre lasers.
5. Metal-vapour lasers.
6. Free electron lasers.
7. Excimer lasers.

11.11 COMPARISON OF SEMICONDUCTOR AND CONVENTIONAL LASERS

Semiconductor lasers are quite different from conventional lasers. In particular:

1. The gain of the semiconductor laser material is very high and is generated by a population inversion between the conduction and valence bands of the semiconductors. In some sense, a semiconductor laser is a two-state laser system.
2. Since the longitudinal mode is on the order of the size of the laser device, the transverse mode of the semiconductor laser is quite different from that of a conventional laser. In particular, the beam is not Gaussian, the beam profile tends to be elliptical, and the beam divergence tends to be large.
3. The gain spectrum is quite large (many THz or hundreds of angstroms).
4. The short cavity length (several hundred microns) means that the longitudinal mode spacing is much larger than that of a conventional gas or solid-state laser (on the order of GHz or angstroms).

5. Due to the small size of semiconductor lasers, they have the potential for mass production, and can be easily integrated on PC boards.

6. The properties of the semiconductor lasers are being rapidly improved. They are becoming more and more powerful and efficient laser sources.

11.12   CLASSIFICATION OF LASERS

Full details on the laser classifications and safety issues are presented in the photonics laboratory safety chapter. This chapter also covers laser devices, safety, operation and utilisation as light sources in lab experiments. Lasers are classified into four basic laser classes, which include Class 1, 2, 2a, 3a, 3b and 4. Higher class numbers reflect an increased potential to harm users. Operating wavelength and mode (CW and pulse) are the criteria used to classify the hazard level of lasers. A summary of the laser classifications and applications is listed below:

1. Class 1 lasers output light in the visible range (450 nm $< \lambda <$ 500 nm) and have very low output power (between 0.04 and 0.40 mW). Some examples of Class I laser devices include the following: CD players, scanners, laser pointers and small measurement equipment.

2. Class 2 lasers output light in the visible range (400 nm $< \lambda <$ 700 nm) and have low output power (must be less than 1 mW of visible CW). Some examples of Class 2 lasers are classroom demonstration laser sources and laser source devices for testing and telecommunications.

3. Class 2a lasers are low output power devices, which are considered to be visible light lasers. A supermarket bar-code scanner is a typical example of a Class 2a laser device. Never view a Class 2a laser beam directly.

4. Class 3 lasers are divided into two subgroups (Class 3a and Class 3b lasers), so there is no plain Class 3.

5. Class 3a lasers are intermediate power devices. Class 3a lasers are allowed to exceed the output power limit of Class 2 lasers by no more than a factor of 5 or produce visible light power less than 5 mW. They are considered to be CW lasers. Some examples of Class 3a laser devices are laser scanners, laser printers and laser source devices for testing and telecommunications.

6. Class 3b lasers are intermediate power devices; they output between 5 and 500 mW of C, or else pulsed 10 J/cm^2 power. They are considered to be CW lasers. Some examples of Class 3b lasers are laser source devices for testing.

7. Class 4 lasers are high power devices; they either output more than 500 mW of CW power or else pulsed 10 J/cm^2 power. They are considered to be very high power lasers. Some applications of Class 4 laser devices include the following: surgery, drilling, cutting, welding and micro-machining.

11.13   LASER-BEAM ALIGNMENT

Researchers and users of commercial laser-based systems need to align the laser beam source to an optical axis of an optical component. In the laboratory, operational alignment flexibility is key when set-ups are modified and re-configured routinely. In commercial products, the critical aspects are quick alignment during assembly, and simple realignment to accommodate efficient replacement or repair of the laser or optics.

The advantage of using mirrors for alignment is that they provide uniform performance over multiple laser wavelengths. Metallic mirrors offer the broadest wavelength performance. Dielectric mirrors offer a narrower reflectivity range but can provide higher-peak reflectivity.

To define an optical axis, it is necessary to fix the *X–Y* location of the beam at some arbitrary point along the *Z*-axis, and to set the angular direction of the beam at this location (Ø, ß). The *X–Y* plane is perpendicular to the direction of beam propagation, which is the *Z*-axis. There are two common ways of establishing such a reference:

1. *The easy way*: Pass the laser beam through two apertures, adjusting the beam angle and position until it goes through the centre of each. The first aperture sets the *X–Y–Z* coordinates, and the second determines the angle of propagation. If using adjustable irises rather than pinholes, open the apertures during initial alignment to save considerable effort finding the axis.
2. *The accurate way*: Replace the apertures with position-sensitive detectors. Quantitative lateral and angular alignment is achieved by centring the beam on each detector or, in the case of a quad cell, by balancing the signals. As these detectors are not transmissive, the first detector must be mounted on a removable base or placed off-axis in conjunction with an on-axis beam splitter.

There are several techniques to accurately perform the positional and angular alignment of a laser beam. They can be divided into two categories. In one category, the laser source device is aligned using a one, two or multi-axis stages. In the second category, the laser beam exiting from the laser source device is aligned directly to an optical application. Some of these techniques are listed below:

11.13.1   ALIGNMENT USING ONE, TWO AND MULTI-AXIS POSITIONERS

Using stages or positioners, as shown in Figure 11.28, is a simple, easy and low cost way of aligning a small laser device like a He–Ne laser tube. Alignment performance parameters, such as specifications for positioner accuracy, precision and resolution are the most quoted. These parameters refer to how well a component performs along the desired axis of travel, and are determined primarily by the drive mechanism.

For the one-axis stage, performance is additionally measured by the deviation from the desired axis of travel. For the single-axis translation stage, as shown in Figure 11.28a, one coordinate of translation is designed to vary. The position of any optical mount can be defined uniquely in terms of six independent coordinates, three translational and three rotational, with respect to some arbitrary fixed coordinate system. The error in motion, then, is defined in terms of the remaining five independent coordinates, which ideally stay fixed in space. Similarly, the two-axis translation stage, as shown in Figure 11.28b, provides linear motion in two directions. Three or multi-axis stages overcome the limitations of one and two-stages. Some of the multi-axis positioners and stages are shown in Figure 11.28c and d.

Figure 11.29a illustrates a schematic diagram for a set-up to align a He–Ne laser tub, so that the laser beam is reflected by mirror M1 towards the target card. Figure 11.29b illustrates a set-up to align a He–Ne laser tube so that the laser beam bounces between two mirrors M1 and M2 and is captured on a target card.

11.13.2   ALIGNMENT USING TWO MIRROR MOUNTS

The most common method of beam steering is to use flat mirrors on tip/tilt adjustable multi-axis mounts. At least two mirrors are needed to deliver the four required degrees of freedom for the alignment steps.

Figure 11.30 shows aligning a laser beam to the two irises. The first step is to roughly position the mirrors so that the beam falls on Iris 1. Then the beam on Iris 1 is centred, using fine adjustments on mirror M_1 only. This step is verified by reducing the aperture of Iris 1 until it is just slightly smaller than the laser beam. When the beam is properly centred, only a halo of light is

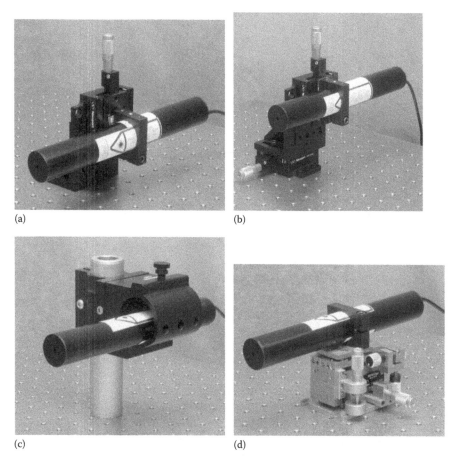

FIGURE 11.28 Laser holder/positioners: (a) one-axis laser holder/stage, (b) two-axis laser holder/stage, (c) multi-directions laser holder/positioner and (d) multi-axis laser holder/stage.

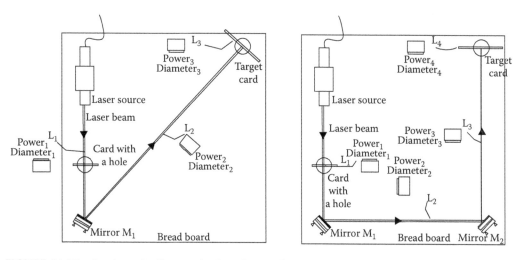

FIGURE 11.29 A schematic diagram for laser-beam alignment: (a) using one mirror and (b) using two mirrors.

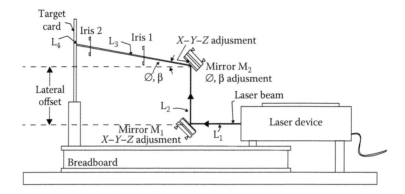

FIGURE 11.30 Two adjustable mirrors aligned to the optical axis of two irises.

seen around the aperture. Next, Iris 1 is re-open and the beam is centred on Iris 2, by adjusting mirror M_2 so that only a halo of light is seen on Iris 2.

Iris 1 is re-checked for beam centration using the halo effect. Iterative adjustment of mirror M_1 on Iris 1 and mirror M_2 on Iris 2 will quickly result in precise alignment. However, if final alignment leaves the beam near the edge of one of the mirrors then the mirror should be grossly repositioned accordingly and the alignment process repeated. Placing mirror M_2 as close as possible to Iris 1 reduces the number of iterations necessary to reach alignment by reducing crosstalk between the degrees of freedom. As the separation approaches zero, adjustments to mirror M_2 have no effect on the position of the beam at Iris 1.

There are several reasons for the popularity of this technique. First, it uses simple, inexpensive components available in most laboratories. But just as important, it is a quickly converging and straightforward technique. The intuitive feel is further enhanced by the fact that the beam's angular displacement is essentially linear over the entire adjustment range of the mounts.

By using closely spaced, large mirrors (compared to the beam diameter), this approach offers fairly large translational and angular adjustment ranges. One potential limitation of a two-mirror mount is that it produces a lateral offset. In many applications, the overall optical layout can accommodate this offset. In others, a zero offset can be achieved by using three or more mirrors.

11.13.3 ALIGNMENT USING THREE AND FOUR MIRROR MOUNTS

A zero offset alignment can be achieved by using three or more mirrors, as shown in Figure 11.31. In a zero-offset layout, the mirrors are roughly aligned, and then the final alignment is performed by adjusting only the first mirror (for Iris 1) and last mirror (for Iris 2). The three-mirror system

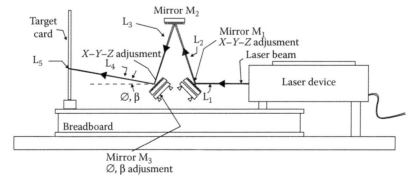

FIGURE 11.31 Three mirrors provide beam alignment with no lateral offset.

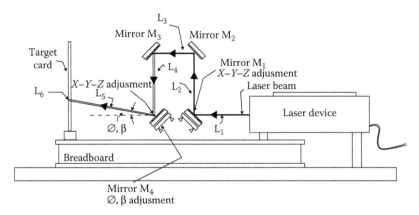

FIGURE 11.32 Four mirrors provide beam alignment with no lateral offset.

is simple but yields incident angles greater than 45 degree. The four-mirror system, as shown in Figure 11.32, produces incidence angle close to 45 degree, which matches well with standard mirror coatings and leads to higher reflectivity.

The choice of mirror mount and actuator depends on the exact dimensions of the set-up and the angular tolerance of the application. Increasing mirror separation produces larger X–Y displacements at the second mirror, for a given actuator adjustment. This can increase the lateral adjustment range of the system but also amplifies the effects of drift or vibration. Also, as mirror separation increases, maintaining a given linear adjustment resolution at the second mirror requires greater angular resolution in the actuator of the first mirror.

11.13.4 ALIGNMENT USING A RISLEY PRISM PAIR

An alternative alignment technique is based on the Risley prism pair. The Risley prism pair consists of two round, paired flats with identical wedge angles. A pair of Risley prisms is wedged and circular windows mounted in independent rotation stages. Each window acts as a thin, refractive prism. The window rotation tilts the beam, as shown in Figure 11.33. By adjusting the two matched prisms, the beam can be deflected at any angle (Ø, ß) within a cone (set by the wedge angle and index of refraction of the prisms). As with mirrors, two sets of the prisms are needed for adjustments in four degrees of freedom (X, Y, Ø, ß). All surfaces in a Risley prism pair system need anti-reflective coating and/or the air gap to be filled in with a refractive-index-matching gel.

An advantage of Risley prisms over two-mirror systems is that they do not introduce a significant lateral beam offset. Also, they offer excellent mechanical stability because of their independent rigid mountings. In a typical rotation stage, the locking mechanism acts orthogonal to the adjustment mechanism, allowing rigid locking without affecting the precise alignment. In contrast, spring tension maintains mirror mount alignment, making the mounts more susceptible to vibration and long-term drift. Consequently, Risley prisms are often found in commercial and aerospace applications where vibration and shock are common.

One significant limitation of the Risley prisms is their lack of orthogonal adjustment. Indeed, their operation is non-intuitive. Rotating them produces tip/tilt re-direction. Furthermore, the mechanical adjustment is non-linear. The result is that it may take more time to adjust precisely. Other drawbacks are the increased optical/mechanical complexity and cost.

In addition, the prisms operate by refraction, with dispersion effects making beam deflection angle a function of wavelength. Unlike mirrors, this can preclude their use in multi-wavelength applications. Prisms also have an inherent trade-off between adjustment range and resolution because the maximum possible deflection depends on the prism wedge angle. Increasing the wedge

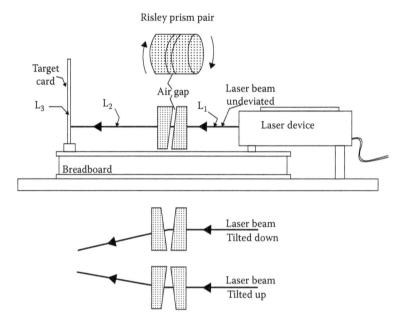

FIGURE 11.33 Risley prism pair provide fine alignment of the beam.

angle increases the adjustment range but reduces the minimum resolution provided by a given rotation mount. The Risley prism pair is better suited to fine-tuning than gross alignment.

11.13.5 Alignment Using an Adjustable Wedge

A simple way of aligning a beam with no lateral offset is to use a single wedge with a continuous angle adjustment, as shown in Figure 11.34. A matching lens pair is made by mating the curved surfaces of a plano-convex lens with a plano-concave lens that has the same radius of curvature.

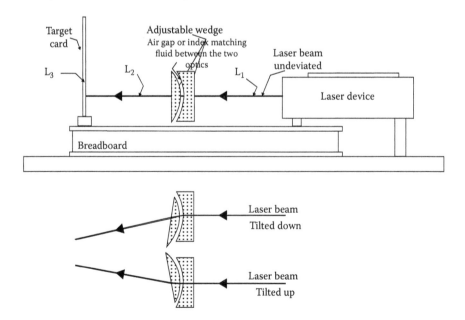

FIGURE 11.34 Matched lenses provide fine alignment of the beam.

Sliding the convex lens relative to the concave surface adjusts the wedge angle. Like the Risley prisms, the adjustable wedge is better suited to fine adjustments than to gross alignment. Two sets of optics are required to provide the full four degrees of freedom. Besides zero offset, adjustment of the variable wedge is intuitive; pushing the plano-convex lens in one direction move the beam in the same direction. Also, for small adjustments beam deflection is fairly linear.

The thin gap between any lenses can contain air or a refraction-index-matching fluid. The advantage of an air gap is a higher damage threshold. But beam-focusing errors can arise if a constant, minimal air gap is not maintained. To maintain throughput efficiency, air/glass surfaces must be anti-reflection coated, whereas surfaces that contact index-matching fluid do not require coating. Although it lowers the damage threshold, an index-matching fluid can have two advantages over an air gap. This fluid reduces the number of air/glass interfaces, which reduces reflection between parallel surfaces of the lenses, and the fluid lubricates the interface while holding the lenses together with surface tension.

Regardless of which alignment method is used, keep some of the following practical tips in mind:

1. The height of the optical axis and laser beam above the mounting surface should be minimised. Using low-profile mounts and posts limits the effects of vibration and thermal drift. If vibration is a problem, consider using damped mounting posts.
2. Each mirror has a thickness when the back is silver coated. The thickness of the mirror refracts the laser beam and increases laser-beam losses. It is better to use other types of mirrors, such as metallic or dielectric mirrors. They are ideal for sensitive applications, such as imaging with no chromatic aberration.
3. For each lens there are two reflections, one from each surface. When the centres of the two reflections are at the height of the laser beam, the height of the lens is properly adjusted. When they are overlapping, the beam is at the centre of the lens. And when they are centred about the laser output, the lens is not tilted with respect to the beam.
4. For each optical component there are many types of losses. The main losses are caused by absorption, dispersion, scattering and reflection of the light.
5. Plan before attempting to align the system. Consider all of the angular tolerances required and the long-term stability and vibration sensitivity of the system should be considered, then specify components accordingly. As with any optical set-up, too much performance adds unnecessary cost, whereas too little presents the risk that the system will not work.

11.14   LASER-BEAM EXPANSION

Often when a laser is used in an optical system, there is a requirement for either a larger beam or a small-beam divergence. In some cases, the size of the beam becomes critical. For example, when measuring the distance from the Earth to the Moon, a beam one metre in diameter travelling to the Moon expands to several hundreds of metres in diameter and returns to the Earth's surface with a diameter of several kilometres. The signal returned from this expansion is millions of times smaller than the original signal, so the divergence of a laser beam needs to be reduced to produce strong and detectable signals. Reduction of the beam divergence is called beam collimation. Even in earthbound experiments, higher beam collimation is required for many applications.

As stated before, light of a laser source is very different from most other light sources. After a description of the simplest type of beam, the transverse mode TEM_{00} Gaussian beam and its parameters will be discussed below. This beam type should be understood before examining ways of collimating the beam.

11.14.1 CHARACTERISTICS OF A GAUSSIAN BEAM

For a laser beam, the term *Gaussian* describes the variation in the irradiance along a line perpendicular to the direction of propagation and through the centre of the beam, as illustrated in Figure 11.35.

The irradiance I is symmetric about the beam axis, and varies radially outward from the propagation axis with the form given by

$$I(r) = I_o e^{-2r^2/r_1^2} \tag{11.25}$$

This equation is illustrated in Figure 11.36, and expressed in terms of a beam diameter as

$$I(d) = I_o e^{-2d^2/d_1^2} \tag{11.26}$$

where:
r_1 is the radius of the beam
d_1 is the diameter of the beam

By definition, r_1 and d_1 occur when I equals $1/e^2$ of I_0.

Figure 11.36 assumes a beam of parallel rays. In reality, a Gaussian beam either diverges from a region where the beam is smallest, called the beam waist, or converges, as illustrated in Figure 11.37. The amount of divergence or convergence is measured by the full angle beam divergence θ, which is the angle subtended by the $1/e^2$ diameter points for distances far from the beam waist, as illustrated in Figure 11.37. In some laser references, the half angle is measured from the beam axis to the $1/e^2$ asymptote. However, it is the full-angle divergence, as defined here that is usually given in the laser

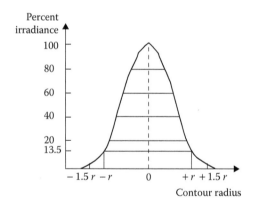

FIGURE 11.35 Gaussian beam profile.

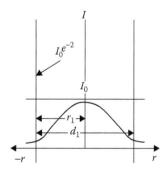

FIGURE 11.36 Gaussian beam waist.

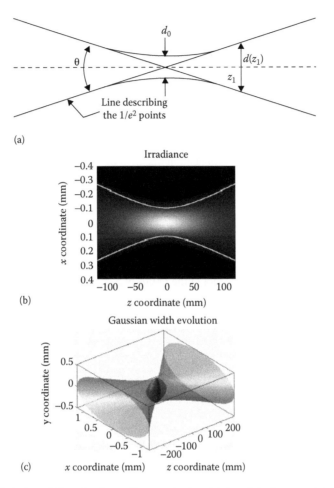

FIGURE 11.37 Variation of Gaussian beam diameter in the vicinity of the beam waist: (a) beam waist and beam divergence, (b) two-dimensional Gaussian beam and (c) three-dimensional Gaussian beam.

specifications. On account of symmetry on either side of the beam waist, the convergence angle is equal to the divergence angle.

According to the laws of geometrical optics, a Gaussian beam converging at an angle of θ should collapse to a point. On account of diffraction, this does not occur. However, at the intersection of the asymptotes that define θ (radians), the beam does reach a minimum beam waist diameter d_o. It can be shown that for a TEM_{00} mode beam, d_o depends on the beam divergence angle as

$$d_o = \frac{4\lambda}{\pi\theta} \tag{11.27}$$

where λ is the wavelength of the beam.

Note that for a Gaussian beam of a particular wavelength, the product $d_o\theta$ is constant. Therefore, for a very small beam waist, the divergence must be large; for a highly collimated beam (small θ), the beam waist must be large. The variation of the beam diameter in the vicinity of the beam waist is illustrated in Figure 11.37, and given as

$$d^2 = d_o^2 + \theta^2 z^2 \tag{11.28}$$

where d is the diameter at a distance $\pm z$ from the waist along the beam axis.

11.14.2 The Rayleigh Range

It is useful to characterise the extent of the beam waist region with a parameter called the Rayleigh range. In other descriptions of Gaussian beams, this extent is sometimes characterised by the confocal beam parameter and is equal to twice the Rayleigh range. Rewriting Equation 11.28 gives

$$d = d_o\sqrt{1 + \left(\frac{\theta z}{d_o}\right)^2} \tag{11.29}$$

The Rayleigh range is defined as the propagation distance from the beam waist to where the diameter has increased to $d_o\sqrt{2}$. Obviously, this occurs when the second term under the radical is unity, when

$$z = z_R = \frac{d_o}{\theta} \tag{11.30}$$

Although the definition of z_R might seem rather arbitrary, this particular choice offers more than just convenience. Figure 11.38 illustrates a plot of the radius of curvature of the wavefronts in a Gaussian beam as a function of z. For large distances from the beam waist, the wavefronts are nearly planar, and the radius tends towards infinity. At the beam waist wavefronts are also planar. Therefore, the absolute value of the radius of curvature of the wavefronts must go from infinity at large distances, through a minimum, and return to infinity at the beam waist. This is also true on the other side of the beam waist but with the opposite sign. It can be shown that the minimum in the absolute value of the radius of curvature occurs at $z = \pm z_R$, that is, at a distance of one Rayleigh range on either side of the beam waist. From Figure 11.38, the collimated region of Gaussian beam waist can be taken as $2z_R$. The point z_R can be considered as the transition of the beam from the near field to the far field. At the focus of the beam using a lens, the Rayleigh range transforms to the depth of the focus.

The Rayleigh range can be expressed in a number of ways:

$$z_R = \frac{d_o}{\theta} = \frac{4\lambda}{\pi\theta^2} = \frac{\pi d_o^2}{4\lambda} \tag{11.31}$$

This shows that all three characteristics of a Gaussian beam are dependent on each other. Given any one of the three quantities, d_o, θ, z_R and the wavelength of the radiation, the behaviour of the beam is completely described.

Figure 11.39 illustrates an example of a helium–neon laser ($\lambda = 632.5$ nm) transverse mode TEM$_{00}$, with a beam waist diameter of 1 mm. The beam divergence angle θ from Equation 11.31 is

$$\theta = \frac{4\lambda}{\pi d_o} = \frac{4 \times 632.5 \times 10^{-9}}{3.141 \times 1 \times 10^{-3}} = 0.805 \times 10^{-3} \text{ rad}$$

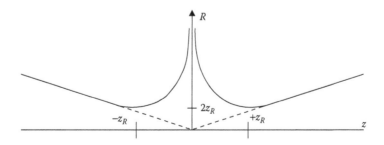

FIGURE 11.38 Radius of curvature (r) versus distance (z) from the beam waist.

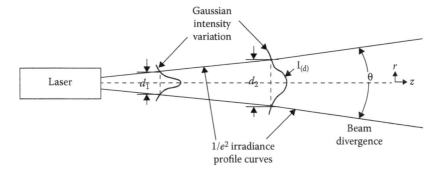

FIGURE 11.39 A Gaussian TEM$_{00}$ laser beam.

The Rayleigh range calculated using Equation 11.30 is

$$z_R = \frac{d_o}{\theta} = \frac{1 \times 10^{-3}}{0.805 \times 10^{-3}} = 1.242 \,\text{m}$$

11.14.3 EXPANSION AND COLLIMATION OF A LASER BEAM

Through the use of lenses, the divergence beam waist and Rayleigh range of the Gaussian beam can be changed. However, from the above discussion it is clear that the relations between the various beam parameters cannot be changed. Thus, to increase the collimation of a beam by reducing the divergence requires that the beam waist diameter be increased, as the beam waist diameter-divergence product is constant.

There are two ways of collimating a Gaussian beam. One method uses a Galilean telescope, which consists of a negative eyepiece lens of short focal length and a positive objective lens of long focal length, as shown in Figure 11.40. Collimation can be done by first creating a beam with a strong divergence and small beam waist, using the negative lens. Then by putting the beam waist at the focal point of the long focal length lens. This amounts to putting the beam through a telescope backwards. The laser beam goes in the eyepiece lens and comes out of the objective lens. Thus, the diverging beam is collimated with a large beam waist and small divergence. It can be shown that the decrease in the divergence is equal to the original divergence divided by the magnification of the telescope. The magnification of the telescope is equal to the ratio of the focal length of the objective and the focal length of the eyepiece.

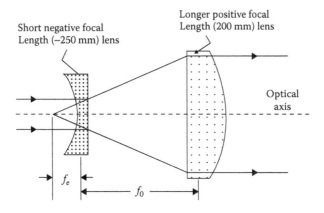

FIGURE 11.40 Gaussian beam collimation by Galilean telescope.

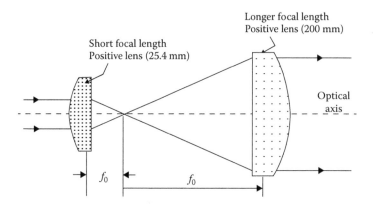

FIGURE 11.41　Gaussian beam collimation by Keplerian telescope.

The second method uses a Keplerian telescope, as illustrated in Figure 11.41. The eyepiece lens is a positive lens of short focal length. The beam comes to a focus and then diverges to be collimated by the objective lens of long focal length.

Each telescope has distinct advantages for beam expansion. The advantage of the Galilean type of beam expander is well utilised in high-power or pulsed-laser systems. As the beam does not come to a focus anywhere inside the beam expander's optical path, the power density of the beam decreases. Thus, if the lenses and environment can survive the initial beam, they can survive the beam anywhere in the optical path. Although the Keplerian beam expander can give similar ratios of beam expansion, the power density at the focus of the first lens is very large. In fact, when using a high power pulsed laser it is possible to cause a breakdown of the air in the space between the lenses. This breakdown is caused by the very strong electrical field that results from focusing the beam to a small diameter creating miniature lightning bolts.

11.15　LASER APPLICATIONS

The unique feature of light from a laser, as mentioned before, is that it is a coherent narrow beam of a single frequency (or several distinct frequencies). On account of this feature, the laser has found many applications. The main applications are as follows:

1. *Industrial applications*: Industry accepted the laser as a tool soon after the laser was invented in 1960. At first, the laser was used for alignment and measurements, but with time, applications using high power laser beams became more common. The intense heat produced in a small area by a laser beam is also used for welding and machining metals and for drilling tiny holes in hard materials. The beam of a laser is narrow in itself (typically a few mm). But because the beam is coherent, monochromatic and essentially parallel and narrow, lenses can be used to focus the light into incredibly small areas without the usual aberration problems. The limiting factor thus becomes diffraction. Also, the energy intensity can be very large. The precise straightness of a laser beam is also useful to surveyors for lining up equipment precisely, especially in inaccessible places.

2. *Applications in chemistry*: There are variety of applications of lasers in chemistry, such as excitation of molecules to specific levels, examination of the emitted radiation, measurements of the relaxation time of specific excited levels of molecules, disruption of chemical bonds in molecules at specific region, Raman spectroscopy and Raman scattering, which is a process of inelastic scattering of a photon by a molecule.

3. *Medical applications*: Lasers are useful surgical tools. The narrow intense beam can be used to destroy tissue in a localised area or to break up gallstones and kidney stones. On account of the heat produced, a laser beam can be used to weld broken tissue, such as a detached retina. For some types of internal surgery, the laser beam can be carried by an optical fibre to the surgical point and sometimes has an additional fibre-optic path on an endoscope for tissue treatment. An example is the removal of plaque that clogs human arteries. Tiny organelles within a living cell have been destroyed using lasers to study how the absence of that organelle affects the behaviour of the cell. Laser beams have been used to destroy cancerous and pre-cancerous cells; at the same time, the heat seals off capillaries and lymph vessels, thus cauterising the wound in the process. Single and multi-coloured tattoos can be removed by Q-switched lasers.

4. *Military applications*: Since the invention of the laser, its potential military uses were exploited. A large number of projects on lasers were done in secret laboratories, and many new types of equipment were put in service. Lasers are used in range finders, detecting laser signals and laser weapons, tracking systems and so on. Directed energy weapons (DEW) include lasers as well as microwaves, and practical beams are still being actively developed by the U.S. military.

5. *Daily applications*: Lasers are used as bar-code readers, compact disc (CD) players and CD-ROMs. The laser beam reflects off the stripes and spaces of a bar-code, and off the tiny pits of a CD. Lasers are used in remote controls and laser printers.

6. *Laboratory applications*: Lasers are used in spectroscopy, advanced microscopes, laser fusion and laser cooling of atoms.

7. *Imaging applications*: One of the most interesting applications of laser light is the production of three-dimensional images called holograms. In an ordinary photograph, the film simply records the intensity of light reaching it at each point. When, the photograph or transparency is viewed, light reflecting or passing through it gives us a two-dimensional picture. In holography, the images are formed by interference. When a laser hologram is made on film, a broadened laser beam is split into two parts by a half-silvered mirror, as shown in Figure 11.42. One part goes directly to the film; the rest passes to the object to be photographed, from which it is reflected to the film. Light from every point on the object reaches each point on the film, and the interference of the two beams allows the film to record both the intensity and relative phase of the light at each point. It is crucial that the light be coherent, which is why a laser is used. After the film is developed, it is placed

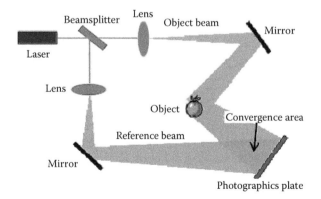

FIGURE 11.42 Production of three-dimensional images by holography.

again in a laser beam and a three-dimensional image of the object is created. When viewed from different sides, the hologram image looks like the original object; but when the image is touched, there is no material present.

8. *Special applications*: The number and variety of laser applications are expanding rapidly, and many new special applications have been discovered in fields, such as energy transmission from space to Earth, optical gyroscopes and fibre lasers.

11.16   EXPERIMENTAL WORK

The experimental works are conducted for the following cases:

11.16.1   EXPERIMENT ONE: LASER-BEAM ALIGNMENT

Students will practice some alignment techniques for a laser beam using optical components and measuring the power and diameter of a laser beam at several distances from the output of the laser source in the following experimental cases:

1. Alignment using one, two and multi-axis positioners
2. Alignment using two mirror mounts
3. Alignment using three and four mirror mounts
4. Alignment using a Risley prism pair
5. Alignment using an adjustable wedge

11.16.2   EXPERIMENT TWO: LASER-BEAM EXPANSION

This experiment demonstrates the design of two types of laser beam expanders: the Galilean telescope and the Keplerian telescope. From the cases of this experiment, students will gain experience in aligning, expanding and measuring the power and diameter of the laser beam. Students will also calculate the diameter of a converging Gaussian laser beam, at various distances from the output of the laser.

11.16.3   TECHNIQUE AND APPARATUS

Appendix A presents the details of the devices, components, tools and parts.

1. 2×2 ft optical breadboard.
2. Laser light source (He–Ne laser light source and power supply).
3. Laser clamp/laser mount assembly.
4. Laser light detector.
5. Laser light power meter.
6. Hardware assembly (clamps, posts, screw kits, screwdriver kits, positioners, post holders, laser holder/clamp, etc.).
7. Two or four mirrors, as shown in Figure 11.43.
8. Mirror holders/positioners, as shown in Figure 11.43.
9. Four mirror holder/rotator/stage assemblies, as shown in Figure 11.44.
10. Lenses (200 mm, −25 mm and 25.4 mm focal lens), as shown in Figure 11.45.
11. Two lens holders/positioners, as shown in Figure 11.45.
12. Card with a hole and card holder, as shown in Figure 11.46.
13. Risley prism pair and prism holder/rotator assemblies, as shown in Figure 11.47.
14. Adjustable wedge and wedge holder/rotator assemblies, as shown in Figure 11.48.
15. Target card and cardholder, as shown in Figure 11.49.
16. Ruler/tape measure.

FIGURE 11.43   Mirror and mirror holder/positioners.

FIGURE 11.44   Mirror and mirror holder/rotating stage assembly.

FIGURE 11.45   Two lenses and holders/positioners.

FIGURE 11.46   Card with a hole and card holder.

FIGURE 11.47   Risley prism pair and holder.

FIGURE 11.48   Adjustable wedge and holder.

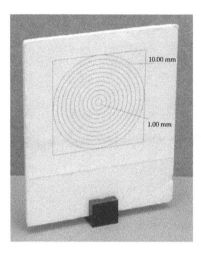

10.00 mm

1.00 mm

FIGURE 11.49 Target card and cardholder.

11.16.4 PROCEDURE

Follow the laboratory procedures and instructions given by the professor and/or instructor.

11.16.5 SAFETY PROCEDURE

Follow all safety procedures and regulations regarding the use of optical instruments and measurements, and light source devices.

11.16.6 APPARATUS SET-UP

11.16.6.1 Experiment One: Laser-Beam Alignment

The following cases are designed for aligning the laser beam towards many directions in space:

11.16.6.1.1 Alignment Using One, Two and Multi-Axis Positioners

This case has two set-ups: section laser-beam alignment using one mirror describes the use of one mirror to align the laser beam from the laser source, and section laser-beam alignment using two mirrors describes the use of two mirrors to direct the laser beam in a two-stage alignment process.

11.16.6.1.1.1 Laser-Beam Alignment Using One Mirror
1. Figure 11.50 shows the experimental apparatus set-up.
2. Bolt the laser short rod to the breadboard.
3. Bolt the laser mount to the clamp using bolts from the screw kit.
4. Put the clamp on the short rod.
5. Place the He–Ne laser into the laser mount and tighten the screw. Turn on the laser device. Follow the operation and safety procedures of the laser device in use.
6. Align the laser beam to be parallel to the edge of the breadboard.
7. Mount a mirror and mirror holder assembly (M_1) to the breadboard at the corner facing the laser beam.

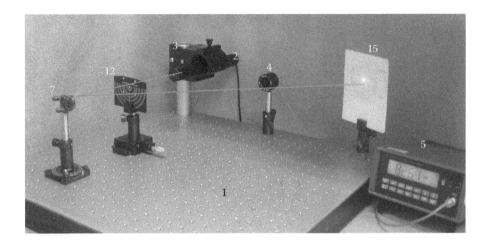

FIGURE 11.50 Laser beam aligned using one mirror.

8. Mount a card with a hole and card-holder assembly between the laser assembly and mirror M_1. Make sure that the hole diameter is slightly larger than the laser beam diameter (about 2 mm), so that the laser beam will pass through, and back reflections from the mirrors can be easily seen.

9. Adjust the position of the laser assembly such that the laser beam passes through the hole and is parallel to the edge of the breadboard.

10. Adjust the height of mirror M_1 until the laser beam intersects the centre of mirror M_1. Finely rotate mirror M_1 post to make sure the laser beam is perpendicular to mirror M_1. In this case, the laser beam reflects back from the mirror, through the hole, to the laser source. Rotate mirror M_1 at an angle of 30° away from the laser beam towards the right corner of the breadboard.

11. Place the target card at the corner of the breadboard that is facing mirror M_1. Adjust the position of the target card so that the laser beam is incident on the centre of the target card, as shown in Figure 11.51.

12. Measure the laser beam power and diameter at several locations between each optical component along the beam path. Fill out Table 11.1 for the listed locations.

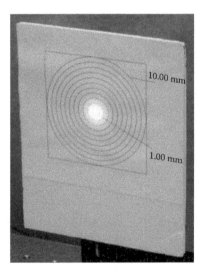

FIGURE 11.51 Incident laser beam on a target card.

TABLE 11.1

Laser Beam Power and Diameter Measurements Using One Mirror

Location	Laser Beam Power P (unit)	Laser Beam Diameter d (unit)
Between laser source and M_1 L_1		
Between M_1 and target card L_2		

11.16.6.1.1.2 Laser-Beam Alignment Using Two Mirrors   Using the same technique as in laser-beam alignment using one mirror, the laser beam can be aligned between two mirrors, as illustrated in Figure 11.29b. Figure 11.52 shows the two-mirror experimental set-up. Start with the following steps to perform alignment of the laser beam using two mirrors:

1. As explained in Section 11.16.1, align the laser beam. The laser beam returning back to mirror M_1 should go back into the laser source.
2. Rotate mirror M_1 at an angle of 45 degree away from the laser beam, towards the right corner of the breadboard.
3. Place a second mirror and mirror holder assembly (M_2) on the breadboard at the adjacent corner.
4. Rotate mirror M_2 at an angle of 45 degree away from mirror M_1, towards the right corner of the breadboard. After this step, the laser beam is successfully aligned, from the laser source to mirror M_1 and mirror M_2. The laser beam is now of the same height and is parallel to the breadboard, as shown in Figure 11.52.
5. Place the target card at the corner of the breadboard that is facing mirror M_2. Adjust the position of the target card so that the laser beam is incident on the centre of the target card, as shown in Figure 11.52.
6. Measure the laser beam power and diameter at several locations between each optical component along the beam path. Fill out Table 11.2 for the listed locations.

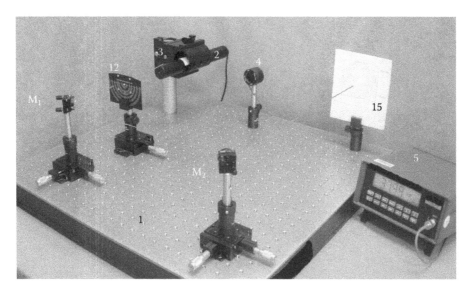

FIGURE 11.52   Laser beam aligned using two mirrors.

TABLE 11.2

Laser Beam Power and Diameter Measurements for Laser-Beam Alignment Using Two Mirrors

Location	Laser Beam Power P (unit)	Laser Beam Diameter d (unit)
Between laser source and M_1 L_1		
Between M_1 and M_2 L_2		
Between M_2 and target card L_3		
At target card L_4		

11.16.6.1.2   Alignment Using Two Mirror Mounts

Using the same technique as in Section 11.16.1, the laser beam can be aligned using two mirror mounts, as illustrated in Figure 11.30. Figure 11.53 shows the experimental set-up for Section 11.16.1, Start with the following steps:

1. Mount a laser device to the middle of the edge of the breadboard, as shown in Figure 11.53.
2. Mount two mirror assemblies, as shown in Figures 11.30 and 11.53, facing each other in the middle of the breadboard in front of the laser device.
3. Align mirror M_1 to reflect the laser beam towards mirror M_2. Adjust M_2 to reflect the laser beam towards the target card, as shown in Figure 11.53.
4. Measure the laser beam power and diameter at several locations between each optical component along the beam path. Fill out Table 11.3 for the listed locations.

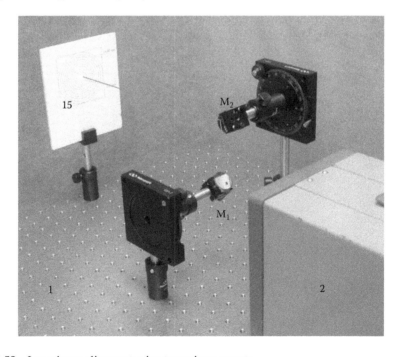

FIGURE 11.53 Laser-beam alignment using two mirror mounts.

TABLE 11.3

Laser Beam Power and Diameter Measurements for Laser Beam Alignment Using Two Mirrors Mounts

Location	Laser Beam Power P (unit)	Laser Beam Diameter d (unit)
Between laser source and M_1 L_1		
Between M_1 and M_2 L_2		
Between M_2 and target card L_3		
At target card L_4		

11.16.6.1.3 Alignment Using Three and Four Mirror Mounts

Using the same technique as in Section 11.16.1, the laser beam can be aligned using three and four mirror mounts, as illustrated in Figures 11.31 and 11.32. Figures 11.54 and 11.55 show the three and four mirror mounts experimental set-ups. Start with the following steps to perform alignment of the laser beam using the three mirror mounts:

11.16.6.1.3.1 Laser-Beam Alignment Using Three Mirror Mounts

1. Mount a laser device near the middle edge of the breadboard, as shown in Figure 11.54.
2. Mount three mirror assemblies, as shown in Figures 11.31 and 11.54, one mirror facing the other two mirrors, in the middle of the breadboard in front of the laser device.

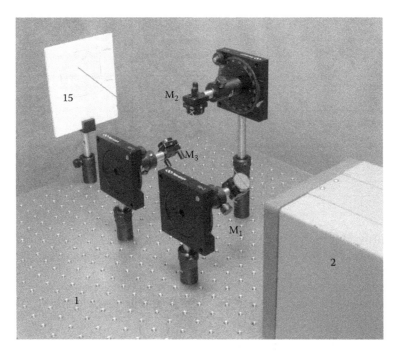

FIGURE 11.54 Laser-beam alignment using three mirror mounts.

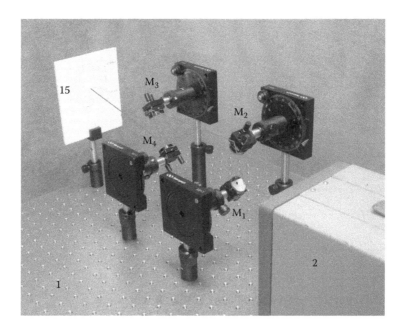

FIGURE 11.55 Laser-beam alignment using four mirror mounts.

TABLE 11.4

Laser Beam Power and Diameter Measurements for Laser-Beam Alignment Using Three Mirror Mounts

Location	Laser Beam Power P (unit)	Laser Beam Diameter d (unit)
Between laser source and M_1 L_1		
Between M_1 and M_2 L_2		
Between M_2 and M_3 L_3		
Between M_3 and target card L_4		
At target card L_5		

---

3. Align mirror M_1 to reflect the laser beam towards mirror M_2. Likewise from mirror M_2 towards mirror M_3. Adjust M_3 to reflect the laser beam towards the target card, as shown in Figure 11.54.
4. Measure the laser beam power and diameter at several locations between each optical component along the beam path. Fill out Table 11.4 for the listed locations.

11.16.6.1.3.2 Laser-Beam Alignment Using Four Mirror Mounts   Using the same technique as in laser-beam alignment using three mirror mounts, the laser beam can be aligned using four mirror

TABLE 11.5

Laser Beam Power and Diameter Measurements for Laser-Beam Alignment Using Four Mirror Mounts

Location	Laser Beam Power P (unit)	Laser Beam Diameter d (unit)
Between laser source and M_1 L_1		
Between M_1 and M_2 L_2		
Between M_2 and M_3 L_3		
Between M_3 and M_4 L_4		
Between M_4 and target card L_5		
At target card L_6		

mounts, as illustrated in Figure 11.32. Figure 11.55 shows the four-mirror mounts experimental set-up. Add the following steps to perform alignment of the laser beam using the following four mirror mounts:

1. Continue the procedure as explained in the three-mirror mounts in laser-beam alignment using three mirror mounts.
2. Mount the fourth mirror assembly, as shown in Figure 11.55, two mirrors facing the other two mirrors, at a location in the middle of the breadboard in front of the laser device.
3. Align mirror M_1 to reflect the laser beam towards mirror M_2. Do likewise from mirror M_2 towards mirror M_3, then towards mirror M_4. Adjust M_4 to reflect the laser beam towards the target card, as shown in Figure 11.55.
4. Measure the laser beam power and diameter at several locations between each optical component along the beam path. Fill out Table 11.5 for the listed locations.

11.16.6.1.4   Alignment Using a Risley Prism Pair

As explained in the theory, the laser beam can be aligned using a Risley prism pair and mount assemblies, as illustrated in Figure 11.33. Figure 11.56 shows the experimental set-up for Section 11.16.1.

1. Mount a laser device near the middle edge of the breadboard, as shown in Figure 11.56.
2. Mount the Risley prism pair and mount assemblies, as shown in Figures 11.47 and 11.56, in the middle of the breadboard in front of the laser device.
3. Try to adjust the Risley prism pair to reflect the laser beam towards different spots on the target card, as shown in Figure 11.56.
4. Align the Risley prism pair to reflect laser beam towards the target card.
5. Measure the laser beam power and diameter at different locations: between the laser device and the Risley prism pair, between the Risley prism pair and the target card and at the target card. Fill out Table 11.6.

FIGURE 11.56   Laser-beam alignment using a Risley prism pair.

TABLE 11.6
Laser Beam Power and Diameter Measurements Using Risley Prism

Location	Laser Beam Power P (unit)	Laser Beam Diameter d (unit)
Between laser device and Risley prism pair L_1		
Between Risley prism pair and target card L_2		
At target card L_3		

11.16.6.1.5   Alignment Using an Adjustable Wedge

As explained in the theory, the laser beam can be aligned using an adjustable wedge and mount assembly, as illustrated in Figure 11.34. Figure 11.57 shows the experimental set-up for Section 11.16.1.

1. Mount a laser device near the middle side of the breadboard, as shown in Figure 11.57.
2. Mount the adjustable wedge and mount assembly, as shown in Figures 11.48 and 11.57, in the middle of the breadboard in front of the laser device.
3. Try to adjust the adjustable wedge to reflect the laser beam towards different spots on the target card, as shown in Figure 11.57.
4. Align the adjustable wedge to reflect the laser beam towards the target card.
5. Measure the laser beam power and diameter at different locations: between the laser device and the adjustable wedge, between the adjustable wedge and the target card and at the target card. Fill out Table 11.7.

FIGURE 11.57 Laser-beam alignment using an adjustable wedge.

TABLE 11.7
Laser Beam Power and Diameter Measurements Using Adjustable Wedge

Location	Laser Beam Power P (unit)	Laser Beam Diameter d (unit)
Between laser device and adjustable wedge L_1		
Between adjustable wedge and target card L_2		
At target card L_3		

11.16.6.2 Experiment Two: Laser-Beam Expansion

The laser beam needs to be aligned, as explained in Experiment One: Laser-Beam Alignment. The following cases are designed to study laser-beam expansion.

11.16.6.2.1 Laser-Beam Expansion by Galilean Telescope

Once the laser beam is aligned as in Section 11.16.1, the laser beam can be expanded by a Galilean telescope arrangement, illustrated in Figure 11.40, by adding the following steps. Figure 11.58 shows the experimental set-up for this case.

1. Insert a short focal length (−25.0 mm) negative lens (Lens 1) into a lens holder/positioner assembly and mount it 127 mm from mirror M_1. Align the lens height in the lens holder and slide the lens holder/positioner so that the diverging beam is centred on mirror M_2.
2. Insert a longer focal length (200 mm) positive lens (Lens 2) into a holder/positioner and place it about 175 mm (the sum of the focal lengths of the two lenses, remembering that the first lens is a negative lens) from the first lens in the diverging laser beam path.

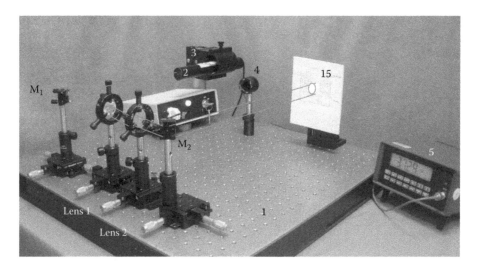

FIGURE 11.58   Laser-beam expansion by a Galilean telescope.

3. Align lens (Lens 2) height in the lens holder and slide the lens holder/positioner, so that the diverging beam is centred on mirror M_2.
4. Carefully adjust the position of Lens 2, by moving it back and forth along the beam, until the laser beam is expanded and incident on the target card, as shown in Figure 11.58.
5. Measure the power and diameter of the laser beam close to the laser source. Fill out Table 11.8.
6. Measure the power and diameter of the laser beam at several distances, two metres apart, from two to ten metres away from mirror M_2. Fill out Table 11.9.

TABLE 11.8
Galilean Telescope Laser Beam Measurements and Calculations Close to the Laser Source

Laser Beam Power P (unit)	Laser Beam Diameter d_o (unit)	Wavelength λ (unit)	Divergence Angle θ (unit)	Rayleigh Range z_R (unit)

TABLE 11.9
Galilean Telescope Laser Beam Measurements and Calculations at Various Distances from M_2

At Distance z (unit)	Laser Beam Power P (unit)	Laser Beam Diameter d (unit)	Divergence Angle θ (unit)	Laser Beam Diameter at z dz (unit)	Rayleigh Range z_R (unit)
z_1					
z_2					
z_3					
z_4					
z_5					

11.16.6.2.2 Laser-Beam Expansion by Keplerian Telescope

Once the laser beam is aligned as in Section 11.16.1, the laser beam can be expanded by a Keplerian telescope arrangement illustrated in Figure 11.41, by adding the following steps. Figure 11.59 shows the experimental set-up for this case.

1. Replace the negative lens with a short focal length positive lens (25.4 mm) and use the same adjustments to centre the beams in the lenses (Lens 1 and Lens 2). Adjust the distance between the two lenses to be the sum of their focal lengths.
2. Carefully adjust the position of Lens 1 and Lens 2, by moving them back and forth along the beam, until the laser beam is expanded and incident on the target card, as shown in Figure 11.59.
3. Measure the power and diameter of the laser beam close to the laser source. Fill out Table 11.10.
4. Measure the power and diameter of the laser beam at several distances, two metres apart, from 2 to 10 m away from mirror M_2. Fill out Table 11.11.

FIGURE 11.59 Laser-beam expansion by a Keplerian telescope.

TABLE 11.10

Keplerian Telescope Laser Beam Measurements and Calculations Close to the Laser Source

Laser Beam Power P (unit)	Laser Beam Diameter d_o (unit)	Wavelength λ (unit)	Divergence Angle θ (unit)	Rayleigh Range z_R (unit)

TABLE 11.11

Keplerian Telescope Laser Beam Measurements and Calculations at Various Distances from M$_2$

At Distance z (unit)	Laser Beam Power P (unit)	Laser Beam Diameter d (unit)	Divergence Angle θ (unit)	Laser Beam Diameter at z dz (unit)	Rayleigh Range z$_R$ (unit)
z$_1$					
z$_2$					
z$_3$					
z$_4$					
z$_5$					

11.16.7 DATA COLLECTION

11.16.7.1 Experiment One: Laser-Beam Alignment

11.16.7.1.1 Alignment Using One, Two and Multi-Axis Positioners

11.16.7.1.1.1 Laser-Beam Alignment Using One Mirror Measure the laser beam power and diameter at several locations between each optical component along the beam path. Fill out Table 11.1.

11.16.7.1.1.2 Laser-Beam Alignment Using Two Mirrors Measure the laser beam power and diameter at several locations between each optical component along the beam path. Fill out Table 11.2.

11.16.7.1.2 Alignment Using Two Mirror Mounts

Measure the laser beam power and diameter at several locations between each optical component along the beam path. Fill out Table 11.3.

11.16.7.1.3 Alignment Using Three and Four Mirror Mounts

11.16.7.1.3.1 Laser-Beam Alignment Using Three Mirror Mounts Measure the laser beam power and diameter at several locations between each optical component along the beam path. Fill out Table 11.4.

11.16.7.1.3.2 Laser-Beam Alignment Using Four Mirror Mounts Measure the laser beam power and diameter at several locations between each optical component along the beam path. Fill out Table 11.5.

11.16.7.1.4 Alignment Using a Risley Prism Pair

Measure the laser beam power and diameter at several locations: between the laser device and the Risley prism pair, between the Risley prism pair and the target card and at the target card. Fill out Table 11.6.

11.16.7.1.5 Alignment Using an Adjustable Wedge

Measure the laser beam power and diameter at several locations: between the laser device and the adjustable wedge, between the adjustable wedge and the target card and at the target card. Fill out Table 11.7.

11.16.7.2 Experiment Two: Laser-Beam Expansion

11.16.7.2.1 Laser-Beam Expansion by Galilean Telescope
1. Measure the laser beam power (P) and diameter (d_o) close to the laser source.
2. Record the measured data in Table 11.8.
3. Measure the laser beam power (P) and diameter (d) at several distances, two metres apart, from 2 to 10 m away from mirror M_2.
4. Record the measured data in Table 11.9.

11.16.7.2.2 Laser-Beam Expansion by Keplerian Telescope
1. Measure the laser beam power (P) and diameter (d_o) close to the laser source.
2. Record the measured data in Table 11.10.
3. Measure the laser beam power (P) and diameter (d) at several distances 2 m apart, from 2 to 10 m away from mirror M_2.
4. Record the measured data in Table 11.11.

11.16.8 CALCULATIONS AND ANALYSIS

11.16.8.1 Experiment One: Laser-Beam Alignment

11.16.8.1.1 Alignment Using One, Two and Multi-Axis Positioners
1. There are no calculations and analysis required for this case.
2. Plot laser beam diameter along the path between the laser source and the target card.
3. Plot laser beam power along the path between the laser source and the target card.

11.16.8.1.2 Alignment Using Two Mirror Mounts
Repeat the steps as explained in Section 11.16.8.1.1.

11.16.8.1.3 Alignment Using Three and Four Mirror Mounts
Repeat the steps as explained in Section 11.16.8.1.1.

11.16.8.1.4 Alignment Using a Risley Prism Pair
Repeat the steps as explained in Section 11.16.8.1.1.

11.16.8.1.5 Alignment using an Adjustable Wedge
Repeat the steps as explained in Section 11.16.8.1.1.

11.16.8.2 Experiment Two: Laser-Beam Expansion

11.16.8.2.1 Laser-Beam Alignment
1. Calculations and analysis are not required for this part.
2. Plot a graph of the power (P) of the laser beam versus distance at various locations (L).
3. Plot a graph of the diameter (d) of the laser beam versus distance at various locations (L).

11.16.8.2.2 Laser-Beam Expansion by Galilean Telescope
1. Calculate the divergence angle (θ) and Rayleigh range (z_R) for the laser beam close to the laser source.
2. Fill out Table 11.8 with the calculated data.
3. Calculate the divergence angle (θ), diameter (d_z) and Rayleigh range (z_R) for the laser beam at various distances (z) away from the mirror M_2.
4. Fill out Table 11.9 with the calculated data.

5. Plot a graph of the power (P) of the laser beam versus distance at various distances (z).
6. Plot a graph of the measured (d) and calculated (d_z) diameters of the laser beam versus various distances (z).
7. Plot a graph of the calculated divergence angle (θ) versus various distances (z).
8. Divergence of the beam diameter along the principle axis of the laser beam can be called the beam profile. To map the beam profile, plot a graph of the beam radius ($\pm\, d_z/2$) versus distance (z) along the axis. The plot of the laser profile can be mapped from the mirror M_2, up to 10 m away.

11.16.8.2.3   Laser-Beam Expansion by Keplerian Telescope

Repeat the steps as explained in Section 11.16.8.2.2. Fill out Tables 11.10 and 11.11.

11.16.9   Results and Discussions

11.16.9.1   Experiment One: Laser-Beam Alignment

11.16.9.1.1   Alignment Using One, Two and Multi-Axis Positioners

1. Report the graph for the laser beam diameter along the path between the laser device and the target card.
2. Report the graph for the laser beam power along the path between the laser device and the target card.
3. Discuss the graphs for the laser beam diameter and power along the path between the laser device and the target card.

11.16.9.1.2   Alignment Using Two Mirror Mounts

Repeat the steps as explained in Section 11.16.9.1.1.

11.16.9.1.3   Alignment Using Three and Four Mirror Mounts

Repeat the steps as explained in Section 11.16.9.1.1.

11.16.9.1.4   Alignment Using a Risley Prism Pair

Repeat the steps as explained in Section 11.16.9.1.1.

11.16.9.1.5   Alignment Using an Adjustable Wedge

Repeat the steps as explained in Section 11.16.9.1.1.

11.16.9.2   Experiment Two: Laser-Beam Expansion

11.16.9.2.1   Laser-Beam Expansion by Galilean Telescope

Repeat the steps as explained in Section 11.16.9.1.1.

11.16.9.2.2   Laser-Beam Expansion by Keplerian Telescope

Repeat the steps as explained in Section 11.16.9.1.1.

11.16.10   Conclusion

Summarise the important observations and findings obtained in this lab experiment.

11.16.11   Suggestions for Future Lab Work

List any suggestions for improvements using different experimental equipment, procedures and techniques for any future lab work. These suggestions should be theoretically justified and technically feasible.

11.17   LIST OF REFERENCES

List any references that were used in the report. Use one format in writing the references. Never mix reference formats in a report.

11.18   APPENDIX

List all of the materials and information that are too detailed to be included in the body of the report.

FURTHER READING

Agrawal, G. P., *Fiber-Optic Communication Systems*, 2nd ed., John Wiley & Sons, New York, 1997.

Agrawal, G. P. and Dutta, N. K., *Long Wavelength Semiconductor Lasers*, Van Nostrand, New York, 1986.

Agrawal, G. P. and Dutta, N. K., *Semiconductor Lasers*, 2nd ed., Van Nostrand, New York, 1993.

Al-Azzawi, A., *Photonics Principles and Practices*, CRC Press, Boca Raton, FL, 2007.

Alda, J., Laser and Gaussian beam propagation and transformation, In *Encyclopedia of Optical Engineering*, Barry Johnson, R. et al., Eds., Marcel Dekker, Inc., New York, 2002.

Black, E., *An Introduction to Pound-Drever-Hall Laser Frequency Stabilization*, California Institute of Technology, LIGO, Pasadena, CA, 2000.

Buus, J., *Tunable Laser Diodes and Related Optical Sources*, 2nd ed., Wiley, Santa Barbara, CA, 2005.

Charschan, S., *Lasers in Industry*, Van Nostrand, New York, 1972.

Chee, J. K. and Liu, J. M., Polarization-dependent parametric and Raman processes in a birefringent optical fiber, *IEEE J. Quantum Elect.*, 26, 541–549, 1990.

Chen, J.-H., Su, D.-C., and Su, J.-C., Holographic spatial walk-off polarizer and its application to a 4-port polarization independent optical circulator, *Opt. Express*, 11, 2001–2006, 2003.

Csele, M., *Fundamentals of Light Sources and Lasers*, Wiley-Interscience, New York, 2004.

Davis, C. C., *Lasers and Electro-Optics, Fundamental and Engineering*, Cambridge University Press, New York, 1996.

Derickson, D., *Fiber Optic Test and Measurement*, Prentice Hall PTR, Upper Saddle River, NJ, 1998.

Duarte, F. J., Note on achromatic multi-prism beam expanders, *Opt. Commun.*, 53, 259–262, 1985.

Duarte, F. J., *Narrow-Lindwidth Pulse Dye Laser Oscillators, in Dye Laser Principles*, F. J. Duarte and L. W. Hillman, Academic, New York, 1990.

Duarte, F. J., *Tunable Lasers Handbook*, Elsevier, New York, 1999.

Duarte, F. J. and Piper, J. A., Dispersion theory of multiple-prisms beam expanders for pulsed dye lasers, *Opt. Commun.*, 43, 303–307, 1982.

Duarte, F. J. and Piper, J. A., Multi-pass dispersion theory of prismatic pulsed dye lasers, *Optica Acta.*, 33, 331–335, 1984.

Duarte, F. J. and Piper, J. A., Narrow-linewidth, high prf copper laser-pumped dye laser oscillators, *Appl. Opt.*, 23, 1391–1394, 1984.

Dutton, H. J. R., *Understanding Optical Communications*, IBM, Prentice Hall, Englewood Cliffs, NJ, 1998.

Fedder, G. K., Iyer, S., and Mukherjee, T., Automated optimal synthesis of microresonators, *Technical Digest of the IEEE International Conference on Solid-State Sensors and Actuators (Transducers'97)*, Chicago, IL, Vol. 2, pp. 1109–1112, 1997.

Griffel, G., Abeles, J. H., Menna, R. J., Braun, A. M., Connolly, J. C., and King, M., Low-threshold InGaAsP ring lasers fabricated using bi-level dryetching, *IEEE Photon. Technol. Lett.*, 12, 146–148, 2000.

Hecht, J., *Understanding Fiber Optics*, 3rd ed., Prentice Hall, Upper Saddle River, NJ, 1999.

Hine, T. J., Cook, M., and Rogers, G. T., An illusion of relative motion dependent upon spatial frequency and orientation, *Vision Res.*, 35, 3093–3102, 1995.

Hoss, R. J., *Fiber Optic Communications—Design Handbook*, Prentice Hall, Englewood Cliffs, NJ, 1990.

Iovine, J., *Homemade Holograms—The Complete Guide to Inexpensive, Do-It-Yourself Holography*, Tab Books, Division of McGraw-Hill, Pennsylvania, 1990.

Jackson, R. A., The laser as a light source for the mach-zehnder interferometer, *J. Sci. Instrum.*, 42, 282–283, 1965.

Kao, C. K., *Optical Fiber Systems: Technology, Design, and Applications*, McGraw-Hill, New York, 1982.

Kasap, S. O., *Optoelectronics and Photonics Principles and Practices*, Prentice Hall PTR, IN, New Jersey, 2001.

Keiser, G., *Optical Communications Essentials*, 1st ed., McGraw-Hill, New York, 2003.

Keiser, G., *Optical Fiber Communications*, 3rd ed., McGraw-Hill, New York, 2000.

Kolimbiris, H., *Fiber Optics Communications*, Prentice Hall, Englewood Cliffs, NJ, 2004.

Kuhn, K., *Laser Engineering*, Prentice Hall, Englewood Cliffs, NJ, 1998.

Lengyel, B., *Lasers*, Wiley, New York, 1971.

Litchinitser, N. M., Dunn, S., Steinvurzel, P., Eggleton, B., White, T., McPhedran, R., and de Sterke, C., Application of an arrow model for designing tunable photonic devices, *Opt. Express*, 12 (8), 1540–1550, 2004.

McComb, G., *The Laser Cookbook – 88 Practical Projects*, Tab Book, Division of McGraw-Hill, Pennsylvania, 1988.

Mouthaan, T., *Semiconductor Devices Explained Using Active Simulation*, Wiley, New Jersey, 1999.

Nanni, C. A. and Alster, T. S., Laser-assisted hair removal: Side effects of Q-switched Nd:YAG, long-pulsed ruby, and alexandrite lasers, *J. Am. Acad. Dermatol.*, 41 (2:1), 165–171, 1999.

Ralston, J. M. and Chang, R. K., Spontaneous-raman-scattering efficiency and stimulated scattering in silicon, *Phys. Rev. B*, 2, 1858–1862, 1970.

Razavi, B., *Design of Integrated Circuits for Optical Communications*, McGraw-Hill, OH, 2003.

Senior, J. M., *Optical Fiber Communications: Principle and Practices*, 3rd ed., Person Education Limited, Essex, England, 2009.

Shashidhar, N., Lensing technology, corning incorporated, *Fiber Product News*, pp. 14–15, 2004.

Simin, G., *High Speed Semiconductor Devices, Basics of Heterostructures*, University of South Carolina, Columbia, SC, Online Course 02/ELCT882, 2005.

Thompson, G. H. B., *Physics of Semiconductor Laser Device*, Wiley, Chichester, 1980.

Topping, A., Linge, C., Gault, D., Grobbelaar, A., and Sanders, R., A review of the ruby laser with reference to hair depilation, *Ann. Plastic Surg.*, 44 (6), 668–674, 2000.

Venkataramanan, V., *Introduction to Laser Safety*, Photonics Research, Ontario, Canada, 2002.

Yeh, C., *Handbook of Fiber Optics. Theory and Applications*, Academic Press, San Diego, CA, 1990.

12 Optical-Fibre Communications

12.1 INTRODUCTION

The availability of information depends on the transmission speed of data, voice and multi-media across telecommunication networks. Despite new technologies that enable legacy of copper telephone lines to carry information more efficiently, optical networks remain the most ideal medium for high-bandwidth communications. There are two distinct modes of optical communications: (1) fibre optics (fibre-optic cable) and (2) optical wireless, based on free-space optics technology. For long-distance network deployments, nothing is better than fibre. When coupled with the wavelength division multiplexing (WDM) technologies, fibre-optic cables are capable of carrying information more densely across the globe. Optical-fibre communications technology became available to bring the connection to the residential areas.

The advantages of optical communications systems for many applications are well recognised: expanded data handling capacity, immunity to electrical noise, electrical isolation, enhanced safety and data security. Many large telecommunication systems are using fibre-optic components, a trend certain to continue for more applications and innovative fields.

Fibre-optic cost in communications systems is often high, the process long and the investment irreversible, but it is profitable over a long period. Optical wireless comes into play because it complements fibre optics in metro area networks and local area networks, with considerably less expense and faster deployment.

12.2   THE EVOLUTION OF COMMUNICATION SYSTEMS

Light has always surrounded us since the creation of the universe. Earth cannot live without light. Communications using light occurred when human beings first communicated by hand signals for short distance under sunlight. This is one of the earlier methods of communication between people, and it is actually a form of optic communication. In daylight, the sun is the source of light that enables a person to see another's hand signals. The information is carried from the sender to the receiver by the sun's light. Hand motion acts as signal generator or modulator. The eye is the message detector device; the brain processes this message and converts it to an action understood by the receiver. Information transfer for such a system is slow. The transmission distance is limited and dependent on light, and the chance of error is great.

Another optical communication system was developed later that used smoke to signal across long distances. The message was sent by varying the pattern and colour of smoke from a fire. The pattern and colour were carried by light. This system required a coding method known to both the sender and the receiver of the message and has several advantages. It is useful for long distances and is accurate, but it also depends on light. This is comparable to modern digital systems that use pulse codes.

Before the nineteenth century, communication was slow and used either optical or acoustic means of sending information. The telegraph was introduced in 1838 by Samuel Morse; this was the start of electrical communications. In 1880, the Canadian scientist Alexander Graham Bell (1847–1922) invented the telephone. The telephone was the first basic system that used cable networks for telecommunications.

The development of light bulbs allowed the construction of a simple communication system. This type of system is widely used in ship-to ship and ship-to-shore links.

In the 1900s, radio, television and microwave used electronic communications by superimposing data onto a carrier, using frequency modulation (FM) or amplitude modulation (AM).

All these communication systems have low rates of information transmission. In 1960, the invention of the laser was a major breakthrough that led to the rapid development of communication systems. The laser provided a coherent and narrow bandwidth beam at specific wavelengths that opened the door to optical communication. Optical communication uses lasers to superimpose electronic data onto an optical carrier using AM. The development of laser diodes along with the advancement of the manufacturability of fibre-optic cables paved the way for the mass deployment of optical telecommunication systems. In the late 1980s, the erbium-doped fibre amplifier (EDFA) was developed and used in long-distance optical communication networks and signal modulation.

12.3   ELECTROMAGNETIC SPECTRUM OVERVIEW

We are surrounded by waves of visible and invisible energy. These waves are created in many different ways. Waves are generated from natural sources, such as sunlight and plants, and from man-made waves, such as telecommunication and radio/television signals, microwaves, remote control infrared rays and x-rays. All these types of energy waves are collectively known as the electromagnetic spectrum.

In order to understand the spectrum and radiation, it is necessary to study two concepts. Wavelength refers to the length of the energy waves (between its peaks). Frequency refers to the number of times, or cycles, per second that a wave cycle occurs; it is measured in hertz (Hz). Radio frequency (RF) is the generic name given to electromagnetic waves that can be used for communications. RFs are at the bottom of the electromagnetic spectrum, having the lowest frequencies, and commonly range from 9 kHz to 30 GHz. A higher frequency wave has a shorter range of travel, and vice versa. Activities with RF of higher frequencies (shorter wavelengths) in electrical communications led to the birth of radio, television, radar and microwave links.

In optical communications, it is customary to specify the band in terms of wavelength instead of frequency. However, due to the high speed of multiple wavelength systems in the middle of the 1990s, the optical source started to be specified in terms of optical frequency.

The optical spectrum ranges from about 50 to about 100 mm, and from ultraviolet to far infrared waves, respectively. In optical-fibre communication systems, the focus will be on the wavelengths ranging from 800 to 1600 nm. The electromagnetic spectrum covers a wide range of wavelengths and photon energies. Light used to see an object must have a wavelength about the same size as or smaller than the object. For example, a source that generates light in the far ultraviolet and soft x-ray regions spans the wavelengths suited to studying molecules and atoms. Table 12.1 presents the wavelength bands used in optical-communication systems. Figure 12.1 shows the applications of the electromagnetic spectrum regions.

TABLE 12.1
Working Bands in Communication Systems

Band	Wavelength (nm)
Original band (O-band)	1260–1360
Extended band (E-band)	1360–1460
Short band (S-band)	1460–1530
Conventional band (C-band)	1530–1565
Long band (L-band)	1565–1625
Ultra long band (U-band)	1625–1675

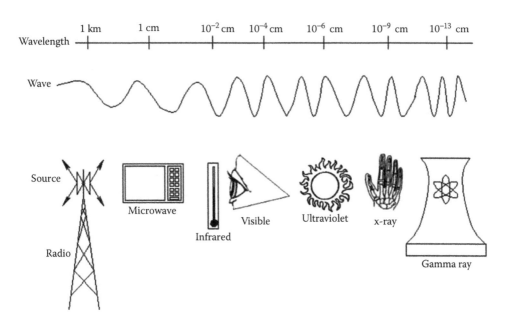

FIGURE 12.1 Applications of the electromagnetic spectrum regions.

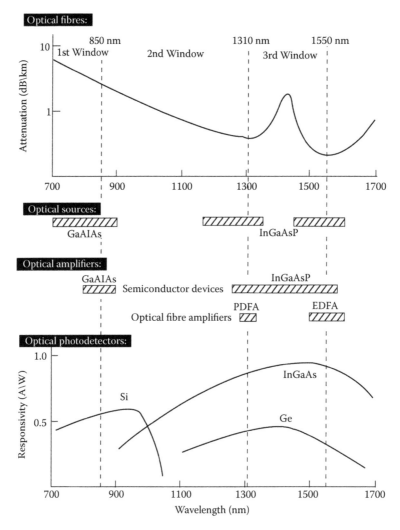

FIGURE 12.2 The evolution of fibre-optic systems.

12.4   THE EVOLUTION OF FIBRE-OPTIC SYSTEMS

Figure 12.2 shows the evolution of operating ranges of optical-fibre systems. The system contains the major key link components: light sources (transmitters), optical fibres, optical amplifiers (and other optical devices) and photodetectors (receivers). The vertical dashed lines indicate the three main generations of operating windows for optical-fibre systems (850, 1310 and 1550 nm). Optical transmission rates are expressed in bits per second (b/s). More details on the three operating windows follow.

12.4.1   THE FIRST GENERATION

This generation encompasses fibre links operated around 850 nm (early silica fibres). These links used gallium–aresenide (GaAs-based) optical sources, silicon photodetectors and multimode fibres. Intercity transmission ranged from 45 to 140 Mb/s, with repeater spacing of around 10 km.

12.4.2 THE SECOND GENERATION

The development of the optical sources and photodetectors allowed the shift from 850 to 1310 nm. Bit rates for long-distance links range are between 155 Mb/s optical carrier (OC-3) and 622 Mb/s optical carrier (OC-12). In some cases transmission is up to 2.5 Gb/s (OC-48) over repeater spacing of 40 km on a single-mode fibre. Bit rates for local area network (LAN) range from 10 to 100 Mb/s, over a distance of 0.5–10 km.

12.4.3 THE THIRD GENERATION

Systems operating at 1550 nm are used for high-capacity long-span terrestrial and undersea transmission links, which carry traffic over a 90 km repeaterless distance at 2.5 Gb/s (OC-48) and 10 Gb/s (OC-192). The introduction of optical amplifiers (OAs) gave a major boost to fibre transmission capacity. Common OAs include GaAlAs-based OAs, praseodymium-doped fibre amplifiers (PDFAs) operating at 1310 nm and EDFAs operating at 1550 nm.

12.5 UNDERSEA DWDM CABLE NETWORK (SEA-ME-WE-3)

The use of the WDM offers a further boost in transmission capacity. The basic principle of WDM is to use multiple sources operating at slightly different wavelengths to send several independent packets of information data over the same fibre cable. For example, one of many of the world's WDM optical networks is the SEA-ME-WE 3-cable system, as shown in Figure 12.3. This undersea network runs from Germany to Singapore, connecting many countries in between; hence the name SEA-ME-WE, which refers to Southeast Asia (SEA), the Middle East (ME) and the West Europe (WE). The network has two pairs of undersea fibres with a capacity of eight STM-16 wavelengths per fibre (equivalent to eight OC-48 that is 8×2.5 Gb/s).

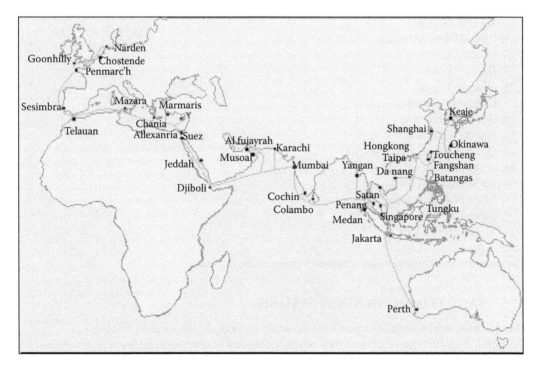

FIGURE 12.3   The SEA-ME-WE WDM cable network connects countries between Germany and Singapore.

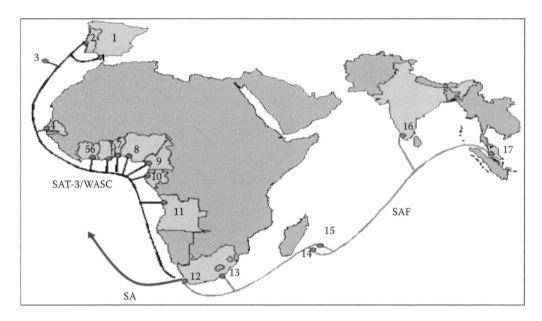

FIGURE 12.4 The SAT-3/WASC-SAF cable network connects countries between Portugal and Malaysia.

Another submarine cable system, the undersea network (see Figure 12.4), connects many countries spanning between

1. Portugal (Sesimbra)
2. Spain (Chipiona)
3. Spain (Altavista)
4. Senegal (Dakar)
5. Côte d'Ivoire (Abidjan)
6. Ghana (Accra)
7. Benin (Cotonou)
8. Nigeria (Lagos)
9. Cameroon (Douala)
10. Gabon (Libreville)
11. Angola (Cacuaco)
12. South Africa (Melkbosstrand)
13. South Africa (Mtunzini)
14. La Reunion (St. Paul)
15. Mauritius (Baie Jacotet)
16. India (Cochin)
17. Malaysia (Penang)

The route distance between Sesimbra and Penang is 23,455 km.

12.6   BASIC COMMUNICATION SYSTEMS

The following are basic definitions for the networks used in communication systems:

Station is a collection of devices with which users wish to communicate, such as computers, terminals, telephones and videos. Stations are also called data terminal equipment (DTE) in network systems, and they can be connected directly to a transmission line.

Network is a group of two or more stations linked or interconnected by a transmission medium, such as a fibre-optic cable or co-axial cable.

Topology is the logical manner or structure in which nodes/devices are linked together via information–transmission channels to form a network.

Switching is the transfer of information from source to destination through a series of inter-mediate nodes. A switch is a device that filters and forwards packets between network segments. Switches operate at the data link layer (layer 2) and sometimes at the network layer (layer 3) of the open-system interconnection (OSI) reference model. A description of the layers is presented in Section 12.10. Switches support any packet protocol. More details on optical mechanical switches are presented in this book.

Routing is the process of moving a packet of data from the source to the destination by the selection of a suitable path through a network. Routing is usually performed by a dedicated device called a router.

Protocol is an agreed-upon format for transmitting data between two devices. The protocol determines the following items:

- Type of error checking to be used
- Data compression method, if any
- Hand shaking, how the device indicates when finished sending a message

A popular protocol used in optical LANs is the fibre-distributed data interface (FDDI) protocol; SONET/SDH protocols are used in optical networks in metro or wider areas. Logical topologies are bound to the network protocols that direct how the data moves across a network. The Ethernet protocol is a common logical bus topology protocol.

12.7   TYPES OF TOPOLOGIES

Topology refers to the shape of a network, or the network's layout. There are different nodes in a network that communicate by the network's topology. Topologies are either physical or logical. Connections between the nodes are via optical couplers. The five most common network topologies are (1) bus topology, (2) ring topology, (3) star topology, (4) mesh topology and (5) tree topology.

12.7.1   Bus Topology

In a bus topology, all stations are connected to a central cable that is called the bus or backbone, as shown in Figure 12.5.

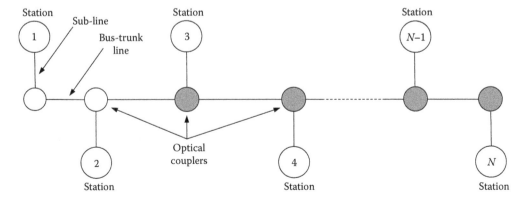

FIGURE 12.5   Bus topology.

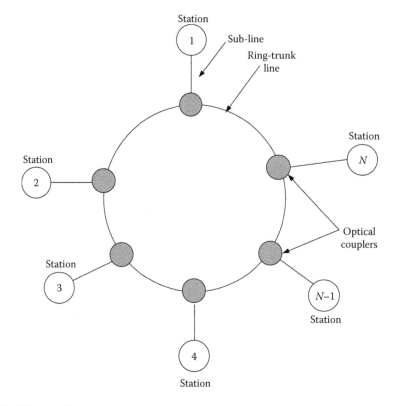

FIGURE 12.6   Ring topology.

12.7.2   RING TOPOLOGY

In a ring topology, all stations are connected to one another in the shape of a closed loop, so that each station is connected directly to two other stations on either side, as shown in Figure 12.6.

12.7.3   STAR TOPOLOGY

In a star topology, all stations are connected to a central hub, which is an optical star coupler. Stations communicate across the network by passing data through the hub, as shown in Figure 12.7.

12.7.4   MESH TOPOLOGY

In a mesh topology, all stations are connected via many redundant interconnections, as shown in Figure 12.8. In a true mesh topology every station has a connection to every other station in the network.

12.7.5   TREE TOPOLOGY

In a tree (hybrid) topology, all stations are connected by various topologies, as shown in Figure 12.9. Groups of star-configured networks are connected to a linear bus backbone.

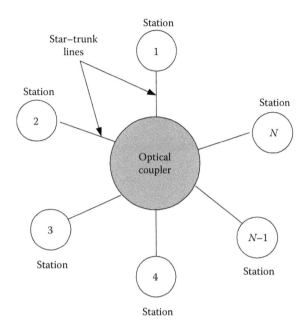

FIGURE 12.7 Star topology.

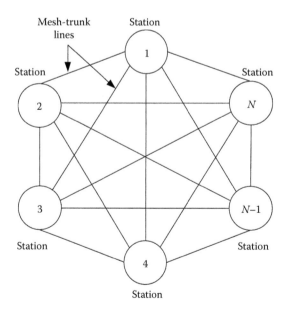

FIGURE 12.8 Mesh topology.

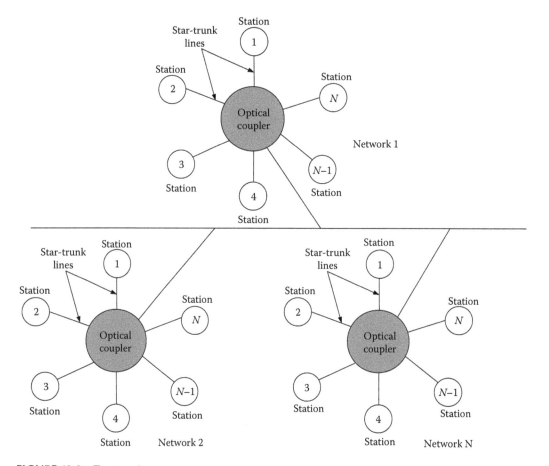

FIGURE 12.9   Tree topology.

12.8   TYPES OF NETWORKS

Networks are divided into the following five types based on the size of the area that the network covers: (1) home area networks (HANs), (2) local area networks (LANs), (3) campus area networks (CANs), (4) metropolitan area networks (MANs) and (5) wide area networks (WANs).

12.8.1   HOME AREA NETWORKS

As shown in Figure 12.10a through c, a HAN is a network that connects the digital devices to computers, which are contained within an individual user's home.

12.8.2   LOCAL AREA NETWORKS

A LAN interconnects users in a localised area such as a department, a small group of building or a factory complex. Figure 12.11 shows a LAN connected to the Internet (through a firewall). This figure shows how workstations are connected to a hub and router. The router provides connections to file servers and printers.

12.8.3   CAMPUS AREA NETWORKS

As shown in Figure 12.12, a CAN is a computer network made up of an interconnection of LANs that are located within a limited geographical area, such as a university or college campus. In the

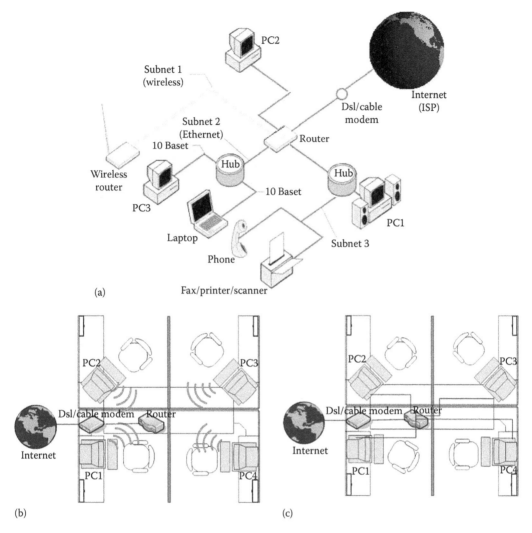

FIGURE 12.10 (a) Home-Area Networks (HANs), (b and c) (Small and Home Offices SOHO).

case of a university campus-based CAN, the network is likely to link a variety of campus buildings, including academic departments, the university library and student halls of residence. A CAN is larger than a LAN, but smaller than a MAN. In addition, CAN stands for corporate area network.

12.8.4 METROPOLITAN AREA NETWORKS

As shown in Figure 12.13 a MAN provides interconnections within a city or in a metropolitan area surrounding a city. MANs typically use wireless infrastructure or optical-fibre connections to link their sites (nodes).

For instance, a university or college may have a MAN that joins together many of their LANs, which are situated in a site whose area is less than a square kilometre. Beyond that, the MAN could have several WAN links to other universities or the Internet.

Some of the technologies used for this purpose are asynchronous transfer mode (ATM) and FDDI. These older technologies are in the process of being displaced by Ethernet-based MANs (e.g. Metro Ethernet) in most areas. MAN links between LANs have been built without cables, by using microwave, radio or infra-red free-space optical communication links.

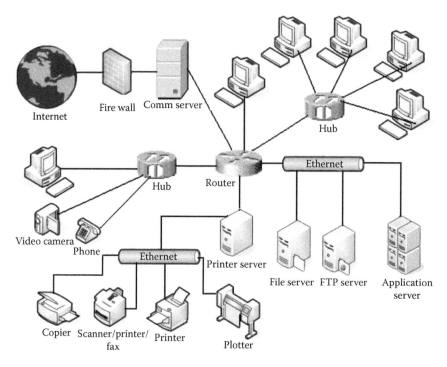

FIGURE 12.11 Local area network (LAN).

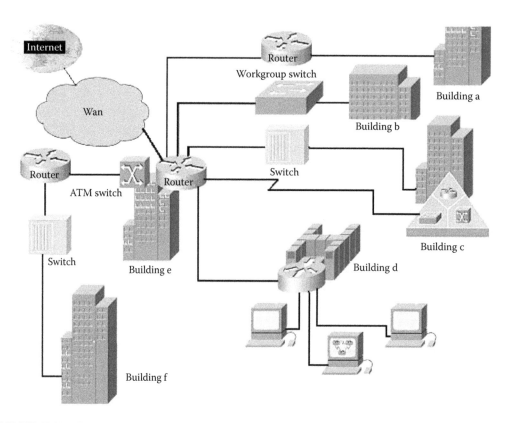

FIGURE 12.12 Campus area network (CAN).

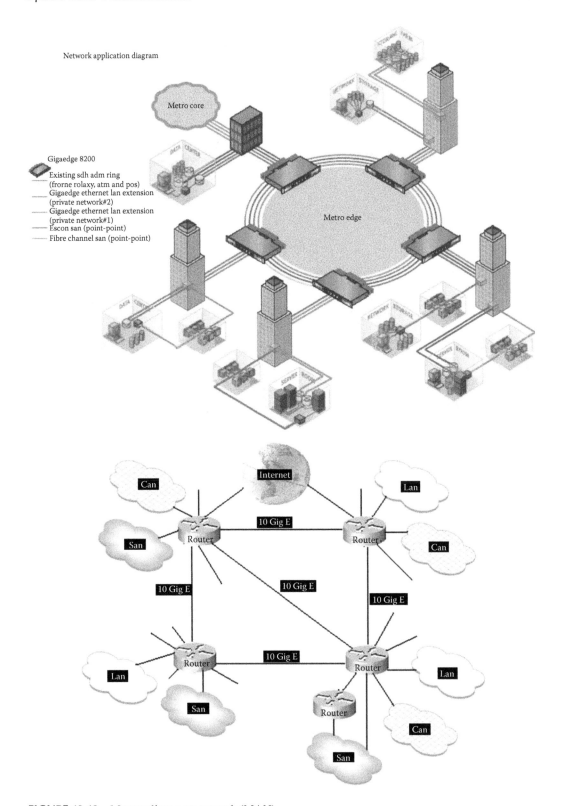

FIGURE 12.13 Metropolitan area network (MAN).

12.8.5 WIDE AREA NETWORKS

As shown in Figure 12.14a and b, a WAN covers a large geographical area connecting cities, countries and continents. The best example of a WAN is the Internet. WANs are used to connect LANs

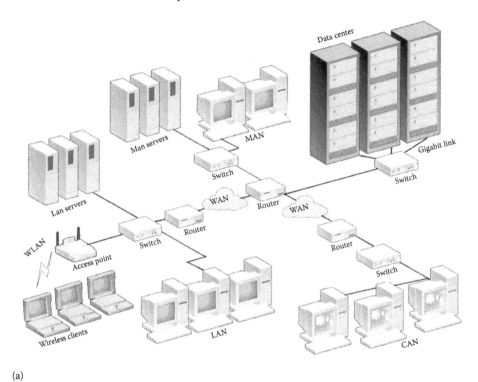

(a)

(b)

FIGURE 12.14 Wide area network (WAN).

or MANs together, so that users and computers in one location can communicate with users and computers in other locations. Many WANs are built for one particular organisation and are private. Others, built by Internet service providers, provide connections from an organisation's LAN to the Internet. A router connects to the LAN on one side of the line, and a hub within the WAN on the other. Network protocols, including TCP/IP, deliver transport and addressing functions. Protocols include Packet-over-SONET/SDH, ATM and Frame Relay. These protocols are often used by service providers to deliver the links used in WANs. X.25 was an important early WAN protocol, and is often considered to be the grandfather of Frame Relay, as many of the underlying protocols and functions of X.25 are still in use today (with upgrades) by Frame Relay.

12.9 SUBMARINE CABLES

Oceans cover more than 70% of our planet, separating the continents and people. People rely on submarine cable networks for voice, data and the Internet communication. Extreme demands between continents for network reliability and high capacity is achieved by using submarine networks that are reliable and well designed. Figure 12.15 shows the map of submarine cable systems across the globe. Submarine cables are sometimes known as underwater cables. The world-leading supplier of submarine networks connected every continent from Europe to Japan, the length of the Americas and across the Pacific. For high capacity, dense wavelength division multiplexer (DWDM) technology is used in telecommunication systems.

12.10 OPEN-SYSTEM INTERCONNECTION

An open-system interconnection (OSI) is a model that defines a networking framework used for implementing protocols in seven layers of communications. Control is passed from one layer to the next, starting at the application layer in one station, and proceeding through to the optical bottom layer. Control passes over the channel (in optical pulses) to the next station and backup the hierarchy. Figure 12.16a shows the seven layers of OSI and data process, whereas Figure 12.16b shows the host and media layers located in the OSI model.

12.10.1 PHYSICAL (LAYER 1)

This layer conveys the bit stream (electrical impulse, light or radio signal) through the network at the electrical and mechanical level. This is transmission of raw data over a communication medium. It provides the hardware means of sending and receiving data on a carrier that includes defining cables, cards and physical aspects. SONET/SDH, Fast Ethernet, RS232 and ATM are protocols with physical layer components.

12.10.2 DATA LINK (LAYER 2)

Layer 2 includes transfer of data frames/packets, addressing and error correction. At this layer, data packets are encoded and decoded into bits. Layer 2 furnishes transmission protocol knowledge and management, and handles errors in the physical layer, flow control and frame synchronisation. The society of electrical and electronic engineers formed the 802 committee that was responsible for dividing the data link layer into two sub-layers: the media access control (MAC) layer and the 802.2 logical link control (LLC) layer. The MAC sub-layer controls how a computer on the network gains access to the data and permission to transmit it. The LLC layer controls frame synchronisation, flow control and error checking.

12.10.3 NETWORK (LAYER 3)

This layer provides switching and routing functions, creating logical paths, known as virtual circuits (e.g. X.25 connection), for transmitting data from node to node. Other functions of this layer

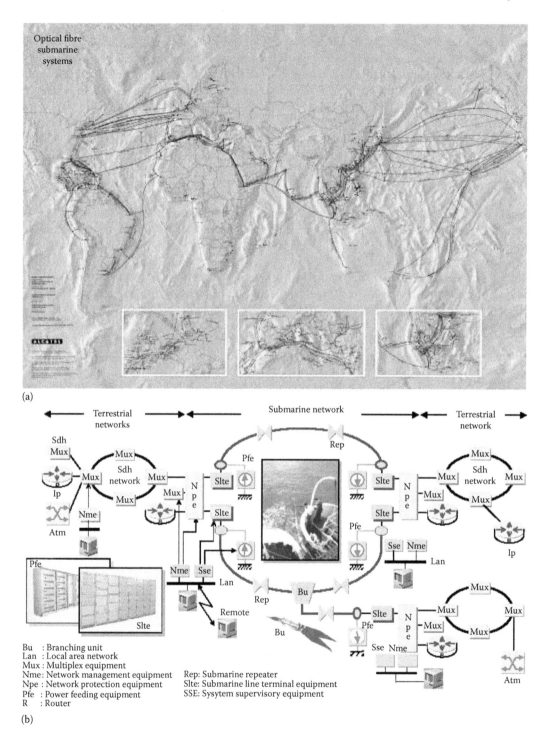

(a)

(b)

Bu   : Branching unit
Lan : Local area network
Mux : Multiplex equipment
Nme: Network management equipment   Rep: Submarine repeater
Npe : Network protection equipment   Slte: Submarine line terminal equipment
Pfe : Power feeding equipment   SSE: Sysytem supervisory equipment
R   : Router

FIGURE 12.15   Map of submarine-cable systems. (a) Map of submarine-cable systems across the globe (Courtesy of Alcatel). (b) Map of submarine configuration (Courtesy of Fujitsu).

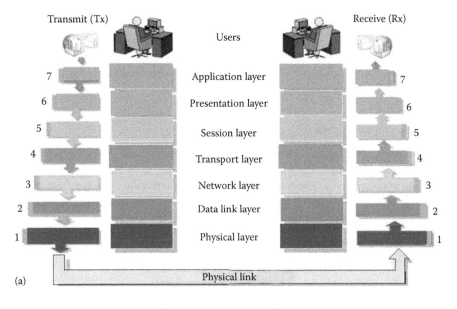

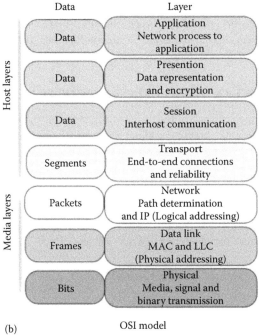

FIGURE 12.16 Open system interconnection (OSI). (a) OSI and data process. (b) OSI model.

are routing and forwarding, as well as addressing, internetworking, error handling, congestion control and packet sequencing. Routing of data packets across networks provides software interface between the physical and data link layers.

12.10.4 TRANSPORT (LAYER 4)

This layer provides transparent transfer of data between end systems, or hosts, and is responsible for end-to-end error recovery and flow control. It ensures complete data transfer.

12.10.5   SESSION (LAYER 5)

This layer establishes, manages and terminates internode connections between applications and uses standards to move data between the applications. The session layer sets up, coordinates and terminates conversations, exchanges and dialogues between the applications at each end. It deals with session and connection coordination.

12.10.6   PRESENTATION (LAYER 6)

This layer involves data formatting, character conversion, security and coding. It provides independence from differences in data representation (e.g. encryption) by translating from application to network format, and vice versa. The presentation layer transforms data into the form that the application layer can accept. This layer formats and encrypts data to be sent across a network, providing freedom from compatibility problems. It is sometimes called the syntax layer.

12.10.7   APPLICATION (LAYER 7)

This layer supports application and end-user processes. Communication partners and quality of service are identified, user authentication and privacy are considered, and any constraints on data syntax are recognised. Everything at this layer is application-specific. This layer provides application services for file transfers, e-mail, network operating systems, application programmes and other network software services. Telnet and FTP are applications that exist entirely in the application level. Tiered application architectures are part of this layer.

12.11   PERFORMANCE OF PASSIVE LINEAR OPTICAL NETWORKS

To evaluate the performance of passive linear networks, consider the fraction of optical power (F_C) lost at a particular interface or component along the transmission path, as shown in Figure 12.17. The power ratio (A_0) over an optical fibre of length (x) will be

$$A_0 = \frac{P_{(x)}}{P_{(0)}} = 10^{-\alpha x/10} \tag{12.1}$$

where $P_{(x)}$ is the power received, $P_{(0)}$ is power transmitted and α is the fibre attenuation (dB/km). If F_C is lost at each port of the coupler, then the connecting loss (L_C) will be

$$L_C = 10 \log\left(1 - F_C\right) \tag{12.2}$$

For example, if $F_C = 20\%$, then $L_C = -0.9691$ dB. The optical power gets reduced by the L_C of 1 dB at any connection junction.

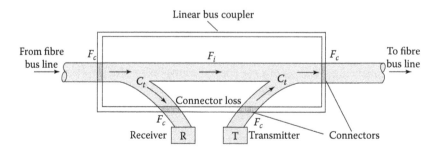

FIGURE 12.17   Losses in a passive linear-bus coupler consisting of two-cascaded directional couplers.

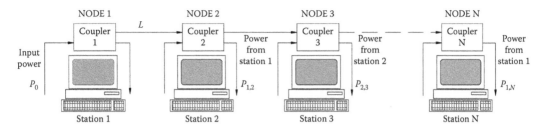

FIGURE 12.18   Topology of a simplex linear bus consisting of N uniformly spaced stations.

The power extracted from the bus is called tap loss (L_{tap}), and is given by

$$L_{tap} = 10 \log C_T \tag{12.3}$$

where C_T is the fraction power removed from the bus and delivered to the detected port. Then the throughput coupling loss (L_{thru}) is given by

$$L_{thru} = 10 \log(1 - C_T)^2 = 20 \log(1 - C_T) \tag{12.4}$$

In addition to the losses L_C and L_{tap}, there is an intrinsic loss (L_i) associated with each bus coupler. If the fraction of power lost in the coupler is F_i, then

$$L_i = 10 \log(1 - F_i) \tag{12.5}$$

All losses are measured in decibels (dB).

A linear bus configuration, as shown in Figure 12.18, consisted of a number of stations (N) separated by various lengths. For simplicity, assume a constant distance L.

The fibre attenuation between two adjacent stations is given by

$$L_{fibre} = 10 \log A_0 = \alpha L \tag{12.6}$$

12.11.1   POWER BUDGET CALCULATION

To calculate the power budget of a fibre link consisting of N stations, as shown in Figure 12.18, the fractional power losses F_C should be examined first.

12.11.2   NEAREST-DISTANCE POWER BUDGET

If P_0 is the optical power launched from the optical source at station 1, then the optical power detected $P_{1,2}$ at the station 2 is given by

$$P_{1,2} \; A_0 C^2{}_T (1 - F_C)^4 (1 - F_i)^2 P_0 \tag{12.7}$$

Then the power budget (considering all losses) between stations 1 and 2 is

$$10 \log\left(\frac{P_0}{P_{1,2}}\right) = \alpha L + 2L_{tap} + 4L_c + 2L_i \tag{12.8}$$

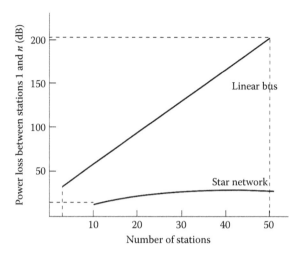

FIGURE 12.19   Total loss versus number of station in linear-bus and star networks.

12.11.3   Largest-Distance Power Budget

If P_0 is the optical power launched from the optical source at station 1, then the optical power detected $P_{1,N}$ at the station N is given by

$$P_{1,N} = A_0^{N-1}(1-C_T)^{2(N-2)}C_T^2(1-F_C)^{2N}(1-F_i)^N P_0 \qquad (12.9)$$

Then the power budget (considering all losses) between stations 1 and N is

$$10 \log\left(\frac{P_0}{P_{1,N}}\right) = N(\alpha L + 2L_c + L_{\text{thru}} + L_i) - \alpha L - 2L_{\text{thru}} + 2L_{\text{tap}}$$

$$= [\text{fibre} + \text{connector} + \text{coupler throughput} + \text{ingress/egress} \qquad (12.10)$$

$$+ \text{coupler intrinsic}] \text{ losses}$$

As shown in Figure 12.19, the losses (in dB) of a linear bus configuration in Figure 12.18 increase linearly with the number of stations N.

12.12   PERFORMANCE OF STAR OPTICAL NETWORKS

The optical input power is evenly divided amongst the output ports in an ideal star fibre coupler, as shown in Figure 12.20.

The total optical-power loss of the coupler consists of its splitting loss and the excess loss in each path through the star configuration. The splitting loss (L_{split}) amongst the N stations is given by

$$L_{\text{split}} = 10 \log\left(\frac{1}{N}\right) = 10 \log N \qquad (12.11)$$

The star fibre coupler excess loss (L_{excess}) is defined as the ratio of the single input power (P_{in}) to the total output power ($P_{\text{out},i}$) of the N stations ($i = 1...N$).

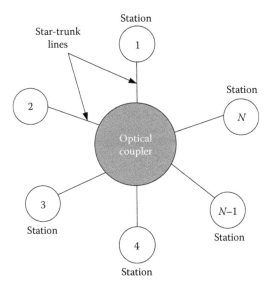

FIGURE 12.20 A star coupler.

$$L_{\text{excess}} = 10 \log \left(\frac{P_{\text{in}}}{\sum\limits_{i=1}^{N} P_{\text{out},i}} \right) \tag{12.12}$$

The total loss within the fibre star coupler is given by

$$\text{Total Loss} = L_{\text{Spilt}} + L_{\text{excess}} \tag{12.13}$$

The optical power balance equation between any two stations in a star network, includes all losses, and is defined as

$$P_S - P_R = L_{\text{excess}} + \alpha(2L) + 2L_c + L_{\text{split}} = L_{\text{excess}} + \alpha(2L) + 2L_c + 10 \log N \tag{12.14}$$

where P_S is the fibre-coupled output power from source in dBm; P_R is the minimum optical power required at the receiver; L_{excess} is the star fibre coupler excess loss; L_{split} is the splitting loss; L_c is the connector loss; L is the distance from the star coupler, assuming all stations at the same distance; and α is the fibre attenuation.

As more stations are added, the loss in a star network increases much slower than the loss in a linear bus network, as shown in Figure 12.19.

12.13 TRANSMISSION LINKS

Transmission links in an optical communication system are the fibre-optic cables that carry light signals between the senders and receivers, and between the communication system's components and peripherals. Signals sent by sender sources over a transmission link or channel can be classified into two formats: analogue and digital signals. Each format has its advantages and disadvantages.

12.13.1 ANALOGUE SIGNALS

An analogue signal carries information through a continuous vibration (waves) of particles in time. Waves can be optical, electrical or acoustical, and have intensities and frequencies, as shown in

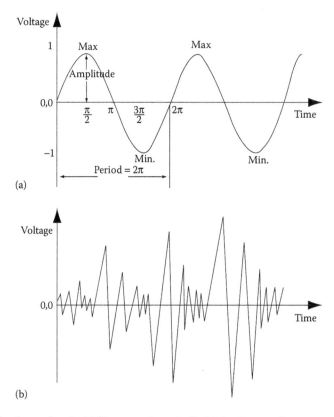

(a)

(b)

FIGURE 12.21   Analogue signals. (a) Sine wave ($y = \sin \theta$). (b) Analogue voice wave.

Figure 12.21. Audio (sound) and video messages are examples of analogue signals. Due to enormous capabilities of computers, the trend in communications has been a fast conversion from analogue to digital transmission format.

12.13.2   DIGITAL SIGNALS

A digital transmission system transmits signals that are in digital form. A digital signal is an ordered sequence of discrete symbols selected from a finite set of elements. The digital signals typically represent information that is alphabetical, numerical and other symbols, such as @, # or %. These discrete symbols are normally represented by unique patterns of pulses of electric voltages or optical intensity, which can take on two or more levels. Figure 12.22 shows common digital

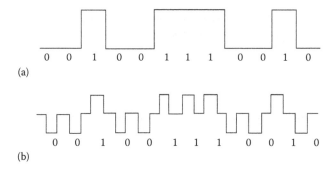

(a)

(b)

FIGURE 12.22   Common digital signal configurations. (a) Digital pulse stream. (b) Dipolar return-to-zero pulse.

signal configurations in the binary waveform. A binary waveform is represented by a sequence of two types of pulses of known shape. The information contained in a digital signal is given by the particular sequence of a presence or high (a binary one, or simply either one or 1), and an absence or low (a binary of zero or simply either zero or 0) of these pulses. These pulses are commonly known as bits. The bit is derived from binary digits. As digital logic is used in the generation and processing of 1 and 0 bits, these bits often are referred to as a logic one (or logic 1) and a logic zero (or logic 0), respectively. Sometimes two voltages are used as high or low level.

The time slot T in which a bit occurs is called the bit interval, bit period or bit time. (Note that this T is different from T used for designating the period of a wave). The bit intervals are regularly spaced, and occur every 1/R seconds (s) at a rate of R bits per second, where R is the rate that data bits are sent, called the bit rate.

Digital signals are easy to process with electronics and optics. It is simple to design a circuit to detect whether a signal is high or low level (on or off).

Fibre-optic cables work very well with digital signals, and were used initially mainly for digital systems. Fibre cables have the high transmission capacity needed for digital transmission with a wide range of wavelengths.

12.13.3 Converting Analogue Signal to Digital Signal

The term digital is used frequently because digital circuits are becoming so widely used in computation, robotics, medical science and technology, communications, transportation and so on. Digital electronics developed from the principle that the circuitry of a transistor could be designed and easily fabricated to have an output of one or two voltage levels, based on its input voltage.

Figure 12.23 shows an analogue-to-digital converter (ADC) used to convert an analogue signal into a digital signal. The ADC can be a single chip, or can be one circuit within a chip. The two voltage levels are usually 5 V (high) and 0 V (low); the levels can be represented by 1 and 0.

The binary numbering system (base-2 numbering system) is the main numbering system used in digital electronics. A digital value is represented by a combination of on and off voltage levels, and expressed as a string of 1s and 0s. Signals that have a theoretically infinite number of levels are converted into signals that have two defined levels.

To convert an analogue signal to a digital form, one starts by taking instantaneous measures of the height of the analogue signal wave at regular intervals, this is called sampling the signal. One way to convert these analogue samples to a digital format is to simply divide the amplitude of the analogue signal into N equally spaced levels, which are designated by integers, and to assign values to one of these integers. This process is called quantisation. Since the signal varies continuously in time, this process generates a sequence of real numbers.

Figure 12.24 shows the equally spaced levels that are the simplest method of quantisation produced by a uniform quantiser. If the digitisation samples are taken frequently enough relative to the rate at which the signal varies, then a good approximation of the signal can be recovered from the samples by drawing a straight line between the sample points. The resemblance of the

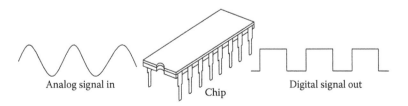

FIGURE 12.23 A chip used to convert an analogue to digital signal.

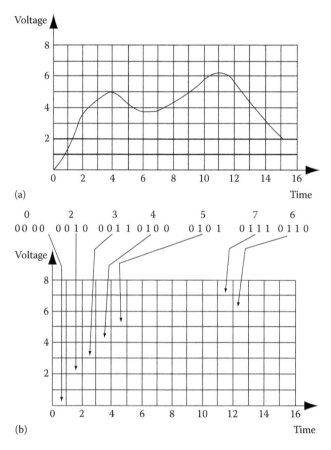

FIGURE 12.24 Digitisation of analogue waveforms. (a) Analogue signal varying between 0 and V volts. (b) Quantised and sampled digital signal.

reproduced signal to the original signal depends on the fineness of the quantising process, and on the effect of noise and distortion added into the transmission system. If the sampling rate is at least two times the highest frequency, then the receiving device can easily reconstruct the analogue signal. Thus, if a signal is limited to bandwidth of B Hz, then the signal can be reproduced without distortion if it is sampled at a rate of 2B times per second. These data samples are represented by a binary code.

As shown in Figure 12.24, eight quantised levels having upper bounds, V_1, V_2, V_3, ..., V_n, can be described by 4 binary digits (e.g. 8 in binary becomes $2^3 = 2^3 \ 2^2 \ 2^1 \ 2^0 = 1 \ 0 \ 0 \ 0$). More digits can be used to give finer sampling levels. Thus, if n binary digits represent each sample, then one can have 2^n quantisation levels. Figure 12.25 shows a conversion process of an analogue signal to digital form, and then to bar code.

12.13.4 BIT ERROR RATE

The performance of a digital communication system is measured by the probability of an error occurring in a data bit; a bit error rate (BER) equal to 0 is ideal. Define p_1 as the probability of misinterpreting a 1 bit as a 0, and p_0 as the probability of misinterpreting 0 as a 1. If the 0 or 1 bits are equally likely to be transmitted, then BER = $1/2 \ p_1 + 1/2 \ p_0$. In telecommunication systems, a typical acceptable BER is 10^{-9} (i.e., an average of one error every 10^9 bit).

FIGURE 12.25   Conversion process of an analogue signal to digital and to barcode.

12.13.5   Fibre-Optic Telecommunication Equipment and Devices

Optical-fibre devices used in any communication system, such as light sources (transmitters), photodetectors (receivers), fibre-optic cables, multiplexers, de-multiplexers, OAs, isolators, circulators and optical switches are explained in detail throughout the book.

12.14   SONET/SDH

12.14.1   Definition of SONET and SDH

With the advent and development *of fibre-optic cables and optical fibre amplifiers, the next important* evolution of the digital time-division *multiplexing (TDM) scheme was a standard signal format.* This format is called *Synchronous Optical NETwork (SONET) in North America, and synchronous digital hierarchy (SDH) in other parts of the world. SONET and SDH are both optical interface* standards that allow internetworking *of services from multiple service providers. SONET and SDH* are almost identical standards dedicated *to transporting data, voice imaging and video over optical* networks. Figure 12.26 illustrates a *standard optical interface between SONET and SDH network* elements (NE) from vendors *A, B and C.*

SONET was specified primarily by Bellcore (now Telcordia) in the late 1980s. It was submitted to the international standards bodies as a proposed international standard. After some negotiation, SDH emerged as the international standard, of which SONET can be considered a complete subset. The standards define a hierarchy of high-speed transmission rates, which currently range from 51.84 Mb/s (SDH starts at 155 Mb/s) up to 40 Gb/s.

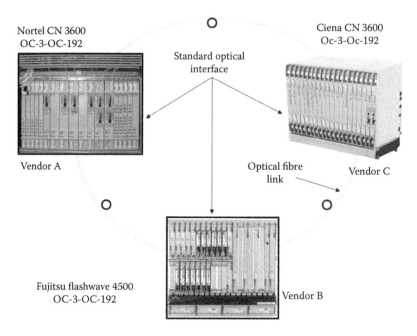

Vendor A

Vendor C

Vendor B

FIGURE 12.26 SONET/SDH.

The standard specifies the physical interfaces, such as the optical wavelengths, pulse shapes and link budgets. For some of the lower rates in the hierarchy, electrical interfaces are specified as well, including line coding and electrical pulse shapes. The organisation of the digital information crossing the interface is defined by specifications for frame structures, payload mappings and overhead assignments. In addition, special signals are specified that allow communication between the two optical network elements for the purpose of operations, administration, maintenance and provisioning (OAM&P). The OAM&P of SONET/SDH networks rely on the Telecommunication Management Network architecture (TMN) as defined by the ITU. SONET/SDH network elements are managed by using translation language 1 (TL1) protocol.

12.14.2 SONET/SDH Purposes and Features

There are several purposes for implementing SONET or SDH. Some of the purposes and features are further explained in the following descriptions.

12.14.2.1 Multi-Vendor Networks

The primary motivation for developing the SONET standard was to allow the deployment of multi-vendor optical networks. Prior to SONET, fibre-optic transport systems could only be deployed in point-to-point configurations, resulting in a lot of unnecessary equipment to terminate the proprietary interfaces. With SONET, a completely optical transport network is envisaged whereby high-capacity fibres interconnect network elements that provide access to the transport network and manage the fibre bandwidth.

12.14.2.2 Cost Reduction

It was also assumed that the resulting network would be cheaper due to the consolidation of network functions, elimination of unnecessary functions and increased competition between vendors. Network providers feel that they no longer have to lock themselves into a single vendor's proprietary (non-standard) solution.

12.14.2.3 Survivability and Availability

With the increasing concern over network reliability and robustness, SONET also provides the ability to build survivable networks; networks that can restore traffic within 50 ms, in the event of fibre cuts and equipment failure.

12.14.2.4 New High-Speed Services

The demand for higher bandwidth pipes continues to increase. SONET provided the opportunity to define an interface, and therefore a network that was capable of carrying a variety of payload types with differing bandwidth requirements, including broadband payloads beyond 50 Mb/s.

12.14.2.5 Bandwidth Management

SONET manages bandwidth by introducing the concept of the payload envelope, into which all payloads are mapped as they enter the SONET network. This allows the infrastructure to be concerned only with transporting the envelopes, regardless of their contents. A limited number of envelopes are defined, with payload capacities ranging from 1.5 Mb/s to 10 GB/s. In addition, different envelopes may be combined on the same fibre.

12.14.2.6 Network Management/Single-Ended Operations

Each rate in the SONET/SDH hierarchy is an integer multiple of a basic rate. For SONET, this basic rate is 51.84 Mb/s; for SDH, it is 155.52 Mb/s. The signal format at each level is created by synchronously multiplexing the basic format. This format is called the synchronous transport signal-level 1 (STS-1) in SONET and the synchronous transport module-level (STM-1) in SDH. Synchronous multiplexing simplifies bandwidth management, allowing access to individual tributaries within the fibre signal without having to completely de-multiplex the fibre signal. The creation of an all-optical network enables management of the network bandwidth using automated techniques. Spare bandwidth on one route may be re-allocated to another route, and new connections between end offices can be created quickly when required. SONET includes overhead allocations for a variety of OAM&P functionality. Examples include a data communications channel to allow network elements to be monitored from a central operations system, and integrity checks to allow single-ended performance monitoring of fibre systems and end-to-end networks.

12.15 MULTIPLEXING TERMINOLOGY AND SIGNALLING HIERARCHY

12.15.1 Existing Multiplexing Terminology and Digital Signalling Hierarchy

In order to understand the role that the SONET standard plays, it is first necessary to understand what interface standards existed previously. In North America, the standard pre-SONET digital hierarchy consisted mainly of digital signals of several levels. The digital signal-level 1 (DS1), is capable of transmitting and receiving data at a bit rate of 1.544 Mb/s (150 Mb/s) and the DS3 has the capability of 44.736 Mb/s, as shown in Figure 12.27. The DS2 really exists only as an intermediate step in the DS1–DS3 multiplex. The DS1 is the main interface to digital voice switches and channel banks, whereas the DS3 is the main interface to fibre-optic transmission systems. The M13 multiplexer links the two together, and is named for its ability to multiplex DS1s into a DS3. This hierarchy is based on TDM.

A similar hierarchy, called electrical signal E, exists in most of the rest of the world. The 2.048 Mb/s interface is the key digital switch interface, whereas most pre-standard fibre-optic transmission systems carry the 139 Mb/s signal. 2.048 Mb/s is called E1, and the hierarchy is based on multiples of 4 E1s:

$$E2 = 4 \times E1 = 8 \text{ Mb/s}$$
$$E3 = 4 \times E2 = 34 \text{ Mb/s}$$
$$E4 = 4 \times E3 = 140 \text{ Mb/s}$$
$$E5 = 4 \times E4 = 565 \text{ Mb/s}$$

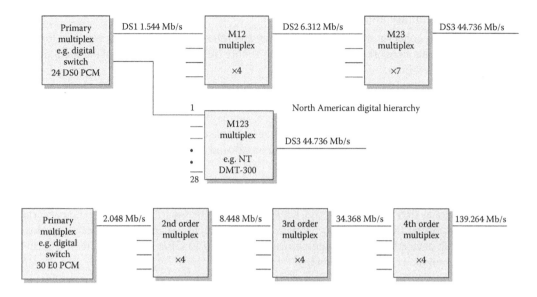

FIGURE 12.27 Existing digital/electrical hierarchies.

The E3 tributaries are faster than the E2 tributaries, whereas the E2 tributaries are faster than the E1 tributaries, and so forth. To synchronise with other tributaries, extra bits, called justification bits, are added. These tell the multiplexers which bits are data and which are spare. Multiplexers on the same level of the hierarchy remove the spare bits, and are synchronised with each other at that level only. Multiplexers on one level operate on a different timing from multiplexers on another level. For instance, the timing between primary rate muxes (that combine 30×64 Kb/s channels into 2.048 Mb/s E1) will be different from the timing between 8 Mbit muxes (that combine up to 4×2 Mb/s into 8 Mb/s).

DS1 is sometimes called transport level-1 (T1). T1 is the optimal rate for accessing low-level devices. It is a type of telephone service capable of transporting the equivalent of 24 conventional telephone lines, using only two pairs of wires. T1 uses two pairs of copper wires (four individual wires) to carry up to 24 simultaneous conversations (channels) that would normally need one pair of wires each. Each 64 Kbit/s channel can be configured to carry voice or data traffic. Most telephone companies allow customers to buy just some of these individual channels, a service called fractional T1. Typically, fractional T1 lines are sold in increments of 56 Kbps (the extra 8 Kbps per channel are used for administration purposes). One of the most common uses of a T1 line is an Internet T1. This connection is used to provide the Internet access to businesses of all sizes, assisting these businesses to meet the challenges of e-commerce. A T1 line can transmit large amounts of data at speeds of 256 Kbit/s, 512 Kbit/s, 1.544 Mbit/s and sometimes 3 Mbit/s.

12.15.2 SONET MULTIPLEXING TERMINOLOGY AND OPTICAL SIGNALLING HIERARCHY

Table 12.2 presents the optically transmitted SONET signal that is referred to as an optical carrier-level N (OC-N). The OC-N is essentially the optical equivalent of the STS-N; however, the STS-N terminology is used when referring to the SONET format. As shown in Figure 12.28, the STS-N consists of a synchronous multiplex of N STS-1s. The STS-1 has a bit rate of 51.84 Mb/s, therefore the STS-N and the OC-N have a bit rate of N times 51.84 Mb/s.

In SONET, all payloads are mapped into synchronous payload envelopes (SPE) at the edge of the SONET network, as shown in Figure 12.27. The core of the SONET network transports the

TABLE 12.2
SONET Terminology

Optical carrier-level *N* (OC-N)	Optical SONET signal at *N* times the basic rate of 51.84 Mb/s
Synchronous transport signal-level *N* (STS-N)	The electrical SONET signal, or SONET format, at *N* times the basic rate of 51.84 Mb/s, consists of a multiplex of N STS-1s
Synchronous transport signal-level 1 (STS-1)	Electrical SONET signal at 51.84 Mb/s, also used to refer to the SONET formal
Synchronous payload envelope (SPE)	In SONET, all payloads are mapped into several types: the VT SPEs carry 1.5–Mb/s payloads, the STS-1 SPE carries 50 Mb/s payloads and the STS-Nc SPE carries 150 Mb/s and higher payloads
Virtual tributary group (VTG)	A logical grouping of VTs prior to multiplexing into the STS-1 SPE
Virtual tributary (VT)	The unit into which the STS-1 SPE can be sub-divided to carry payloads that require much less than 51.84 Mb/s

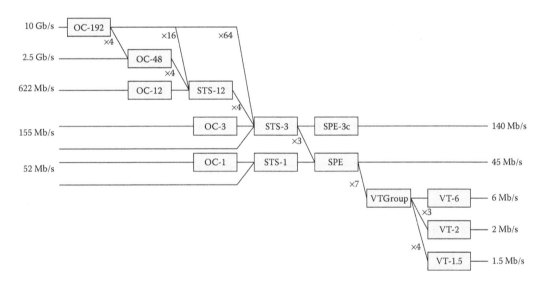

FIGURE 12.28   SONET multiplexing hierarchies.

envelopes. Carried within the STS-1 is the STS-1 SPE, which has a payload capacity of approximately 51.84 Mb/s. N STS-1s may be concatenated to carry an STS-Nc SPE, which has a payload capacity of N × 51.84 Mb/s, as shown in Table 12.2.

In Figure 12.28, the STS-1 SPE can be subdivided into virtual tributary (VT). There are four sizes of VT: VT1.5, VT2, VT3 and VT6. The VT1.5 carries a VT1.5 SPE that has a payload capacity of approximately 1.5 Mb/s (Note: VT1.5 bit rate is equivalent to the bit rate of DS1 or T1). Similarly, there are VT2, VT3 and VT6 SPEs with capacities of about 2, 3 and 6 Mb/s, respectively. Within the STS-1 SPE, the VTs are grouped by types into VT Groups. Within a VT Group, there may be four VT1.5s, three VT2s, two VT3s or one VT6. There are seven VT Groups within an STS-1 SPE, and different groups may contain different VT sizes within the same STS-1 SPE.

12.15.3   SDH MULTIPLEXING TERMINOLOGY AND OPTICAL SIGNALLING HIERARCHY

As discussed previously, a European standard, SDH, was developed parallel to the SONET standard. SDH uses a different terminology, as shown in Table 12.3 and Figure 12.29. Aside from the terminology, most other features of SONET can be extended to SDH.

TABLE 12.3
SDH Terminology

Synchronous transport module-level *N* (STM-N)	A synchronous multiplex of *N* STM-1s
Synchronous transport module-level 1 (STM-1)	The basic rate (155.52 Mb/s) and format of the SDH hierarchy; also refers to the optical signal
Administrative unit group (AUG)	A logical grouping of like AUs
Administrative unit (AU)	Similar to the TU; consists of a higher order VC and a payload pointer
Tributary unit group (TUG)	A logical grouping of like TUs
Tributary unit (TU)	A logical element consisting of a lower order VC and a payload pointer
Virtual container (VC)	The SDH structure into which all payloads are mapped

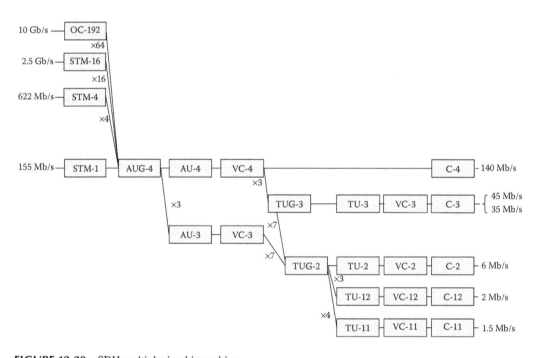

FIGURE 12.29   SDH multiplexing hierarchies.

12.16   SONET AND SDH TRANSMISSION RATES

Although the SONET multiplexing scheme would in theory allow any multiple of STS-1s, only certain rates are defined as standard transmission rates. Physical (photonic) interfaces are specified for these rates. Table 12.4 lists the most commonly supported SONET rates with their SDH equivalents.

12.17   NORTH AMERICAN OPTICAL AND DIGITAL SIGNAL DESIGNATION

Table 12.5 shows parts of the optical signal and carrier designations, as well as the complete digital signal and carrier designations that are used in North America. The T designation refers to the bit rate and the copper-transmission system, and the DS designation refers to the bit format and framing. However, often the terms are used interchangeably. A single 64 kbps channel is called a DS0.

TABLE 12.4
SONET/SDH Rates

SONET	SDH	Rate (Mb/s)
OC-1	STM-0	51.84
OC-3	STM-1	155.52
OC-12	STM-4	622.08
OC-48	STM-16	2488.32
OC-192	STM-64	9953.28
OC-768	STM-256	39813.12

TABLE 12.5
Optical and Digital Signal Designation in North America

Digital Signal Designation	Transmission Rate	Carrier Designation	Number of Channels
DS-0	64 Kbps	—	1
DS-1	1.544 Mbps	T1	24
DS-2	6.312 Mbps	T2	96
DS-3	44.736 Mbps	T3	672
DS-4	274.186 Mbps	T4	4032
0C-1	51.840 Mbps	STS-1	—
OC-3	155.52 Mbps	STS-3	—
OC-12	622.08 Mbps	STS-12	—

The T1 rate of 1.544 Mbps for 24 channels of 64 kbps each is referred to as a DS1. The T3 rate of 44.736 Mbps for 28 T1s is referred to as a DS3.

The optical carrier (OC) is the fundamental unit used in SONET. OC indicates an optical signal and the number following OC represents increments of 51.84 Mbps, the minimum transmission rate. The standard SONET frame format for 51.84 Mbps is called STS-1; the equivalent optical transmission rate is called OC-1. SONET standardises higher transmission bit rates, OC-N, as OC-3, OC-12, OC-48 and OC-192, which are exact multiples of OC-1 (N × 51.84 Mbps). SONET also standardises the overhead formats and other details of optical transmission to implement mid-span links between different vendors' equipment.

12.18   SONET SYSTEMS

SONET network elements are combined to create systems that are classified based on the mechanism used to provide traffic protection and survivability. The two main types of protection provided in SONET are linear and ring.

12.18.1   LINEAR SYSTEMS

Linear systems transport traffic along a single route that may consist of one or more working fibres. A one plus one (1 + 1) system has one protection fibre and one working fibre, as shown in Figure 12.30. The traffic is permanently bridged onto both fibres, so that the receiving end can autonomously choose the fibre that is operating better.

FIGURE 12.30   Linear SONET system.

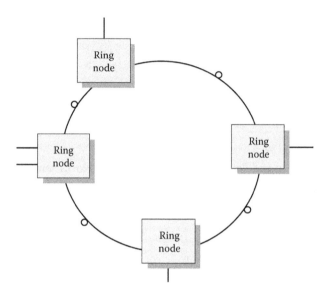

FIGURE 12.31   SONET ring system.

12.18.2   RING SYSTEMS

Ring systems transport traffic around a ring, allowing traffic to be added and dropped anywhere along the ring, as shown in Figure 12.31. Spare capacity is allocated around the ring so that when a failure occurs at any one point; the affected traffic can be restored using the spare capacity. In a unidirectional path switched ring (UPSR), traffic added to the ring is bridged onto both directions, such that the drop node can autonomously select the better path. In a bi-directional line switched ring (BLSR), the traffic affected by a failure at any point on the ring is re-routed the other way around the ring. A protocol operating around the ring coordinates this action.

12.19   STS-1 FRAME STRUCTURE

The 810 bytes of an STS-1 frame are most conveniently represented as a matrix consisting of 9 rows and 90 columns, as shown in Figure 12.32. The intersection of a row and a column is one byte. The order of transmission is from left to right, and from top to bottom. The first three columns carry transport overhead, which consists of section overhead and line overhead. The remaining 87 columns carry the STS-1 SPE. The STS-1 SPE has its own frame consisting of 9 rows and 87 columns. Although the SPE fits within the STS-1 envelope capacity, the first byte of the SPE does not necessarily occupy the first byte position within the envelope capacity. Thus the SPE is represented as being offset from the STS-1 frame. The STS payload pointer is a line overhead function that locates the start of the SPE within the envelope capacity.

12.19.1   SERIAL TRANSMISSION

A SONET signal is transmitted as a continuous serial bit stream, as shown in Figure 12.33. The bit stream is organised into bytes, and the bytes are grouped in 125 μs blocks called frames. At the

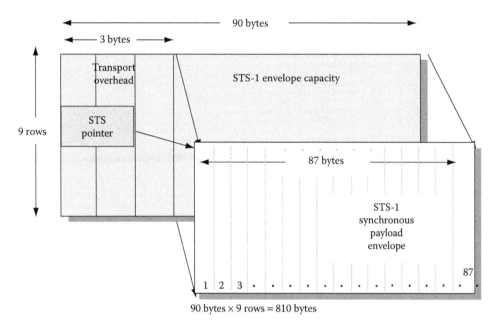

FIGURE 12.32   STS-1 frame structure.

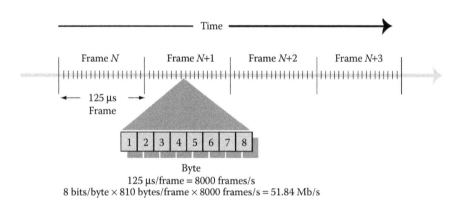

FIGURE 12.33   Serial transmission.

basic STS-1 rate, there are 810 bytes per frame, which results in a basic rate of 51.84 Mb/s. All STS-N rates are N times the basic rate.

Currently, the common supported rates are OC-1, OC-3, OC-12, OC-48, OC-192 and OC-768.

12.19.2   TRANSPORT OVERHEAD

Transport overhead is divided into two main types—section overhead and line overhead—as shown in Figure 12.34. Section overhead is allocated to the first three rows of the transport overhead. Subsequent slides describe each section overhead byte. Line overhead is allocated to the remaining six rows of the transport overhead. In a detailed treatment of SONET theory, each of the bytes and their functions would be described in detail; this information can be found in many of the SONET references.

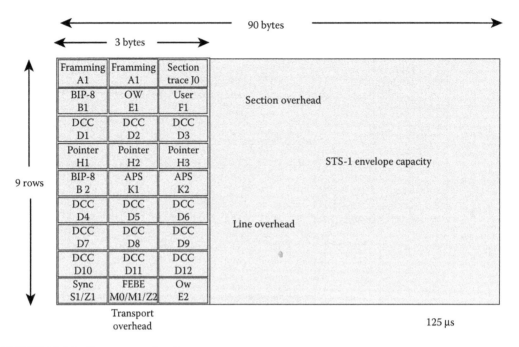

FIGURE 12.34 Transport overhead.

12.19.3 STS-1 SPE Path Overhead

Path overhead consists of the first column of the STS SPE, as shown in Figure 12.35. STS path overhead allows for integrity verification of the end-to-end STS path. In general, it is not modified at line terminations.

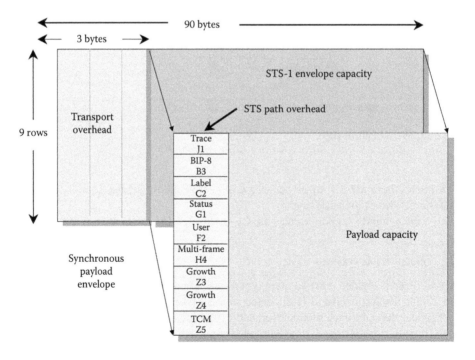

FIGURE 12.35 STS-1 SPE path overhead.

Frame-aligned STS-1s

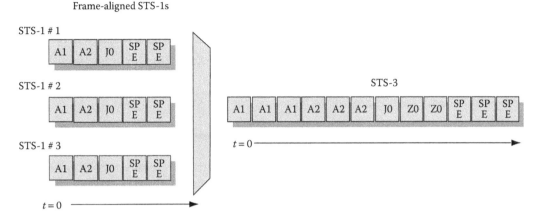

FIGURE 12.36   Multiplexing method.

12.19.4 MULTIPLEXING METHOD

When multiplexing STS-1s, the individual STS-1 frames must first be aligned. The STS payload pointer is used to accomplish alignment. The STS-1s are byte interleaved, as shown in Figure 12.36. In other words, a byte is transmitted from the first STS-1, then a byte from the second STS-1 and so on, until the first byte has been transmitted from every STS-1. Then the second byte is transmitted and so on.

12.20 METRO AND LONG-HAUL OPTICAL NETWORKS

Metro optical networks can be thought of as consisting of core networks and access networks, as shown in Figure 12.37. SONET/SDH metro and long-haul core networks are typically configured as point-to-point or ring connections that are spaced in tens to thousands of kilometres apart.

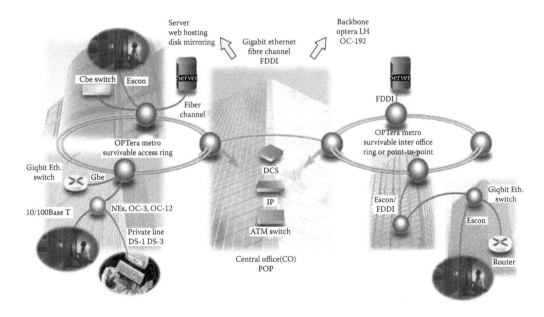

FIGURE 12.37   Metro and long-haul optical networks.

The metro optical network consists of optical links between the end users and a central office (CO). The ring configurations shown in Figure 12.37 contain three or four nodes. Optical add/drop multiplexers provide the capability to add or drop multiple wavelengths to other locations or networks.

12.21 NETWORK CONFIGURATION

Network configuration is explained in the following text.

12.21.1 AUTOMATIC PROTECTION SWITCHING

Ring automatic protection switching (APS) provides an increased level of survivability for SONET/ SDH networks by allowing as much traffic as possible to be restored, even in the event of a cable cut or node failure.

12.21.2 SONET/SDH RING CONFIGURATIONS

SONET/SDH has two main types of ring configurations: (1) a UPSR and (2) a BLSR.

A UPSR consists of two fibres on each span transmitting in opposite directions between adjacent nodes.

There are two types of BLSR: two-fibre BLSR and four-fibre BLSR. In the two-fibre BLSR, there are two fibres transmitting in opposite directions between adjacent nodes. In the four-fibre BLSR, there are four fibres between adjacent nodes, with two fibres transmitting in one direction and other two fibres in the opposite direction. The four-fibre BLSR supports more traffic.

For a UPSR network, the selection of data path is made on a per path basis using the path layer integrity information. Thus, the ring is called path switched. In the case of a BLSR network, the decision to switch to the other path is made by the nodes adjacent to the failure using line-layer integrity information. Thus, the ring is called line switched.

12.21.2.1 Two-Fibre Unidirectional Path Switched Ring Configuration

Figure 12.38 shows a two-fibre UPSR network. By convention, in a unidirectional ring, the normal working traffic travels clockwise around the ring on the primary (working) path. For example, the

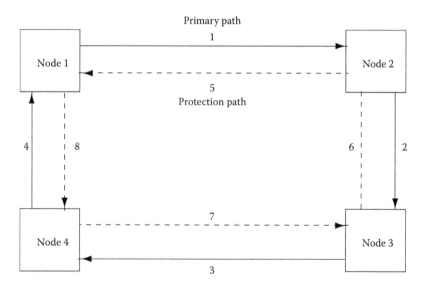

FIGURE 12.38 A two-fibre unidirectional ring with a counter-rotating protection path.

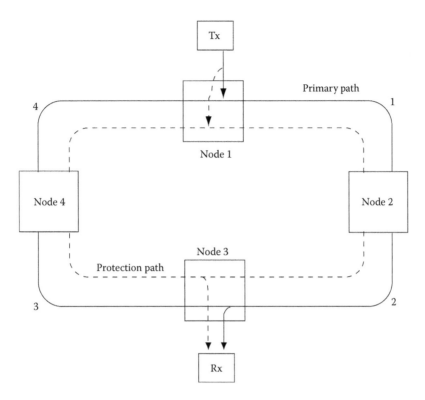

FIGURE 12.39 Flow of primary and protection traffic from Node 1 to Node 3.

connection from the node 1 to node 3 uses links 1 and 2, whereas the traffic from the node 3 to node 1 traverses links 3 and 4. In a UPSR ring, the counterclockwise path is used as an alternate route for protection against link or node failures. This protection path (links 5–8) is indicated by dashed lines. The signal from a transmitting node is dual-fed into both the primary and protection fibres. This establishes a designated protection path on which traffic flows counterclockwise; namely, from node 1 to node 3 via protection links 8 and 7.

12.21.2.1.1 Two-Fibre UPSR Configuration (Traffic Flow)

In Figure 12.39, two signals from node 1 arrive at their destination at node 3 from opposite directions. The receiver at node 3 selects the signal from the primary (working) path. However, if the quality of received signal on the primary path is poor, then it selects the signal from the protection path. In case of any failure in the node 2 equipment or on the primary path 2, node 3 will switch to the protection path via node 4 to receive the signal from node 1.

12.21.2.2 Four-Fibre BLSR Configuration

In Figure 12.40, two primary fibre loops are used for normal bi-directional communication, whereas the other two secondary fibre loops are standby links for protection purposes. The two primary fibre loops have fibre segments labelled 1p through 8p, which provides for an arrangement of 1p, 2p, 3p and 4p in one primary loop, and 8p, 7p, 6p and 5p in the second primary loop. The two secondary fibre loops have fibre segments labelled 1s through 8s, grouping 1s, 2s, 3s and 4s in one secondary loop, and 8s, 7s, 6s and 5s in the second secondary loop.

Consider the connection from node 3 to node 1. The traffic from node 1 to node 3 flows in a clockwise direction along the links 1p and 2p. The traffic in the return path flows counterclockwise from node 3 to node 1, along links 6p and 5p. Thus, the information between node 1 and node 3 does not tie up any of the primary channel bandwidth in the other half of the ring.

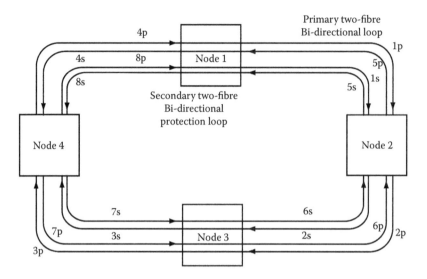

FIGURE 12.40   Four-fibre bi-directional line switched ring network.

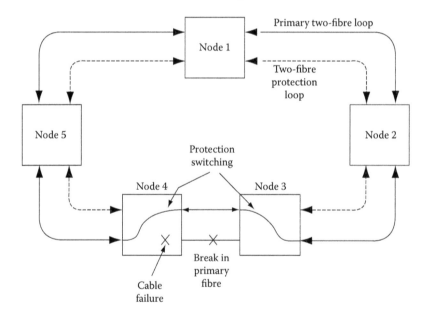

FIGURE 12.41   Re-configuration under transceiver card or line failure.

12.21.2.2.1   Four-Fibre BLSR Reconfiguration (Failure 1)

Consider the scenario shown in Figure 12.41, where a transmitter or receiver circuit card used on the primary ring fails (in either node 3 or 4, in this case). In this case, the affected nodes detect a lose-of signal (LOS) condition and switch both primary fibres, connecting them to the secondary protection pair. The protection between these nodes (3 or 4, in this case) now becomes part of the primary bi-directional loop.

12.21.2.2.2   Four-Fibre BLSR Reconfiguration (Failure 2)

The exact same re-configuration scenario as in failure 1 will occur when the primary fibre connecting two nodes (in this case, nodes 3 and 4) breaks, as in Figure 12.41. Note that in any case, the other links remain unaffected.

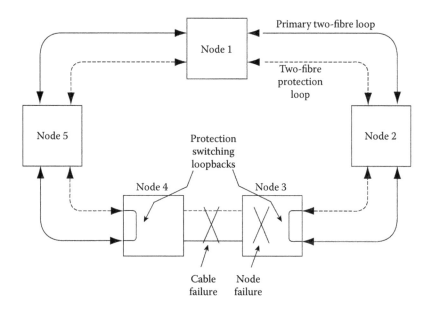

FIGURE 12.42   Re-configuration under node or fibre-cable failure.

12.21.2.2.3   Four-Fibre BLSR Reconfiguration (Failure 3)

In Figure 12.42, consider the scenario where an entire node fails (in this case node 3), or both the primary and the protection fibres in a given span are severed, which could happen if they are in the same cable duct between 2 nodes (in this case nodes 3 and 4). In this scenario, the nodes on either side of the failed internal span will internally switch the primary path connection from their receivers and transmitters to the protection fibres, in order to loop traffic back to the previous node.

12.21.3   Generic SONET Network

SONET/SDH architecture allows the interconnections and interoperability of a variety of network configurations, as shown in Figure 12.43. One can build point-to-point links, linear chains, UPSR,

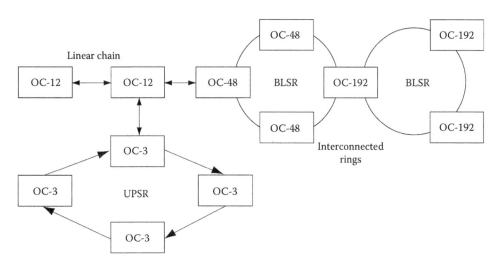

FIGURE 12.43   Generic configuration of large SONET network consisting of various types of interconnected systems.

BLSR and interconnected rings. Each of the individual configurations has their own failure-recovery and protection mechanism, and SONET/SDH network management procedures.

12.21.4 SONET ADD/DROP MULTIPLEXER

One of the important features in SONET/SDH architecture is the add/drop multiplexer (ADM), as shown in Figure 12.44. Various pieces of equipment are fully synchronised, and a byte-oriented multiplexer is used to add and drop sub-channels within an OC-N stream. Here, various OC-12s and OC-3s are multiplexed into OC-48 stream. On entering an ADM, these multiplexed sub-channels can be individually dropped by the ADM, and others can be added.

12.21.5 DENSE WDM DEPLOYMENT

SONET/SDH architectures can also be implemented with wavelengths. Figure 12.45 shows an example of DWDM deployment one an OC-192 trunk ring for n wavelengths (e.g. one could have $n = 16$). The different wavelength outputs from each OC-192 transmitter are first passed through the variable attenuator (VA) to equalise the out powers. These are then fed into a wavelength multiplexer, possibly amplified by a post-optical amplifier, and sent out over the transmission fibre. Additional OAs might be located at intermediate points or at the receiver end.

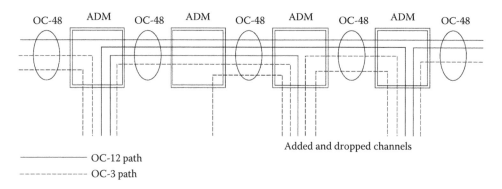

FIGURE 12.44 Functional concept of an add/drop multiplexer (ADM) for SONET applications.

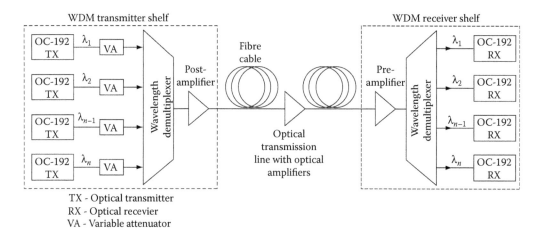

FIGURE 12.45 DWD deployment of n wavelengths in an OC-192 trunk ring.

FURTHER READING

Agrawal, G. P., *Fiber-Optic Communication Systems*, 2nd ed., John Wiley & Sons, New York, 1997.

Bean, J., Optical Wireless: Secure High-Capacity Bridging: The Adoption of Optical Wireless Solutions as High-capacity Links Will See Changes During 2005, (Cover story). In *Fiber Optic Technology*, Gale Group, Farmington Hills, MI, pp. 10–13, 2005.

Black, U., and Sharleen, W., *SONET and T1 Architectures for Digital Transport Networks*, Prentice Hall, Upper Saddle River, NJ, 1997.

Boyd, W. T., *Fiber Optics Communications, Experiments and Projects*, 1st ed., Howard W. Sams & Co., Washington, DC, 1987.

Cole, M., *Telecommunications*, Prentice Hall, Englewood Cliffs, NJ, 1999.

Derfler, F. J., Jr. and Les, F., *How Networks Work*, Millennium Edition, Que Corporation, Indianapolis, IN, 2000.

Derickson, D., *Fiber Optic Test and Measurement*, Prentice Hall PTR, Upper Saddle River, NJ, 1998.

Dutton, H. J. R., *Understanding Optical Communications*, IBM, Prentice Hall, Englewood Cliffs, NJ, 1998.

Goff, D. R., and Hansen, K. S., *Fiber Optic Reference Guide: A Practical Guide to the Technology*, 2nd ed., Butterworth-Heinemann, London, UK, 1999.

Goralski, W. J., *SONET*, 2nd ed., McGraw-Hill, New York, 2000.

Green, P. E., *Fiber Optic Networks*, Prentice Hall, Englewood Cliffs, NJ, 1993.

Hecht, J., *Understanding Fiber Optics*, 3rd ed., Prentice Hall, Englewood Cliffs, NJ, 1999.

Hioki, W., *Telecommunications*, 3rd ed., Prentice Hall, Englewood Cliffs, NJ, 1998.

Hoss, R. J., *Fiber Optic Communications—Design Handbook,* Prentice Hall, Englewood Cliffs, NJ, 1990.

Kao, C. K., *Optical Fiber Systems: Technology, Design, and Applications*, McGraw-Hill, New York, 1982.

Keiser, G., *Optical Fiber Communications*, 3rd ed., McGraw-Hill, Boston, MA, 2000.

Keiser, G., *Optical Communications Essentials*, 1st ed., McGraw-Hill, New York, 2003.

Kolimbiris, H., *Fiber Optics Communications*, Prentice Hall, Englewood Cliffs, NJ, 2004.

Nortel Networks, *Products, Services, and Solutions Change Notification, Nortel Equipment, Nortel Products*, Manual, Ottawa, Canada, 2005.

Palais, J. C., *Fiber Optic Communications*, 4th ed., Prentice Hall, Englewood Cliffs, NJ, 1998.

Razavi, B., *Design of Integrated Circuits for Optical Communications*, McGraw-Hill, New York, 2003.

Senior, J. M., *Optical Fiber Communications: Principle and Practice*, 2nd ed., Prentice Hall, Englewood Cliffs, NJ, 1986.

Shamir, J., *Optical Systems and Processes*, SPIE Optical Engineering Press, Bellingham, WA, 1999.

Ungar, S., *Fiber Optics: Theory and Applications*, Wiley, New York, 1990.

Yeh, C., *Handbook of Fiber Optics: Theory and Applications*, Academic Press, San Diego, CA, 1990.

Yeh, C., *Applied Photonics*, Academic Press, San Diego, CA, 1994.

13 Fibre-Optic Lighting

13.1 INTRODUCTION

This chapter will explain the basic principles of fibre-optic lighting using fibre-optic cables to transfer light from the source to applications. Fibre-optic lighting is used to provide the light power required for residential, commercial and office buildings, sport centres, theatres, retirement homes, swimming pools, backyards, walkways, stairways, show cases, warning signs and advertisement panels. In medical applications, fibre-optic lighting is important for delivering illumination to remote parts of the body and for carrying coherent images.

Fibre-optic lighting transfers light from an electrical or solar light source through a bundle of fibres in a cable to the locations where the light is needed. In this way, light from outside can be transferred to indoor spaces during the daytime. A fibre-optic cable is considered a tool used to transfer light between two points. The use of fibre-optic cable in lighting systems is one of the simple and easy ways to achieve green building conditions, help the environment and reduce the cost of energy consumption in the building sector. This chapter will present multiple types of fibre-optic lighting systems. Particular emphasis will be given to the study of the design of a lighting system and application. Also in this chapter, along with the theoretical presentation, three experimental cases will demonstrate the principles of the fibre-optic lighting system.

13.2 LIGHT

Light waves have a wide range of wavelengths. Lighting applications use wavelengths in the visible region. Light frequently behaves as a particle, and at other times it behaves as a wave. Light makes things visible. Light transmits from a source to a receiver in the form of either pulses or waves. A light source can be an electrical bulb, a laser or solar rays.

Fibre-optic cables are used as a light carrier. The types of fibre-optic cables are explained in the fibre-optic cables principles. If the diameter of the fibre-optic cable is large, then it is suitable for use in fibre-optic lighting system designs.

13.3   ELECTRICAL ENERGY CONSUMPTION BY LIGHTING

With energy rates rising, high operating costs are a concern for building owners. Maintenance costs for items such as light bulbs, which require frequent replacement, must also be considered. Society consumes large amounts of electricity, much of which is used for indoor and outdoor lighting. In the average household, 18%–22% of total energy consumption is used for lighting, as illustrated in Figure 13.1.

One objective of saving energy is to demonstrate that buildings can be attractively and effectively lit while using only a fraction of normal electrical demand. Four lighting strategies can be employed to meet this objective: natural daylight, energy-efficient light fixtures, task lighting and fibre-optic lighting using solar rays.

Natural daylight is the most efficient source of building lighting; however, its full potential has yet to be demonstrated in buildings. Day lighting provides two very different challenges to designers: (1) perimeter spaces in buildings can suffer from over lighting and glare and (2) interior spaces without windows are too dark for normal use. Solutions to these problems depend on the building design. Window size and placement can provide sufficient daylight for a building without increasing heating and cooling loads. Translucent fabric window coverings such as roller blinds and horizontal blinds can diffuse light and reduce glare.

Rapid advances in lighting and appliance technologies have made significant energy reductions possible. Research on typical apartment buildings suggests that savings on lighting systems can average 75% through the use of fluorescent technologies. Even greater savings are possible using induction lamp technologies. Artificially-lit, energy efficient light fixtures are used in buildings to cut energy consumption. Most modern light fixtures have electronic dimmable ballasts in indirect/direct lighting fixtures. These lights use 35% less electricity than 40 W tubes with magnetic ballasts, while providing the same lighting level.

Electric lights are controlled by a modulating dimming system, in order to maintain desired light levels. Lights will dim on bright days, and brighten on dark days. Motion sensors and timers ensure that lights are on only when required. In addition, the lighting design emphasises

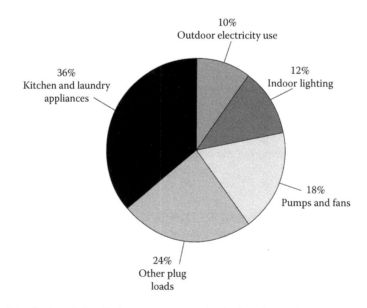

FIGURE 13.1   Distribution of electrical energy consumption in the advanced houses in Canada.

task lighting. Task lighting does not illuminate an entire room; it focuses on a small area, such as a desk top, where illumination is required. Compact fluorescents can be used for task and accent lighting while reserving halogen spotlights with parabolic reflectors for very small task lit areas. The use of automated controls, occupancy sensors and energy management systems can produce additional savings. More efficient lighting systems also reduce building cooling requirements.

Solar technology utilises free energy from the sun. The simplest form of this technology occurs when sunlight passing through windows heats a building and provides light. There are many passive solar heating and lighting designs that provide at least about one-third of the annual space heating and lighting for buildings. In a fibre-optic lighting system, a fibre-optic cable bundle is used to transmit solar rays for indoor lighting. The system can be located on the south-facing roof of the building, driveways or empty spaces to facilitate the collection of solar rays. The system consists of solar collectors, fibre-optic cable bundles and diffusers. One design of a fibre-optic light system is presented in detail in Section 12.4 of this chapter.

Fibre-optic lighting systems can also utilise a hybrid design, in order to provide light during both day and night. At night, an efficient electric light bulb can provide the required lighting. Even during cloudy days, the hybrid system provides lighting, whereas electric light bulbs are used to supplement the daylight. The hybrid system eliminates the need for storage batteries and the resulting ventilation, maintenance and disposal problems associated with some other solar photovoltaic technologies.

Each year, hundreds of dollars are spent to provide lighting in a typical home. A fibre-optic lighting system would provide the same comfort for less energy, which is especially helpful as energy prices go up. It is good for the environment, too. Whenever combustion fuels are burned in a home or in generating stations to produce electricity, carbon dioxide, nitrogen oxides and other emissions are released. In conclusion, by utilising lighting and appliances equipment, and fibre-optic lighting systems, homes will consume less energy and make less of an impact on the environment. Using a fibre-optic lighting system is good for both budget and comfort.

13.4   LIGHT MEASUREMENT

13.4.1   LUMINOUS FLUX OR LIGHT OUTPUT

The study of the measurement of light is called photometry. Two important measurable quantities in photometry are the luminous flux (I), or the light output, and the illumination of a surface (E). Luminous flux measures the brightness of a light source, and it is defined as the total quantity of light emitted per second by a light source. The quantity of light emitted varies with the wavelength, reaching a maximum at the wavelength of 555 nm. The unit for luminous flux (I) is the candle or candela (cd). The early use of certain candles for standards of intensity led to the name of this unit. Currently, a platinum source is used at a specific temperature as the standard for comparison. Another unit, the lumen (lm), is often used to measure the flux of a light source; one candle produces 4π lumens. The lumen is defined as the luminous flux associated with a radiant flux of 1683 W at the wavelength of 555 nm in air. Lumens are equal to the quantity of light emitted by a lamp.

13.4.2   LUMINOUS EFFICACY

The luminous efficacy or efficiency of a light source is defined as the ratio of the light output (lm), to the energy input (watts). The effectiveness of a lighting system is measured in lumens per watt (lm/W), and can vary dramatically—from less than 10 to more than 200 lm/W.

13.4.3 LUMINOUS FLUX DENSITY OF LIGHTING LEVEL

The luminous flux is a density expression that relates the amount of luminous radiative flux to a specific surface area. The luminous flux density is also known as the illuminance, or the quantity of light on a surface or the lighting level. It is assumed that the surface being illuminated is perpendicular to the light source. The unit of the luminous flux density or lighting level is the lux (lx). The relation is given as

$$\text{Illuminance flux} = \frac{\text{Luminous flux (lumens)}}{\text{Area}} \tag{13.1}$$

where:
  lux (lx) = lm/m²
  footcandles (fc) = lm/ft²
  1 fc = 10.76 lux

In specific technical terms:

$$E_{\text{ave}} = \frac{\Phi_S}{A_S} \tag{13.2}$$

where:
  E_{ave} is the average surface illuminance (lx or fc)
  Φ_S is the total luminous flux, or light (lumens), that falls onto the total surface area
  A_S the total surface area (m² or ft²).

The lighting level is measured by a photometer, as shown in Figure 13.2. The minimum required lighting levels for different tasks are given in Table 13.1.

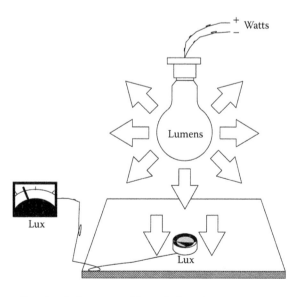

FIGURE 13.2 Luminous flux density is measured by a photometer.

TABLE 13.1

Lighting Levels by Building Area and Task

Building Area and Task	Lighting Level fc	lux	Comments
Auditoriums	15	150	Include provision for higher levels
Banks—tellers' stations	75	750	
Barber shops	75	750	
Bathrooms	30	300	
Building entrances	5	50	
Cashiers	30	300	
Cleaning	15	150	
Conference rooms	30	300	Plus task lighting
Corridors	15	150	
Dance halls	7.5	75	
Drafting—high contrast	75	750	
Drafting—low contrast	150	1500	
Elevators	15	150	
Exhibition halls	15	150	Include provision for higher levels
Floodlighting—bright surroundings	30	300	Less for light surfaces—more for dark
Floodlighting—dark surroundings	15	150	Less for light surfaces—more for dark
Hospital—examination rooms	75	750	High colour rendition Variable
Hospital—operation rooms	150	1500	(dimming or switching)
Kitchen	75	750	
Laundry	30	300	
Lobbies	30	300	
Office—general	75	750	
Parking Areas—covered	5	50	
Parking Areas—open	2	20	
Reading/writing	75	750	
Restaurant—dining	7.5	75	
Restaurant—food display	75	750	
Stairways	15	150	
Stores—sales area	75	750	
Street lighting—highways	1.5	15	
Street lighting—roadways	0.5	5	
Utility rooms	30	300	
Video display terminals	7.5	75	

13.5   ELECTRICAL LIGHTING SYSTEM

Figure 13.3 illustrates basic types of lights in an electrical lighting system, which consists of the following main elements:

1. A *lighting unit*, which serves as a light source. Light source can be any type of lamp(s).
2. A *ballast*, which is a device used with a gas discharge lamp and provides the lamp with the necessary starting and operating electrical conditions.

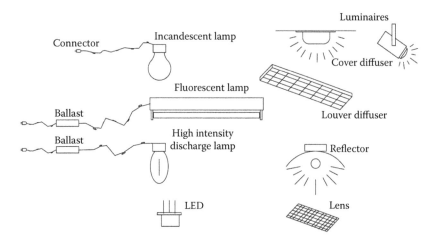

FIGURE 13.3 A basic design of an electrical lighting system.

3. *End fixtures*, which produce the desired quality of light for the task. The end fixtures can be any design of light diffusers that diffuse light uniformly over the illuminated areas. The fixtures are available in different designs. Some fixtures have diffusers or reflectors, or they are covered with a lens to distribute light uniformly over the illuminated areas.
4. *Copper wire*, which carries electrical power from the main power supply to the fixture through the ballast (in the case of high intensity discharge lamp) and to the lamp(s).

13.6   FIBRE-OPTIC LIGHTING SYSTEM

This technology is based on light being transferred from a source to several destination outlets through fibre-optic cables. As no heat or electricity is spent within the system it is more efficient. This technology is not limited in its application around water or in situations involving adverse environments, such as outdoor lighting.

Figure 13.4 illustrates a basic design of a fibre-optic lighting system, which consists of the following main elements:

1. An *illuminator* (or engine), which serves as a light source (or provider or supplier). Light sources can be electric light bulbs, solar ray collectors/concentrators and hybrids (a combination of electric light bulbs and solar collectors/concentrators).
2. A *fibre-optic bundle* carries the light from the illuminator (or engine or solar collector/concentrator) to the task. Fibre-optic bundles can be classified as follows:
   a. Glass fibre bundles
   b. Small core diameter plastic fibre cables
   c. Large core diameter plastic fibre cables
   d. Hollow core fibre tubing
   e. Liquid core tubing
   f. Free air optics
3. The *end fixtures* produce the quality of light for the task. They can be any design of light diffusers that diffuse light uniformly over the illuminated areas. Unlike the standard electric lighting, fibre-optic lighting enables changing of the light colour without changing the light fixture. This can be done by placing a small rotator that carries a wheel with coloured filters. The wheel rotates and changes the colour of the light exiting from the fixture. The user can control the light intensity and select the colour.

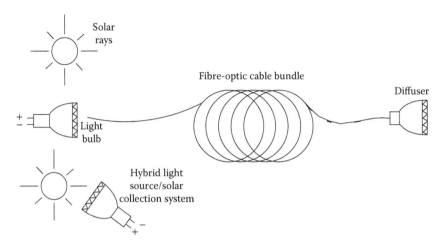

FIGURE 13.4   A basic design of a fibre-optic lighting system.

Standardisation of fibre-optic lighting components and modifications of the building codes will enable the usage of fibre-optic lighting systems in both old and new buildings. These systems are one of many ways to reduce energy consumption in lighting. The systems are commercially available for common lighting applications. Currently, fibre-optic lighting systems are widely used in a variety of applications in buildings around the world.

13.7   ADVANTAGES OF FIBRE-OPTIC LIGHTING

Fibre-optic lighting can significantly reduce electrical lighting requirements and energy costs in many residential, commercial and industrial buildings as well as schools, libraries and hospitals. Using fibre-optic lighting in buildings may also reduce air-conditioning costs, because fibre-optic lighting produces no heat.

Fibre-optic lighting systems using solar rays capture the visible light while excluding the ultraviolet and infrared light, which are separated from the visible wavelengths by the focusing action of lenses and filters. Fibre-optic lighting increases fire and electromagnetic safety by transporting light rather than electricity. This technology eliminates heat transfer and structural support hardware, simplifying light fixtures, installation and architectural requirements. It produces more pleasing light with colour changes and special effect options. It centralises maintenance and cleaning, and reduces maintenance and labour costs.

Some of the advantages of the fibre-optic lighting systems are

1. Their small size and light weight
2. The elimination of hazardous area lighting concerns
3. The ability to select a range of visible colour/specific wavelengths from the light source
4. Their electrical safety
5. Their low maintenance
6. The maintenance cost reductions
7. The elimination of running cost
8. The simplicity of the system
9. Their virtually unbreakable equipment
10. The elimination of vibration

11. The elimination of noise
12. The elimination of heat
13. The elimination of ultraviolet rays
14. Their flexibility and durability
15. The elimination of consumables and replacements
16. Their long-life running systems
17. The variety of system designs and selections
18. Their sound design for the environment
19. The reduction of air-conditioning loads in buildings
20. An increase in human inspiration and comfort

13.8 FIBRE-OPTIC LIGHTING APPLICATIONS

Fibre-optic lighting systems can be applied to the interior and exterior of commercial, retail and residential buildings. New applications are being explored in landscapes, waterscapes, medical lighting instruments and theme parks.

Electric lighting represents a large portion of the total energy consumption in buildings. When power plants generate electrical energy, they produce gas emissions. Such emissions proportionally increase with increased demands for lighting. Fibre-optic lighting using a solar collector is one way to replace conventional electric lighting. Fibre-optic lighting is a tool to be considered in renewable energy applications. It can be coupled with renewable energy sources like solar energy. Figure 13.5 shows a lighting system design using fibre-optic bundles to guide sunlight from a solar concentrator inside a building. This system might be suitable for indoor, outdoor and emergency lighting applications.

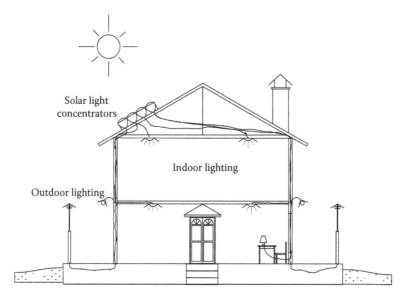

FIGURE 13.5 Proposed fibre-optic lighting system.

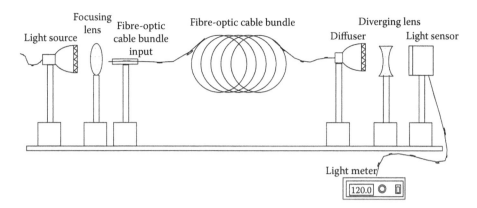

FIGURE 13.6   A fibre-optic lighting system.

13.9   EXPERIMENTAL WORK

This experiment is designed to launch light into a bundle of fibre-optic cables. Figure 13.6 illustrates a schematic diagram for the experimental cases. The experiment can also be conducted using focusing and diverging lenses that are located at the input of the fibre-optic cable bundle and at the diffuser side, respectively. The lenses are used to improve the collection and the diffuser efficiencies. Other types of lenses may also be used in this design. In the fibre-optic lighting system experiment, the student will perform the following cases:

1. Fibre-optic lighting with diffuser (light source, fibre-optic bundle and diffuser).
2. Fibre-optic lighting with lens and diffuser (light source with focusing lens, fibre-optic bundle and diffuser).
3. Fibre-optic lighting with lenses and diffuser (light source with a focusing lens, fibre-optic bundle and diffuser with a diverging lens).

13.9.1   TECHNIQUE AND APPARATUS

Appendix A presents the details of the devices, components, tools and parts.

1. 2 × 2 ft optical breadboard
2. Light source
3. Fibre-optic cable
4. Fibre-cable holder/positioner assembly
5. Diffuser
6. Hardware assembly (clamps, posts, screw kits, screwdriver kits, etc.)
7. Fibre-cable end preparation kit
8. Focusing lens
9. Diverging lens
10. Lens holder/positioner assembly
11. Light sensor
12. Light meter
13. Post holder/positioner assembly
14. Black/white card and cardholder
15. Ruler

13.9.2 PROCEDURE

Follow the laboratory procedures and instructions given by the professor and/or instructor.

13.9.3 SAFETY PROCEDURE

Follow all safety procedures and regulations regarding the use of optical components, electrical and optical devices and optical cleaning chemicals.

13.9.4 APPARATUS SET-UP

13.9.4.1 Fibre-Optic Lighting with Diffuser

1. Figure 13.7 (without lenses) shows the apparatus set-up.
2. Mount a light source on the breadboard.
3. Make a bundle of 10 fibre-optic cables packed together, 6 ft long.
4. Prepare the fibre-optic cable ends with a good cleave at each end, as described in fibre-optic cable end preparation procedure in the Fibre-Optic Chapter.
5. Pack the fibre cables together so that the ends meet at the same position.
6. Insert each end of the fibre-optic cable into a 10 mm long metal tube with a diameter of 2.5 mm.
7. Dip a needle into clear epoxy and apply to each end of the fibre-optic cable bundle filling the space in the metal tube and in between the fibre-optic cables. Place the tube metal about 3 mm from the fibre-optic cable bundle ends. Allow the epoxy to cure according to the epoxy manufacturer's instructions.
8. Insert the fibre-optic cable bundle into a heat shrink tube for mechanical protection of the fibre-optic cables.
9. Hold the bundle end upright, and polish the ends, following the polishing procedure for a fibre-optic connector described in the fibre-optic connectors chapter.
10. Inspect the bundle ends under the microscope to see the condition of the polished end surfaces.

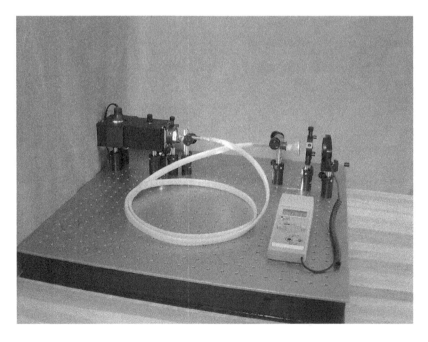

FIGURE 13.7   Fibre-optic lighting apparatus set-ups.

TABLE 13.2
Light Level Adjustment

Factor	Reduce Lighting Level by 30%	Increase Lighting Level by 30%
Reflectance of task background	Greater than 70%	Less than 70%
Speed of accuracy	Not important	Critical
Workers' age (average)	Less than 40 years	More than 55 years

11. Turn on the light source.
12. Mount a fibre-cable holder/positioner onto the breadboard so that the light from the light source passes over the centre hole.
13. Insert one end (input) of the bundle into the fibre-cable holder/positioner.
14. Mount a second fibre-cable holder/positioner on the other side of the breadboard facing the first fibre-cable holder/positioner.
15. Insert the second end (output) of the bundle into the fibre-cable holder/positioner.
16. Extend the end of the bundle so that the bundle input end is at the centre of the fibre-cable holder/positioner. This is a very important step for obtaining maximum light from the light source to the output end.
17. Verify the alignment of your light launching arrangement by ensuring that the input end of the bundle remains at the centre of the light source.
18. Ensure that you have light output from the output end of the bundle. Point the output towards the centre of the light sensor head.
19. Place the light sensor head in front of the output end of the bundle at a distance (L). Remember to keep the same distance in the following experimental cases.
20. Turn off the lights of the lab.
21. Measure the light intensity at the input and output ends. Record data in Table 13.2.
22. Turn on the lights of the lab.

13.9.4.2 Fibre-Optic Lighting with Lens and Diffuser

In addition to what you have learned in Case (a), you can add a focusing lens at the input end of the bundle to focus the light from the light source into a small spot onto the bundle input end.

1. Figure 13.7 (without the diverging lens) shows the apparatus set-up.
2. Mount a lens holder/positioner to the breadboard at the input end of the bundle so that the light passes through the lens.
3. Place a focusing lens into the lens holder/positioner.
4. Verify the alignment of the input end of the bundle with the focusing lens. Ensure that the focal point of the focusing lens remains at the front of the input end of the bundle.
5. Turn off the lights of the lab.
6. Measure the light intensity at the input and output ends. Record date in Table 13.2.
7. Turn on the lights of the lab.

13.9.4.3 Fibre-Optic Lighting with Lenses and Diffuser

In addition to what you have learned in Case (b), you can add a diverging lens at the output end of the bundle to diverge light from the output end of the bundle.

1. Figure 13.7 shows the apparatus set-up.
2. Mount a lens holder/positioner to the breadboard at the output end of the bundle so that the light passes over the centre of the diverging lens.

TABLE 13.3
Light Intensity Measurements

Case	Light Intensity		Calculated Efficiency	L (unit)
	At Input (unit)	At Output (unit)		
(a)				
(b)				
(c)				

3. Place the diverging lens into the lens holder/positioner.
4. Verify the alignment of the output end of the bundle with the diverging lens.
5. Point the light output from the diverging lens towards the light sensor head. Keep the same distance (L) between the diverging lens and the light sensor head as in Case (a).
6. Turn off the lights of the lab.
7. Measure the light intensity at the input and output of the bundle ends. Record data in Table 13.3.
8. Turn on the lights of the lab.

13.9.5   Data Collection

13.9.5.1   Fibre-Optic Lighting with Diffuser
1. Measure the light intensity at the input and output.
2. Record the collected data in Table 13.3.

13.9.5.2   Fibre-Optic Lighting with Lens and Diffuser
1. Measure the light intensity at the input and output.
2. Record the collected data in Table 13.3.

13.9.5.3   Fibre-Optic Lighting with Lenses and Diffuser
1. Measure the light intensity at the input and output.
2. Record the collected data in Table 13.3.

13.9.6   Calculations and Analysis

13.9.6.1   Fibre-Optic Lighting with Diffuser
1. Calculate the efficiency of the fibre-optic lighting system in Case (a).
2. Record the calculated result in Table 13.3.
3. Consider this case to be the basic case for comparison with other cases.

13.9.6.2   Fibre-Optic Lighting with Lens and Diffuser
1. Calculate the efficiency of the fibre-optic lighting system in Case (b).
2. Record the calculated result in Table 13.3.
3. Compare the efficiency of the fibre-optic lighting system against Case (a).

13.9.6.3   Fibre-Optic Lighting with Lenses and Diffuser
1. Calculate the efficiency of the fibre-optic lighting system in Case (c).
2. Record the calculated result in Table 13.3.
3. Compare the efficiency of the fibre-optic lighting system against Cases (a) and (b).

13.9.7 RESULTS AND DISCUSSIONS

13.9.7.1 Fibre-Optic Lighting with Diffuser

1. Report the light intensity.
2. Consider this case to be the basic case for comparison with other cases.

13.9.7.2 Fibre-Optic Lighting with Lens and Diffuser

1. Report the light intensity.
2. Compare the light intensity in this case with Case (a).
3. Discuss the effect of the presence of the focusing lens on the light intensity when the focusing lens is placed at the fibre-cable bundle input.

13.9.7.3 Fibre-Optic Lighting with Lenses and Diffuser

1. Report the light intensity.
2. Compare the light intensity with Cases (a) and (b).
3. Discuss the effect of the presence of the focusing and diverging lenses on the light intensity when the focusing and diverging lenses are placed at the fibre-cable bundle input and output.

13.9.8 CONCLUSION

Summarise the important observations and findings obtained in this lab experiment.

13.9.9 SUGGESTIONS FOR FUTURE LAB WORK

List any suggestions for improvements in using different experimental equipment, procedures and techniques for any future lab work. These suggestions should be theoretically justified and technically feasible.

13.10 LIST OF REFERENCES

List any references that were used in the report. Use one format in writing the references. Never mix reference formats in a report.

13.11 APPENDIX

List all of the materials and information that are too detailed to be included in the body of the report.

FURTHER READING

Al-Azzawi, A. R. and Casey, P., *Fiber Optics Principles and Practices*, Algonquin College, Algonquin Publishing Centre, Ottawa, Canada, 2002.

Beiser, A., *Physics*, 5th ed., Addison-Wesley, Reading, MA, 1991.

Blaze Photonics Limited, *Product Summary of Photonic Crystal Fibers Catalog*, Blazephotonics, Bath, UK, 2003.

Derickson, D., *Fiber Optic Test and Measurement*, Prentice Hall PTR, Upper Saddle River, NJ, 1998.

Duffie, J. A. and Beckman, W. A., *Solar Energy Thermal Processes*, Wiley-Interscience Publication, New York, 1974.

Falk, D., Brill, D., and Stork, D., *Seeing the Light Optics in Nature, Photography, Color, Vision, and Holography*, Wiley, New York, 1986.

Hood, D. C. and Finkelstein, M. A., *Sensitivity to Light Handbook of Perception and Human Performance. Sensory Processes and Perception*, Boff, K. R., Kaufman, L., and Thomas, J. P., Eds., Vol. 1, Wiley, Toronto, Canada, 1986.

IESNA ED-150.5A, *IESNA Lighting Education—Intermediate Level*, Illuminating Engineering Society of North America, New York, Vol. 21, 1993.

Jameson, D. and Hurvich, L. M., Theory of brightness and color contrast in human vision, *Vision Res. J.*, 4 (1), 135–154, 1964.

Kao, C. K., *Optical Fiber Systems: Technology, Design, and Applications*, McGraw-Hill, New York, 1982.

Key, G., *Fiber Optics in Architectural Lighting Methods, Design, and Applications*, McGraw-Hill Book Company, New York, 2000.

Kreith, F. and Kreider, J. K., *Principles of Solar Engineering*, McGraw-Hill Book Company, New York, 1980.

Lerner, R. G. and Trigg, G. L., *Encyclopedia of Physics*, 2nd ed., VCH Publishers, New York, 1991.

Malacara, D., *Geometrical and Instrumental Optics*, Academic Press, Boston, MA, 1988.

McDaniels, D. K., *The Sun: Our Future Energy Source*, 2nd ed., Wiley, New York, 1984.

National Resources Canada, Advanced houses, *Testing New Ideas for Energy-Efficient Environmentally Responsible Homes*, CANMET's Buildings Group, National Resources Canada, Ottawa, Canada, 1993.

National Resources Canada, Household lighting, *Consumer's Guide to Buying and Using Energy Efficient Lighting Products, Energy Guide*, National Resources Canada, Ottawa, Canada, 1993.

Nolan, P. J., *Fundamentals of College Physics*, Wm. C. Brown Publishers, Dubuque, IA, 1993.

Ocean Optics, Inc, *Product Catalog,* Ocean Optics, Inc., Dunedin, FL, 2003.

Overheim, D. R. and Wagner, D. L., *Light and Color*, Wiley, New York, 1982.

Pedrotti, F. L. and Pedrotti, L. S., *Introduction to Optics*, 2nd ed., Prentice Hall, Englewood Cliffs, NJ, 1993.

Pritchard, D. C., *Environmental Physics: Lighting*, Longmans, Green, London, 1969.

Salah, B. E. A. and Teich, M. C., *Fundamentals of Photonics*, Wiley, London, 1991.

SEPA, *Solar Power*, Solar Electric Power Association, Washington, DC, 2002.

Serway, R. A., *Physics for Scientists and Engineers*, Saunders Golden Sunburst Series, London, 1990.

Warren, M. L., *Introduction to Physics*, W.H. Freeman and Company, San Francisco, CA, 1979.

Weisskopf, V. F., How light interacts with matter, *Sci. Am.*, 219 (3), 60–71, 1968.

Wilson, J. D. and Buffa, A. J., *College Physics*, 5th ed., Prentice Hall, Englewood Cliffs, NJ, 2000.

14 Fibre-Optic Testing

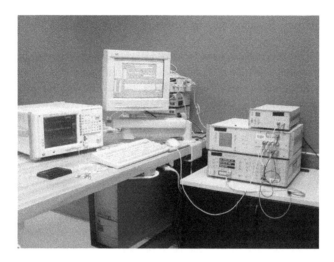

14.1 INTRODUCTION

This chapter describes the basic techniques employed to test optical fibre devices, using optical switches as an example. It presents common testing methods required by industry standards. In these tests, losses and other parameters, such as wavelength, will be measured. Students will become familiar with types of losses and customer requirements. Students will also learn how to use some optical-testing instruments, such as an optical spectrum analyser (OSA). This chapter also presents three experimental cases to demonstrate the principles of testing a fibre-optic device.

14.2 TESTING PHOTONICS COMPONENTS

The testing of optical-fibre devices is based on two different measurement aspects of light output: power and frequency. Power measurements refer to light intensity, as measured by photosensitive devices, usually called photodetectors (e.g. photodiodes and photosensors). The photodetectors generate an electrical signal proportional to the intensity (or power) of the incident light, thus allowing measurement of the light power. Photosensitive devices will not, however, react differently to different light frequencies. They will be more sensitive to certain frequencies while not being able to detect others. On account of this frequency dependence, power meters are calibrated for predetermined frequencies only. To ensure accurate readings, it is very important to verify the calibration of the power meter, before making any power measurements.

Photodetectors also depend on the polarisation of the incident light. Different polarisations will generate different current intensities in the photosensitive devices, thus affecting the power readings during measurement. Polarisation effects can be seen when measuring light power for a fibre-optic device while moving the fibres or optical components. Movements will change the polarisation of the light passing into the device, causing the power readings to fluctuate slightly. For this reason, fibres and/or optical components should not be moved during test measurements.

Polarisation measurements require more complex calculations than power measurement alone. Many methods have been developed over the years to calculate the polarisation-dependent loss (PDL) of optical-fibre devices. For each state of polarisation, a device will absorb varying amounts of light.

Thus, in order to determine the PDL of a device, the device needs to be illuminated with all possible polarisation states while measuring the loss for each. Then the maximum difference in loss for all the polarisation states can be calculated. This process is tedious and time-consuming. However, another calculation method, known as the Mueller matrix, allows the calculation of the PDL with enough accuracy using only four basic polarisation states (linear 0°, 90°, 45° and circular polarisations). By using this matrix calculation method significantly reduces the time required to measure the PDL of any device. This algorithm is used in most PDL meters today to perform measurements.

The frequency of light refers to the colour of the light. Colours cannot be determined simply with the use of a photosensitive cell alone. Instead, comparison methods are used to determine the colour of light. A reference beam of light (a laser with a known and fixed single wavelength, such as a He–Ne laser or another gas laser) is combined with the beam of light to be tested. From the interference of the two beams, the frequency of the incident beam can be determined. This method requires a more complex test circuit and more components (e.g. reference laser) than power measurements. This is why wavelength meters are more expensive and bigger than power meters. Furthermore, the alignment between the test beam and the reference beam is a critical factor in analysing the interference of the two beams. Thus, it is crucial to exercise caution when handling wavelength meters, to avoid misalignment of the internal optical components of the wavelength meter.

Calibration is always an issue in measurement practises for optical devices. The calibration of a power meter involves knowledge of the electrical properties of photodetectors. From this knowledge, calculations can establish the relationship between the current generated by the photodetector and the intensity of the test beam measured at a specific light wavelength. Lasers with adjustable light intensity are used to calibrate power meters with respect to a master power meter. The power measured by the power meter has to be equal to the power detected by the master power meter. Any deviation is recorded, and the values (constants) are adjusted in the calculation algorithms of the power meter. Power meters need to be calibrated on a regular basis, because the properties of the photodetectors are known to change over time. Power meters can be calibrated in a very short time using software.

Wavelength meters also need to be calibrated for power measurements. They too use photodetectors, which measure the power of the interfering beam of light in order to determine the light colour. Improper power calibration could result in incorrect readings. Frequency measurements do not, however, only require power calibration in terms of seconds; they also require proper optical alignment. Thus, full calibration of this equipment requires opening the device box and realigning the internal optical components. Further calibration of a wavelength meter requires the skills of a qualified technician for aligning the optical parts.

As explained earlier, the optical-fibre measurements can be divided into two types:

1. Optical-power measurements (intensity I in Watts)
2. Optical-frequency measurements (wavelength λ in metres)

14.3   OPTICAL-POWER MEASUREMENTS (INTENSITY)

The following sections describe the parameters for optical-power measurements.

14.3.1   OPTICAL-POWER MEASUREMENT UNITS

Three common units are used in optical-power measurements:

1. Linear (mW): The milliwatt is the standard unit of measurement. For example, typical communications lasers used are in the range of 1–10 mW.

2. Logarithmic (absolute) (dBm): In the absolute scale, 0 dBm = 1 mW.

$$P_{\text{dBm}} = 10 \times \log_{10}\left(P_{\text{mW}}\right) \tag{14.1}$$

3. Logarithmic (relative) (dB): Indicates a change in power level independent of the absolute power at the input,

$$L_{\text{dB}} = P_{\text{in}}\left(\text{dBm}\right) - P_{\text{out}}\left(\text{dBm}\right) \tag{14.2}$$

14.3.2 OPTICAL-POWER LOSS MEASUREMENTS

The following sections describe common optical-power loss measurements.

14.3.2.1 Insertion Loss

Insertion loss (IL) occurs in all optical-fibre devices, such as fibre-optic cables, optical-passband filters and optical switches, which are discussed in more detail throughout the text.

Consider an optical switch as an example. An optical switch redirects the optical signal in one or another direction or path, with low attenuation (loss) of the light. Thus, a very important parameter to test for a switch is the IL. Internal components of a switch (such as lenses, mirrors or prisms) will always attenuate the light intensity, because of either absorption, diffraction, diffuse reflection or scattering. Thus, the power transmitted out of the switch is less than the power incident on the switch. Thus, IL is never zero. For best switch performance, the IL should be as low as possible for all the paths of the switch.

A common procedure is used in testing optical-fibre devices for IL. Using a single-wavelength laser source, the IL can be measured with a power meter calibrated for that wavelength. All optical-testing procedures described below are based on this IL test. The procedure remains more or less the same; only the test parameters will change.

The intensity of light at one of the output ports, specified in dB, is attenuated, relative to the input signal power. The general light-loss formula is used to calculate the insertion loss, as given in the following equation:

$$\text{IL(dB)} = -10\ \log_{10}\frac{P_{\text{out}}}{P_{\text{in}}} \tag{14.3}$$

A perfect optical device would have no internal losses and would transmit 100% of the incident light ($P_{\text{in}} = P_{\text{out}}$). In other words, IL = 0 dB in an ideal device.

14.3.2.2 Crosstalk

The purpose of an optical switch, as mentioned above, is to redirect the light in one, and only one, direction (channel). Ideally, when activating a switch, all the incoming light will go through one specific output port (active port), and no light at all will go through any other output port (inactive port). Therefore, in addition to measuring the IL for the active path, the loss for all the inactive paths must be measured, in order to discover if light leaks to other channels. In theory, all light is transmitted through the active channel, and absolutely no light goes through the inactive channels. In practise, however, this is not always the case. Diffuse reflection, diffraction or back reflection of the incident light on an internal component of a switch will redirect the light elsewhere inside the switch. This light can escape from the package by any other input or output port of the switch, depending on where the light is headed. This leakage of light into another channel can be measured in a manner similar to that used for the IL; the IL for the inactive channel (through which no light should be travelling) can be measured. The same procedure as used for the IL test is employed here, except that the optical path chosen is not the active path. The leak towards a wrong or inactive output port is calculated with the same formula used to calculate the IL of the active port, but P_{out} is measured at an inactive output port.

Crosstalk (XT) is the leakage from the active channel into the inactive channel. Crosstalk between channels (ports) occurs in optical devices, such as filters, wavelength separators and switches. Crosstalk is critically detrimental in wavelength division multiplexer systems. When signals from one channel arrive in another undesired channel, they become noise in the other channel. This can have serious effects on the signal-to-noise ratio, and, hence, on the error rate of the communication system.

A perfect device would have no crosstalk (P_{out} of inactive channel $= 0$ mW). In other words, XT $= -\infty$ in a perfect device.

14.3.2.3 Polarisation-Dependent Loss

The polarisation-dependent loss (PDL) is another important parameter for optical-fibre devices, such as an optical switch. An optical switch should not affect or distort the optical signal, only re-route it. Therefore, the PDL should be as low as possible for the active path. The loss in the switch should not vary when the polarisation of the incident light is changed. Any change in loss due to polarisation is called PDL.

PDL can be measured in many ways. One way is to illuminate the device under test (DUT) with laser light in which the polarisation state is changing. Then, the loss for all possible states of polarisation can be measured. However, as mentioned earlier, this method consumes too much time and too many resources to be cost-effective for companies to use routinely.

In order to measure the PDL property of a device, one of the most popular techniques used in manufacturing is the Mueller method. Mueller discovered that, by using the four basic light polarisation states of the light (linear 0°, 45°, 90° and circular polarisation), the PDL can be calculated with great accuracy in a few seconds. Quite complex, the calculations take the form of the Mueller matrix. Most PDL meters use this matrix to calculate the PDL of a device.

PDL meters have a built-in laser source of known polarisation, combined with a power detector, and also include polarising filters, which can be moved in and out of the laser path to vary the polarisation.

As the components included in the PDL meter are expensive and their alignment needs to be precise, the cost of a PDL meter is much higher than that of a power meter. Also, because the PDL meter has to adjust the polarisation of the light to take four measurements and then perform a calculation, it takes approximately 1 s before the PDL value is refreshed on the display. Furthermore, any change in the light path can modify the PDL value, so it is very important that the components and fibres do not move during the measurement. If the fibre-optic cables of the device are moving during testing, the PDL values will change on the display unit and will not be as accurate.

There are three other polarisation effects: polarisation-dependent gain (PDG) in optical amplifiers (OAs); polarisation-mode dispersion in optical fibres; and polarisation-dependent modulation in electro-optic modulators. These polarisation effects occur in almost all other optical-fibre devices transmitting polarised light. Such devices are polarising beam-splitter devices and erbium-doped fibre amplifiers.

14.3.2.4 Return Loss or Back reflection

The return loss (RL) of the DUT should be as high as possible, depending on the definition of loss. This is because having a large amount of incident light reflecting back into the system is not desired. Light reflected back towards the light source can damage the source, especially when laser light is employed.

A small portion of incident light will always be reflected due to a change in the refractive index between two adjacent surfaces. This reflection can be reduced when an anti-reflective coating is applied to one of the surfaces, or by using angled surfaces along the path of light. It is difficult, however, to eliminate any undesired reflection completely. Furthermore, a microbend in a fibre-optic cable, glue joint or anywhere else along the path of light, will create back reflection.

In order to measure the back reflection, the path of light must be terminated. A situation must be created in which all the transmitted light in the device will dissipate or diffract in the fibre and escape the device. To achieve this situation, the fibre needs to be coiled into fibre loops of a very small diameter (usually around 1–2 cm). This is the only situation in which it is acceptable to coil the fibres in very small loops. Careful coiling of the fibres is important so as not to coil part of the fibre onto another fibre loop. This coiling could crush the fibre underneath and create damage.

14.3.2.5 Temperature-Dependent Loss

In most cases, an optical device, such as an optical switch, is used inside a module containing many other devices. These can include optical devices (e.g. couplers and filters) and electrical devices (e.g. switch controller, power supply and fan). On account of the electronic components creating heat inside the module, the temperature can rise and/or fluctuate, so that the switch will almost never operate at ambient temperature. Therefore, the temperature properties of the DUT need to be measured.

The temperature-dependent losses (TDL) are measured through the use of a test called the thermal gradient stability test, which consists of cycling the temperature of the DUT and measuring the optical losses of the DUT over the temperature range.

Usually, the operating temperature range of an optical module is small, varying from −10°C to −40°C, to less than +100°C. Under these small temperature changes, the properties of the materials composing the switch will only change slightly (on the order of microns), and the effect on the optical properties will be small. Such small losses, usually less than 1 dB, are significant for the IL and PDL, but are not significant for XT or RL. This is why usually only the IL and sometimes PDL are measured over the temperature range for a switch.

However, only the steady state losses are measured to calculate the TDL; the transient losses are not considered. Thus, the measurements are performed when the temperature has been achieved and the losses stabilised. Then a simple subtraction (maximum loss value achieved over the temperature range minus the minimum loss value achieved over the temperature range) is calculated.

The parameters most affected by temperature are the IL and PDL, as mentioned above. The variation of IL over temperature is called TDL. However, the PDL difference over the temperature range is not measured; instead, the maximum PDL value achieved over the temperature range is given, and it is called Max PDL over temperature.

14.3.2.6 Wavelength-Dependent Loss

On account of the properties of the anti-reflecting (AR) coating on the lens and other optical components of an optical-fibre device, the IL can vary depending on the wavelength of the incident light. The variation of IL over a wavelength range is called the wavelength-dependent loss (WDL). It is also sometimes called the wavelength flatness or transmission curve. The WDL is a direct consequence of any coating or physical properties of the materials composing the optical-fibre device.

In an optical switch, the AR coating must be able to transmit all the wavelengths with the same efficiency and have a WDL as low as possible. Thus, the difference between the IL measured at 1310, 1480 or 1550 nm should be near zero.

14.3.2.7 Chromatic Dispersion

All forms of dispersion degrade a light wave signal, reducing the data-carrying capacity through pulse broadening. Chromatic dispersion results from a wavelength-dependent variation in propagation delay and is affected by materials and dimensions of the waveguide of an optical-fibre device or the fibre. Chromatic dispersion measurements characterise the way in which the velocity of propagation changes with wavelength, while travelling down the length of the fibre or through the waveguide of optical components.

The concept of optical phase should be considered in the discussion of chromatic dispersion. A mathematical relationship exists between optical phase and chromatic dispersion or group delay.

It is important to mention optical phase before any explanations of chromatic dispersion or group delay. Group delay is defined as the first derivative of optical phase with respect to optical frequency. Chromatic dispersion is the second derivative of optical phase with respect to optical frequency. These quantities are represented as follows:

$$\text{Group delay} = \frac{\partial \varphi}{\partial f} \tag{14.4}$$

$$\text{Chromatic dispersion} = \frac{\partial^2 \varphi}{\partial f^2} \tag{14.5}$$

where optical phase φ is the measured modulation in degrees. Optical frequency f is measured in Hz or THz. For example, a typical communications wavelength is 1550 nm = 193.4 THz.

Both of these phenomena occur because all optical signals have a finite spectral width, and different spectral components propagate at different speeds along the length of the fibre. One cause of this velocity difference is that the index of refraction of the fibre core is slightly different for different wavelengths. This condition is called material dispersion; it is the dominant source of chromatic dispersion in single-mode fibres.

Another cause of dispersion is the wavelength dependence of the cross-sectional distribution of light within the fibre. Shorter wavelengths are more completely confined to the fibre core, whereas a larger portion of the optical power at longer wavelengths propagates in the cladding. As the index of the core is greater than the index of the cladding, the wavelengths in the core travel slightly more slowly. Thus, this difference in spatial distribution causes a change in propagation velocity amongst the wavelengths. This phenomenon is known as waveguide dispersion, which is relatively small compared to material dispersion.

Chromatic dispersion can also cause bit errors in digital communications, distortion and a higher noise in analogue communications. These outcomes can pose a serious problem in high-bitrate systems, if dispersion is not measured accurately, and if some form of dispersion compensation is not employed.

14.4   OPTICAL-FREQUENCY MEASUREMENTS

The relation between light wavelength λ and frequency f is given by the following equation:

$$\lambda = \frac{c}{nf} \tag{14.6}$$

where:
   λ is the light wavelength in nm
   c is the speed of light in a vacuum in m/s
   f is the light frequency in Hz or THz
   n is the index of refraction of an optical material

Optical measurements are typically calibrated to the light wavelength in a vacuum ($n = 1$). Therefore, when $n = 1$, Equation 14.3 can be rewritten as

$$\lambda = \frac{c}{f} \tag{14.7}$$

14.5   TESTING OPTICAL FIBRE SWITCHES

The following are the most common tests on optical fibre switches required by industry standards and consumers. These tests simulate the conditions that may be present during the switch's lifetime operation in the field.

There are few reasons why testing of the optical performance is not the only important factor in testing an optical device. Mechanical, electrical and environmental tests of devices are as important as optical testing, before delivering the devices to the market.

14.5.1 MECHANICAL TESTS

Mechanical tests are carried out to test the durability of the moving mechanical parts inside optical devices, such as switches. Inside an optical switch, there is usually a moving part (either a moving mirror, moving prism or moving lens) that is moved in and out of the optical path; this changes the direction of the optical path (e.g. if the mirror is out of the optical path, the light will pass straight through the device, unaffected, to one output port; if the mirror is placed in the optical path, the light will be reflected by the mirror into a different output port).

Like all other devices, optical components wear out with time. For example, in an optical switch, the physical joint between the mirror or lens, and the moving arm holding them, are affected by mechanical stresses. These stresses are induced by movement, change in ambient temperature and humidity level, and they are due to the different intrinsic physical properties (e.g. thermal coefficient, stiffness and elasticity) of the epoxy, mirror and moving arm. Of course, manufacturers try to choose an epoxy closely matching the physical properties of the parts, but in practise, there is always a slight difference. On account of this difference, a stress situation can occur in which temperature increase will cause the epoxy to expand more (or less) than the mirror. This expansion causes the mirror angle to change, and thus the beam direction will be deviated and no longer be optimised. Therefore, the loss properties of the optical device will change due to mechanical stress.

Another situation can arise if there are internal moving parts in an optical device. The physical joint between moving parts, for example, between a mirror and the arm to which it is attached, can become weak after many movement cycles. This weakness causes the mirror to move a little with each cycle and induces instability in the beam deviation, and this weakness will of course affect the optical properties of the device.

Still another factor that needs to be taken into consideration concerns the device's electrical properties. Although it is not obvious, when dealing with the mechanical properties of a device, electrical signals are often linked to the observed mechanical defects. As an example, take an actuator with a coil creating a magnetic field used to move an arm up and down, depending on the orientation of the electric signal. A mirror is glued to the tip of the moving arm. The mirror can be moved either in or out of the optical path of a beam of light, forming an electro-optical switch. If the amplitude of the electrical current changes, the magnetic field will change accordingly, and thus will make the arm move more quickly or more slowly. This will affect the speed at which the mirror moves in and out of the optical path. Such factors can seem insignificant, but in some applications where the speed of the transmission of a signal is important (and customers always want faster components), this variation in time can become a real problem.

14.5.2 ENVIRONMENTAL TESTS

Industry standards require that a representative number of samples of a product be subjected to a programme of environmental challenges including high and low temperature storage, temperature cycling and humidity. An environmental-testing procedure might consist of three to six stages of temperature and humidity tests on selected devices. The characteristics of a DUT must be measured before and after each test stage. In some cases, continuous or interval testing is conducted during each test stage.

Environmental test systems are usually integrated in an automated test system, which is intended for long-term reliability testing of optical components under environmental stress conditions. Such stress conditions are listed in Telcordia specifications GR-326-CORE, GR1209-CORE and

GR-1221-CORE. The Telcordia GR-xxx-CORE standards are quality standards used in the fibre-optic industry, and they are more complete and restrictive than the ISO quality standards.

Devices under test are subjected to a range of environmental stress conditions in a test chamber, usually over a period of many weeks. Chamber conditions are recorded at specified time intervals during the environmental tests, and the required parameters for each DUT are measured. User selected parameters are calculated from these responses and recorded along with the time and environmental data.

Switches combined with the appropriate source and monitoring hardware along with software can create fully automated measurement systems. An optional polarisation controller is installed when PDL measurements are required. A computer is used to set up the tests, control the measurements and monitor the test conditions and results.

For measuring IL and PDL, the optical component environmental test system (OCETS) uses a combination of up to three internal Fabry–Perot lasers, and a broadband source (BBS) with a filter or an external source. The light from any of these sources can be routed to either end of each DUT. The power meter measures either the IL through the DUT, in either direction, or the back reflection from either end. A polarisation controller enables PDL to be measured. Second and subsequent tests can be added for additional sets of DUTs, up to the switch capacity limit, whereas the first test is running. In this way, an environmental chamber running a long-term test (on a first prototype, for example) could evaluate the performance of product improvements, by installing later devices in the chamber and configuring a second test to run with the same conditions and measurements.

14.5.3   REPEATABILITY TEST

If there are moving parts in the optical path, the optical signal may slightly change each time the parts move from one position to another. This change in optical signal, or loss variation at each cycle, is a measure of repeatability. The difference in performance between test cycles is due to the physical properties of the components in the optical device. Minimal difference between the cycles indicates good repeatability. For example, the repeatability test for an optical switch measures the maximum IL variation, for a given optical path, over number of cycles, at a given constant temperature. The results of repeatability tests are often displayed as a graph. This graph enables a quick view of the reading stability, and it allows determination of the occurrence of spikes or non-regular variations.

The repeatability test is usually very sensitive to external vibrations (such as someone walking near the test station or banging on the table) because such vibrations induce additional movement to the internal parts of the DUT. This test can also be affected by the orientation of the DUT (due to gravitational forces on the DDT's internal moving parts). It is also important to keep the temperature constant during this test, because the physical properties of the materials composing the DUT change with temperature. This induces additional measurement variations, which are not due to the moving parts themselves.

14.5.4   SPEED TEST

Moving parts do not move instantly; they need an amount of time to react to the signal that controls them in order to move. The time elapsed between the application of the control pulse and the moment when the optical signal achieves its steady state is called the switching time for optical switches (also often called speed time for switches).

For an optical switch, the switching time is usually defined as the time between the application of the activation pulse and the time when the optical signal reaches 90% of its steady state, at a given constant temperature. In order to measure the switching time, a very fast optical power meter (OPM) is required. The sampling speed of the OPM should be at least ten times faster than the switching speed to be measured. For example, if a switching time in the order of milliseconds is to be measured, the sampling speed of the OPM should be at least 0.1 ms faster.

The results of this test are often displayed as a graph to see if the steady state of the optical signal is flat and high enough, and if the dynamic range between light-on signal and light-off signal is good.

This test is very sensitive to fluctuations of the laser light source. For example, a variation in optical signal could be perceived as an unwanted movement inside of a switch. Also, when the OPM is set in the fast-reading mode, it is very sensitive to the power level of the laser because the response time of the OPM detector varies with the power of light: the lower the power, the slower the response time. Poor OPM performance results in additional noise in the measurements. However, if the laser power is too high, the detector will saturate and give inaccurate readings. Thus, it is important to find the acceptable power range for optimum response time of the OPM detector for this test.

Temperature is also an important factor for this test, because the properties of the materials will change with temperature (electrical current moves faster at higher temperatures, for example). So the temperature should be kept constant during the speed test.

The orientation of the DUT is also a factor to be considered. If the moving parts of the DUT are heavy, then gravitational forces will not be negligible, and will have an effect on the speed. Thus, it is important to perform some tests to determine if the orientation of the DUT will have an effect on the speed test measurements.

14.6   LIGHT WAVELENGTH MEASUREMENTS

Wavelength measurements provide various information about most optical fibre devices. In these measurements, optical properties, such as IL, are measured over a wide range of wavelengths. For example, WDL is the variation in IL over a specified wavelength range shown in a spectral plot.

From the spectral plot (IL versus λ), many device parameters can be measured that are as follows:

1. Flatness (WDL)
2. Passband width (bandwidth, BW)
3. Full width half maximum (FWHM) (−3 dB BW)
4. Cross-channel isolation to reduce crosstalk
5. Maximum/minimum loss in passband
6. Free spectral range (FSR)

Light wavelength measurements can be achieved using either a tunable laser source with fixed detector or a broadband light source with an OSA. In addition, advanced wavelength measurement systems conduct different measurements, including different types of losses.

14.7   DEVICE POWER HANDLING TESTS

There are two categories of power-handling failures that occur during the power-handling test of an optical-fibre device:

1. Increasing the input optical power until a device failure occurs: this happens in an epoxy joint, when the epoxy ruptures and the surface coating burns: in this case, a graph of loss versus power shows the results for each DUT.
2. Applying a higher than normal input optical power over a prolonged period of time, until a device failure occurs: in this case, a graph of losses versus time for different input powers shows the results for each DUT.

14.8 TROUBLESHOOTING

The main sources of problems that can arise during any of the preceding tests are as follows:

1. Unwanted reflections occur when the refractive index along the light path changes, for example, at a glass-air interface, dirty and/or damaged connectors and terminated reflections from unused fibre ends. Reflection between two parallel optical surfaces can be prevented, by using non-perpendicular surface angles.
2. Unstable laser source occurs if laser temperature is fluctuating, light is reflected back into the laser source (that can be solved by installing an isolator device after the laser source). It also occurs because of dirty and/or damaged connectors, damaged jumpers or a sharply bent or too tightly coiled cable. It can also occur if the power meter is in fast-sampling mode.
3. Inaccurate power measurements occur when the power meter needs calibration, or are caused by stray light (low power measurement can be affected by room light; therefore, connections must be properly shielded). It also occurs due to dirty and/or damaged connectors or detectors or poor switch repeatability. It can also occur because of the connection to the detector differs between the reference and measurement.

When a problem arises during testing, and troubleshooting is required, one must first ensure that these above-mentioned problems are not present at the test station before calling the DUT a failure. Even if none of these situations occurs during testing, there are other sources of errors, intrinsic to any test system that must be taken into account.

14.9 SOURCES OF ERROR DURING FIBRE-OPTIC MEASUREMENTS

The following are the main categories for sources of error in a test system:

14.9.1 Resolution

Resolution is defined as the size of the smallest increment or unit in which the instrument can read. The resolution of the system should be ≤ 1/10 of the total tolerance width. The resolution of equipment is an intrinsic property of the equipment's internal components, which the person performing the test has no control over (unless, of course, that person decides to buy another piece of equipment with a better resolution). Consider a ruler: the smallest divisions are millimetres. The human eye can differentiate, without the use of a magnifier, details as small as one tenth of a millimetre. However, when measuring the length of a line, it would be difficult to say whether the measurement is 11.2 or 11.3 mm. Therefore, the resolution of the human-eye-and-ruler system is ± 0.5 mm. Another example is a digital meter: the resolution of the digital meter is ± the last digit; if the meter gives a reading of 0.67 dB, then the resolution of the measurement is ± 0.01 dB.

14.9.2 Accuracy

Accuracy is defined as the difference between the observed average for a series of measurements and the true value being measured. The true value is measured using a reference measurement system, a master power meter. A master power meter has been referenced and calibrated with the world's master power meter located at the National Institute of Standards and Technology, and used to calibrate all the master power meters in the world (not that many). This reference ensures that any power meter in any facility worldwide, when calibrated with a master meter, will give the same measurement result for a given device. The accuracy of a power meter therefore usually depends on how long ago the meter was calibrated. Likewise, when measuring wavelength, the reference is a master wavelength meter.

14.9.3 STABILITY (DRIFT)

Stability is defined as the change in measurements over time. Measurement readings will vary over time, because the equipment wears out with time, dust can accumulate and temperature and humidity, along with other factors, can affect the readings. If a device remains connected to a power meter for a long time and measurements are repeatedly performed at pre-determined time intervals, the reading will vary over time. Most of the time, the reading variation follows a cycle, such as a sinusoid curve or a series of abrupt changes (steps). This pattern can be caused by instability in, for example, electrical components and heating components. In fact, the instability is closely related to inaccuracy. Stability can be considered a measure of the variation in accuracy over time. The more time that passes, the greater the likelihood that a drift will occur in the readings. This is why periodic calibration has to be performed on any measurement equipment.

14.9.4 LINEARITY

Linearity is defined as the difference in accuracy values over the expected operating range. Some electrical and optical components will not have linear response depending on the intensity of electric current or incident light. Linearity is the linear correlation of an input power or signal to the corresponding optical power output. A mathematical formula is used to calculate the linearity value of a device. The correlation between electric current and light power, for an optical diode, is not perfectly linear; the correlation graph will usually be a curve. The more the curve bends, the worse the linearity is. This means that for a power value near the middle of the measurement range of the apparatus, the readings will be closer to the real value than when the power value is close to one end of the curve. The linearity and the correlation coefficients are adjusted during calibration of the meter or equipment.

14.9.5 REPEATABILITY ERROR

Repeatability error is defined as the consistency of a given system, making repeated measurements of the same part, using the same measurement instrument. It is a measure of the inherent variation in the system. Basically, to determine the repeatability error of a measurement system (a complex automated system or a simple power meter and an operator), a number of devices (between five and thirty, but usually around ten to fifteen) have to be selected and tested. The selected devices should have measured values covering the entire range of probable results. These devices should include those that have both passed and failed testing. All of the devices are tested, and the test results are recorded as the first set of results. The order of the device testing is changed. Then, a new test is performed, and the results are recorded. This procedure can be repeated a number of times (between two times and five times, usually three to five times), and each time the testing order of the devices is changed. After each test, the results are correlated to identify all of the devices throughout the tests. Once all the series of tests are performed, the results can be recorded in a table, wherein each row corresponds to a device and the columns contain the test results for each measurement performed for each device. From the difference in test results for each device, analysis can be performed to determine the repeatability error of the test system.

14.9.6 REPRODUCIBILITY

Reproducibility is defined as consistency of different operators measuring the same part using the same measurement instrument. To determine the reproducibility error of a measurement system, a number of devices (between 5 and 30, usually around 10 to 15) have to be selected and tested. The same devices used for the repeatability-error test can be used for this. Selected devices should have measured values covering the entire range of probable results, and the devices should include those that passed and failed testing. A number of people are then selected (between two and five) to perform the tests. The devices are identified with a code, and one person is asked to test all of the devices and record the test results. This is the first set of results. Then the ID codes of the devices are changed, and another person tests the devices again and records the results. The same procedure continues until all the selected persons have tested the devices. The results can be recorded in a table, wherein each row corresponds to a device, and the columns correspond to the test results obtained for each person. From the difference between person-to-person results for each device, analysis can be performed to determine the reproducibility error of the test system.

The following points should be considered, when examining the sources of measurement errors:

1. The errors mostly stem from the measurement system and not only from the persons performing the test. If the test system were perfect, it would always give exactly the same results regardless of test operator or the number of tests. But systems are never perfect, and there will always be variations in the test results or errors. It is possible to minimise error occurrence by using experienced persons. System error itself can be reduced with more frequent calibration and a more-stable or better-designed system.

2. The repeatability and reproducibility errors will almost always be the major source of error in any test system. It is important to characterise these two sources of error, in order to determine if the results obtained from the test system are reliable or not. The measurement tool is sometimes called gauge. The most common way to determine the repeatability and reproducibility error factors of a system is to perform a gauge capability study (GCS) of the system, also known as gauge repeatability and reproducibility study (Gauge R&R Study). This GCS study includes simultaneously performing repeatability and reproducibility studies.

3. If the gauge R&R results show a systematic error in the measurements (results are always lower/higher than the real value), then this can be reduced by a proper calibration and increasing the frequency of calibrations.

4. If the gauge R&R results show random errors in the measurement results, then it is better to add a guardband to the test results. For example, if the test specification for the IL value is less than 1.0 dB, and the system gives a random error of ± 0.05 dB, then the test must fail if any device that has an IL higher than 0.95 dB.

5. The resolution, accuracy, stability and linearity sources of error are quoted in technical equipment specifications. These values can be improved with the calibration of the equipment.

14.10   EXPERIMENTAL WORK

In this experiment, students will practice testing of optical-fibre devices. Students will work with the following cases:

14.10.1   Testing a Fibre-Optic Device Using an Optical Spectrum Analyser

Case One familiarises students with the use of an OSA for testing optical components. Refer to the manufacturer's operation manual for the OSA in use. As its name implies, an OSA analyses the optical spectrum of a DUT. In other words, an OSA measures the optical power of the DUT relative to a band of frequencies (or wavelengths) selected by the user. As an OSA is quite big, heavy, very expensive and more complicated to use than a power meter, its use is usually restricted to R&D laboratories. It is also used for testing fibre-optic devices, which could not be fully tested using a power meter, such as passband filters, WDM devices and BBNS.

To measure the wavelength of the light coupled to its input port, the OSA uses a free-space optics interferometer. An adjustable power meter measures the optical power of the light for each wavelength in the range specified by the user. The data are stored in memory and displayed as a graph along with the analysed parameters. The user can then see the power distribution of a device for a determined wavelength or frequency range.

Students will learn how to configure the proper settings of an OSA for testing optical devices, and familiarise themselves with its basic functions.

14.10.2 Testing Mechanical Properties of Fibre-Optic Devices

This experiment familiarises students with tests performed on fibre-optic devices, which require mechanical and environmental testing in addition to optical testing. Mechanical testing is required for any device containing moving parts (e.g. switches, tuneable filters and variable attenuators). Environmental testing is usually performed during the R&D stage of a new product, to ensure that the product will perform within normal specifications over time, over different stress conditions and over variable temperature and humidity levels. The students will learn about the most common mechanical tests performed in industry for optical devices, as well as develop intuition in determining the required tests for new devices.

14.10.3 Testing a Fibre-Optic Cable Using an Optical Spectrum Analyser

This experiment introduces students to measuring the losses in a multi-mode fibre cable. Students will also verify the test results with the technical specifications provided by the manufacturer. An OSA is used to obtain accurate measurements for the losses in the fibre-optic cable. Students will also practice alignment, connection and set up of optical tests, as shown in Figure 14.1.

This experiment can be repeated using a fibre-optic cable with a connector on each end, and employing a laser source with output connectors matching the cable connectors. In this case, the cable can be connected directly to the laser source, as shown in Figure 14.2. This experimental set-up gives more precise measurements, because the direct connection between the laser source and the OSA reduces the losses caused by outside light interference, lens coupling and air gap between the connectors.

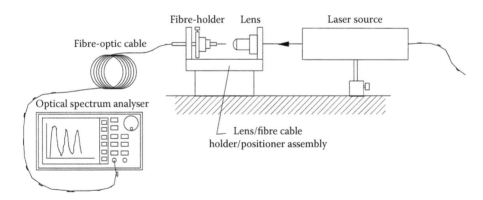

FIGURE 14.1 Fibre-optic cable testing.

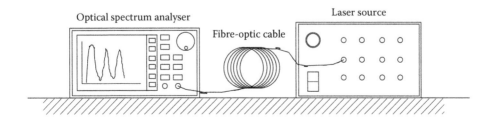

FIGURE 14.2 Fibre-optic cable testing by direct connection.

14.10.4 Technique and Apparatus

Appendix A presents the details of devices, components, tools and parts.

1. C-band BBNS or white light source
2. Optical spectrum analyser (OSA)
3. Single-mode fibre (SMF) cable with an angled connector FC/APC at one end, to connect to the BBNS
4. Single-mode fibre (SMF) cable with a flat connector FC/PC at one end to connect to the OSA
5. Multi-mode fibre (MMF) cable with an angled connector FC/APC at one end, to connect to the BBNS
6. Devices for testing:
   An SMF cable
   An MMF optic cable
   A passband filter
   A WDM device
7. 2 × 2 ft optical breadboard
8. He–Ne laser light source and power supply
9. Laser mount assembly
10. Lens/fibre cable holder/positioner assembly
11. Multi-axis translation stage
12. Hardware assembly (e.g. clamps, posts, screw kits, screwdriver kit, positioners and holder)
13. Optical assembly
14. Objective lens
15. Black/white card and cardholder

14.10.5 Procedure

Follow the laboratory procedures and instructions given by the professor and/or instructor.

14.10.6 Safety Procedure

Follow all safety procedures and regulations regarding the use of optical components, electrical and optical devices, light sources, measurement instruments and optical cleaning chemicals.

14.10.7 Apparatus Set-Up

14.10.7.1 Testing a Fibre-Optic Device Using an Optical Spectrum Analyser

1. Connect the SMF cable with the FC/APC connector to the BBNS output port. Turn the BBNS on. The light source usually needs up to 30 min to warm-up and stabilise. In the meantime, gather all the material and equipment and prepare the set-up for the measurements. If the DUT does not have connectors on input and/or output ports, make a connection to attach the input port to the BBNS light source and the output port to the OSA. Gather all the necessary tools and prepare the fibre tips for connectors. Clean the tips of the fibres properly and set them aside. If the DUTs have connectors on their input/output ports, clean the connectors properly. Find the appropriate mating sleeve to connect the DUT's input port to the BBNS source fibre.
2. Turn on the OSA. The initialisation and warm-up process takes from a few seconds up to a minute.

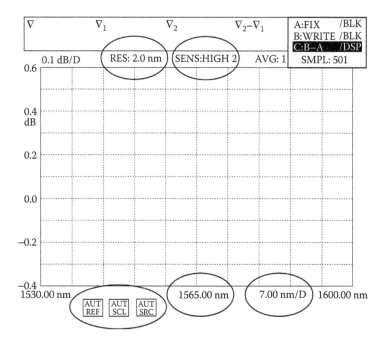

FIGURE 14.3 Test parameters displayed on the optical spectrum analyser.

3. During this time, for non-connectorised DUTs, connect the SMF cable with the FC/PC connector to the input port of the OSA. Connect the BBNS source fibre to the measurement fibre of the OSA.

4. For connectorised DUTs, connect the BBNS source fibre directly to the OSA input port.
   The following is the set-up procedure to prepare the OSA for optical measurements. Figure 14.3 shows the settings for the OSA.

5. Set up the scanning parameters on the OSA to take optical measurements as follows:
   Press the SETUP key.
   Press the RESOLN soft key.
   Turn the rotary knob, until a resolution of 20 nm is displayed.
   Note: the keypad can also be used to enter the resolution. Type "20" and then press the nm/ENTER key.

6. Set the sensitivity required for the measurements as follows:
   Press the SENS soft key.
   Turn the rotary knob, until a sensitivity of HIGH1 is displayed.
   Note: The UP and DOWN arrows on the keypad can also be used to select the proper sensitivity.
   To select the sampling size, press the SAMPLING soft key. Using the keypad, enter 501 and then press the nm/ENTER key.

7. Set the parameters of the vertical axis of the OSA graph as follows:
   Press the LEVEL key.
   Press the AUTO REF LEVEL soft key, until the AUT-REF icon becomes highlighted on the bottom of the OSA screen.
   Press the MORE 1/2 soft key to go to the next page of the LEVEL menu.
   Press the AUTO SUB SCALE soft key, until the icon AUT-SCL becomes highlighted on the bottom of the OSA screen.
   At the bottom of the display, both the AUT-REF and the AUT-SCL should be displayed. If it is not, press the soft keys AUTO REF LEVEL and AUTO SUB SCALE again.

8. Set up the horizontal axis of the graph to measure the proper wavelength range for the experiment as follows:

> Press the CENTER key.
>
> Press the CENTER soft key.
>
> Using the keypad, enter the wavelength 1545 nm, and then press the nm/ENTER key.
>
> Press the SPAN key, and then press the SPAN soft key.
>
> Using the keypad, enter the wavelength range (50 nm for the SMF DUT, 30 nm for the passband filter and 50 nm for the WDM DUT).
>
> *Note:* A reference reading must be taken any time the wavelength range, wavelength span or centre wavelength is changed. The OSA only stores the data for the wavelengths selected and displayed on the screen. Any wavelength outside the range displayed on the screen was neither analysed nor stored in memory by the analyser.

Therefore, take a new set of reference readings every time the horizontal axis is changed, as follows:

> Press the nm/ENTER key.
>
> Press the PEAK SEARCH key.
>
> Press the MORE $^1/_2$ soft key, to display the next page of options for the PEAK SEARCH menu.
>
> Press the AUTO SEARCH soft key. The AUT-SRC icon should appear on the bottom of the screen. If not, press the AUTO SEARCH soft key again.
>
> Press the SWEEP key, for measurement sweeping.
>
> Press the SINGLE soft key. This will initialise the new settings of the OSA for single-sweep measurement.
>
> *Note:* The settings are displayed on the OSA screen.

1. Take the reference reading, with the BBNS source connected to the OSA.
2. Store the reference reading as follows:

   > Press the TRACE key.
   >
   > Activate Trace A, using the TRACE ABC soft key.
   >
   > Press the WRITE A soft key.
   >
   > Press the SWEEP key. Then press the REPEAT soft key, to get continuous measurement.
   >
   > Verify that the source power is stable and does not change with every sweep. When *t* the source power is stable, press the SINGLE soft key.
   >
   > When the single sweep is completed, a power distribution can be seen, which is similar to that shown in the BBNS user's manual.
   >
   > Press the TRACE key again, and then press the FIX A soft key. Doing so will store the reference reading in memory.
   >
   > Press the BLANK A soft key, to hide the reference stored.
   >
   > Activate Trace B and press the WRITE B soft key. Press BLANK B.
   >
   > Activate Trace C.
   >
   > Press the soft key CALCULATE C.
   >
   > Press the B–A (B/A) soft key.
   >
   > Press the DISPLAY C soft key.
   >
   > The following settings should now appear in the top right corner of the OSA screen:
   >
   > A:FIX/BLK, B:WRITE/BLK, C:B/A/DSP.

3. Press the SWEEP key and press the REPEAT soft key. The readings should be a flat line centred at 0 dB.
4. The peak-IL (due to noise in the signal or dirt on the connections and measurement error) will be indicated in the top left corner of the OSA display, under ▼ PEAK. The wavelength at which this loss is located will also be displayed.

> *Note:* The reference reading is now complete. The OSA is ready to start testing the DUT.

5. Record the reference IL and wavelength. If the reference IL is greater than 0.2 dB, re-do the reference set-up procedure, steps 2–10.

It is strongly recommended to print out all the graphs obtained during this experiment. Press the COPY button, and tear off the printed graph from the OSA printer.

To identify the graphs, write a descriptive text on the label field portion of the OSA screen, which will appear on the printout. To add/modify text on the label, press the DISPLAY key, then press the LABEL soft key. Using the rotary knob, the up/down and left/right arrows, and the keypad, write the desired text. When satisfied with the text, press the DONE soft key. Print the identified graph by pressing the COPY key. If there is not enough blank paper below the graph to tear the roll without loosing part of the graph, press the FEED key until there is enough clearance to tear off the chart paper.

14.10.7.1.1   *Measuring the IL of a wavelength-independent DUT*

The following is the testing procedure to measure the IL of a wavelength-independent DUT (such as a fibre-optic cable, switch, circulator and coupler).

1. Disconnect the reference fibre from the OSA and connect it to the input port of the DUT.
2. Connect the OSA to the DUT output port corresponding to the channel to be tested.
3. A line should appear on the OSA screen. The line should be mostly flat, because the DUT does not present any wavelength dependency. The lowest IL value will be indicated in the top left corner of the OSA screen, along with the corresponding wavelength.

Note: If the loss line shows ripples or waves instead of being mostly flat, these problems are probably caused by dirt on connections, damage in the fibre-optic cables causing an etalon effect (two partially reflective surfaces parallel to each other in the optical path cause interference) or by a bad or damaged AR coating of the internal components inside the DUT.

Using an OSA to measure the loss of a wavelength independent DUT does not add significant value to the test results that could be obtained using a regular and less expensive power meter.

14.10.7.1.2   *Measuring the IL of a wavelength-dependent DUT*

The following is the Testing procedure to measure the IL of a wavelength-dependent DUT. This test can use a passband filter, WDM device or Fabry–Perot etalon as a DUT.

1. Disconnect the reference fibre from the OSA, and connect it to the input port of the DUT.
2. Connect the OSA to the DUT output port corresponding to the channel to be tested.
3. The optical spectrum for the DUT should appear on the OSA screen. The peak IL value will be indicated in the top left corner of the OSA screen, along with the corresponding wavelength. Record the peak IL value and associated wavelength. Figure 14.4 shows a chart printout from the OSA for the DUT.
4. Calculate the FWHM, or in other words the −3 dB bandwidth (−3 dB BW), as follows: Press the MARKER key.
   Set the horizontal line marker one at the top of the transmission peak.
   Place the horizontal line marker two to be 3 dB lower than the horizontal line marker one (the difference between the two markers will be indicated on the display).
   Place the vertical line markers three and four at the intersection of the line marker two and the trace.
   The difference between vertical line markers three and four is displayed on the OSA screen. This is the −3 dB BW, in nanometres.
   For the −20 dB BW, follow the same instructions but place the horizontal line marker two at 20 dB below the horizontal line marker one.

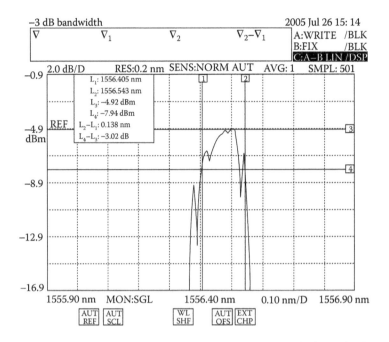

FIGURE 14.4 A chart printout from the optical-spectrum analyser for the device under the test.

Note: It is evident that many valuable optical parameters can be determined from this test for wavelength-dependent DUT. The information given by the power loss versus wavelength graph gives more details about the behaviour of the device tested than what could be obtained from using only a power meter and a single-wavelength laser source. For example, from these graphs one can calculate the FSR of a filter device (distance, in nanometres, between consecutive transmission peaks), or calculate the flatness (difference in IL between the maximum and minimum transmission peaks over a specified wavelength range). In addition, the data illustrated on the OSA display panel can be saved and downloaded to a computer to be analysed in an Excel spreadsheet or similar application, and the graph obtained can be printed either via the OSA or a computer. Customers often like to have these graphs for the devices they purchase, because it gives them more detail about their new device than just data alone.

14.10.7.2 Testing Mechanical Properties of Fibre-Optic Devices

The following is the procedure for testing mechanical properties of fibre-optic devices:

1. Connect the SMF cable with the FC/PC connector to the laser output port, and turn the laser on. The light source usually needs up to 30 min to warm-up and stabilise. In the meantime, gather all the material and equipment, and set up for the measurements.
2. If the DUTs do not have connectors on their input and/or output ports, a connection is needed to attach their input port to the light source and their output port to the power meter. Gather all the necessary tools and prepare the fibre tips for connections. Clean the tips of the fibres properly and set them aside.
3. If the DUTs have connectors on their input/output ports, clean the connectors properly. Find the appropriate mating sleeve to connect the DDT's input port to the laser source fibre.
4. Turn on the power meter; the initialisation and warm-up process will take a few seconds.
5. For non-connectorised DUTs, connect the SMF cable with the FC/PC connector to the input port of the power meter. Connect the laser source fibre to the measurement fibre of the power meter. Take a reference reading.

6. For connectorised DUTs, connect the laser source fibre directly to the power meter input port. Take a reference reading.
7. Repeat the procedure explained in Section 14.10.7.1 to set up the OSA.

14.10.7.3 Testing a Fibre-Optic Cable Using an Optical Spectrum Analyser

The following describes the process employed for fibre-optic cable testing using an optical-spectrum analyser:

1. Set up the experimental apparatus, in the way illustrated in Figure 14.5.
2. Measure the light loss in the optical materials, using the light attenuation equation.
3. Prepare a fibre-optic cable with a connector on one end. Make sure that the connector is the correct type, matching the input terminal connection of the OSA.
4. Prepare the other end of the cable, as explained in the fibre-optic cable end preparation section of the Fibre-Optic Cables Chapter. Cleave the end of the cable at a 90° angle. Clean the fibre end.
5. Mount the stripped end of the fibre into a copper fibre-holder, with the tip of the fibre extending beyond the end of the copper fibre-holder. Now insert the copper fibre-holder into the lens/fibre-cable holder/positioner assembly.
6. Using a He–Ne laser source, project the light into the lens/fibre cable holder/positioner assembly, ensuring that the light is fully coupled into the fibre cable end.
7. Turn the laser on.
8. Turn off the laboratory lights before taking measurements.
9. Using the OSA, measure the laser power and wavelength of the laser source. Figure 14.6 shows a sample of data collected by the OSA printed on a paper chart. Ask the supervisor/instructor for the OSA operational instructions and referencing the source. Record the results for Port 1 in Table 14.1.
10. Turn on the laboratory lights after completing the measurements.

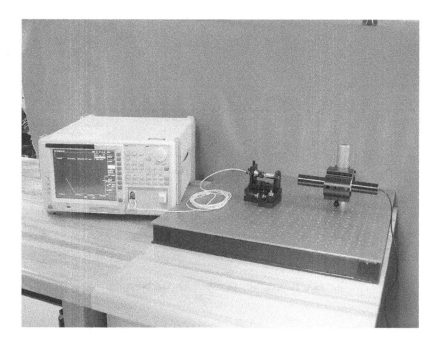

FIGURE 14.5   Fibre-optic cable testing apparatus set-up.

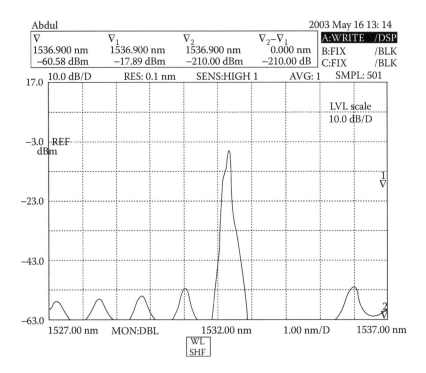

FIGURE 14.6   A sample of collected data printed on a chart from the optical spectrum analyser.

TABLE 14.1
Loss versus Wavelength Measurements Using the OSA

DUT	Reference Max Loss		DUT Peak IL		−3 dB BW (nm)	−20 dB BW (nm)
	Wavelength (nm)	Loss (dB)	Wavelength (nm)	Loss (dB)		
WDM device						
Fabry–perot etalon						
Pass band Filter						
Fiber-Optic Cable						

14.10.8   DATA COLLECTION

14.10.8.1   Testing a Fibre-Optic Device Using an Optical Spectrum Analyser
Fill out Table 14.1 with the test values of the DUT.

14.10.8.2   Testing Mechanical Properties of Fibre-Optic Devices
Fill out Table 14.1 with the test values of the DUT.

14.10.8.3   Testing a Fibre-Optic Cable Using an Optical Spectrum Analyser
Fill out Table 14.1 with the test values of the fibre-optic cable.

14.10.9   CALCULATIONS AND ANALYSIS

14.10.9.1   Testing a Fibre-Optic Device Using an Optical Spectrum Analyser
1. Print the graphs obtained and attach them to the report.
2. Record the maximum loss obtained during the reference measurements.
3. Present the peak IL and wavelength of the tested devices.
4. Calculate the BW as requested and record the results.

14.10.9.2   Testing Mechanical Properties of Fibre-Optic Devices
1. Print the graphs obtained and attach them to the report.
2. Record the maximum loss obtained during the reference measurements.
3. Present the peak IL and wavelength of the tested devices.
4. Calculate the BW as requested and record the results.

14.10.9.3   Testing a Fibre-Optic Cable Using an Optical Spectrum Analyser
1. Print the graph obtained and attach to report.
2. Record the maximum loss obtained during the reference measurements.
3. Present the peak IL and wavelength of the tested fibre-optic cable.
4. Calculate the BW as requested and record the results.

14.10.10   RESULTS AND DISCUSSION

14.10.10.1   Testing a Fibre-Optic Device Using an Optical Spectrum Analyser
1. From the BBNS power distribution illustrated on the OSA screen (obtained before the reference measurements), discuss the importance of taking a reference reading prior to performing any optical measurements.
2. Depending on the optical range available for the BBNS light source, noise in the reference readings is probably visible towards the edges of the graphs, especially for wider ranges. Compare the flatness of the reference line at the centre and the edges of the wavelength range; explain any difference in flatness, called noise.
3. From the manipulations performed during this experiment, and knowing that power meters, OSAs and other testing equipment are usually equipped with parallel, serial or GPIB ports, explain why, from an industrial point of view, some testing procedures should be automated instead of manual.

14.10.10.2   Testing Mechanical Testing Properties of Fibre-Optic Devices
1. As explained in Section 14.10.10.1.
2. List all the items that should appear on a test report to be sent out to a customer for one particular device chosen amongst those tested during this lab. Justify the list of items (why they should appear on the report) and state the reason why some (if any) of the tested parameters should not appear on the customer's report.

14.10.10.3   Testing a Fibre-Optic Cable Using an Optical Spectrum Analyser
1. As explained in Section 14.10.10.1.
2. List all the items that should appear on a test report to be sent out to a customer for the fibre-optic cable tested. Justify the list of items (why they should appear on the report), and state the reason why some (if any) of the tested parameters should not appear on the customer's report.

14.10.11   CONCLUSION

Summarise the important observations and findings obtained in this lab experiment.

14.10.12 SUGGESTIONS FOR FUTURE LAB WORK

List any suggestions for improvements using different experimental equipment, procedures and techniques for any future lab work. These suggestions should be theoretically justified and technically feasible.

14.11   LIST OF REFERENCES

List any references that were used in the report. Use one format in writing the references. Never mix reference formats in a report.

14.12   APPENDIX

List all of the materials and information that are too detailed to be included in the body of the report.

FURTHER READING

Camperi-Ginestet, C., Kim, Y. W., Wilkinson, S., Allen, M., and Jokerset, N. M., Micro-opto-mechanical devices and systems using epitaxial lift off, JPL, in *Proceedings of the Workshop on Microtechnologies and Application to Space Systems*, (SEE N94-29767 08-31), Category Solid-State Physics, Georgia Inst. of Tech., Atlanta, GA, pp. 305–316, June 1993.

Chee, J. K., and Liu, M., Polarization-dependent parametric and raman processes in a birefringent optical fiber, *IEEE J. Quantum Electron.*, 26, 541–549, 1990.

Dennis, D., *Fiber Optic Test and Measurement*, Prentice Hall, Englewood Cliffs, NJ, 1998.

Duton, H. J. R, *Understanding Optical Communications*, IBM, Prentice Hall, Englewood Cliffs, NJ, 1998.

Edmund Industrial Optics, *Optics and Optical Instruments Catalog*, Edmund Industrial Optics, Barrington, NJ, 2004.

Gerd, K., *Optical Fiber Communications*, 3rd ed., McGraw-Hill, UK, 2000.

Goff, D. R., *Fiber Optic Reference Guide: A Practical Guide to the Technology*, 2nd ed., Butterworth-Heinemann, Sudbury, MA, 1999.

Golovchenko, E., Mamyshw, P. V., Pilipetskii, A. N., and Dianiv, E. M., Mutual influence of the parametric effects and stimulated raman scattering in optical fibers, *IEEE J. Quantum Electron.*, 26, 1815–1820, 1990.

Hagness, S. C., Rafizadeh, D., Ho, S. T., and Taflone, A., DTD Microcavity simulations: design and experimental realization of waveguide-coupled single-mode ring and whispering-gallery-mode disk resonators, *IEEE J. Lightwave Technol.*, 15, 2157–2164, 1997.

Hibino, K., Error-compensating phase measuring algorithms in a fizeau Interferometer, *Opt. Rev.*, 6, 529–538, 1999.

Hibino, Y., Recent advances in high-density and large-scale AWG multi/demultiplexers with higher index-contrast silica-based PLCs, *IEEE J. Sel. Top. Quantum Electron.*, 8, 1090–1101, 2002.

Hine, T. J., Cook, M., and Rogers, G. T., An Illusion of relative motion dependent upon spatial frequency and orientation, *Vis. Res.*, 35, 3093–3102, 1995.

Ho, M. C., Vesaka, K., Marhic, M., Akasaka, Y., and Kazovsky, L. G., 200-nm-bandwidth fiber optical amplifier combining parametric and raman gain, *J. Lightwave Technol.*, 19, 977–981, 2001.

Horng, H. E., Chich, J. J., Chao, Y. H., Yang, S. Y., Hony, C. Y., and Yang, H. C., Designing Optical-fiber modulators by using, magnetic fluids, *Opt. Lett.*, 30 (5), 543–545, 2005.

IGI Consulting, Inc., *Optical Amplifiers: Technology and Systems*, Global Information, Boston, MA, 1999.

Jackel, J., GoodMan, M. S., Bron, J. E., Tomlinson, W. B., Chang, G. K., Igbal, M. Z., and Song, G. H., Acousto-optic tunable filters (AOTF' s) for multiwavelength optical cross-connects: Crosstalk considerations, *J. Lightwave Technol.*, 14, 1056–1066, 1996.

Javier, A., Laser and gaussian beam propagation and transformation, in *Encyclopedia of Optical Engineering*, Barry Johnson, R. et al., eds., Marcel Dekker, New York, 2002.

JDS Uniphase Corporation, Opto-Mechanical Switches, SN Series, JDS Uniphase Corporation, San Jose, CA, 2003.

Kao, C. K., *Optical Fiber Systems: Technology, Design and Applications*, McGraw-Hill, New York, 1982.

Kuhn, K., *Laser Engineering*, Prentice Hall, Englewood Cliffs, NJ, 1998.

Laude, J. P., *DWDM Fundamental, Components, and Applications*, Artech House, Boston, MA, 2002.

Li, X., Chem, J., Wu, G., and Ye, A., Digitally tunable optical filter based on DWDM thin film filters and semiconductor optical amplifiers, *Opt. Express*, 13, 1346–1350, 2005.

Parry-Hill, M. J., and Michael W., Davidson, *Fiber optics testing*, National High Magnetic Field Laboratory, Florida State University, Tallashassee, FLSeptember 2006.

Ralston, J. M., and Chang, R. K., Spontaneous-raman-scattering efficiency and stimulated scattering in silicon, *Phys. Rev.*, B2, 1858–1862, 1970.

Robillard, J. J., and Luna-Moreno, D., All-optical switching with fast response variable-index materials, *Opt. Eng.*, 42, 3575–3578, 2003.

Salah, B. E. A., and Teich, M. C., *Fundamentals of Photonics*, John Wiley and Sons, New York, 1991.

Sato, Y., and Aoyama, K., OTDR in optical transmission sdystems using Er-Doped fiber amplifiers containing optical circulators, *IEEE Photon. Technol. Lett.*, 3, 1001–1003, 1991.

Shen, Y. R., and Bloembergen, N., Theory of stimulated brillouin and Raman scattering, *Phys. Rev.*, A137, A1787–A1805, 1964.

Shen, L. P., Hvang, W.-P., Chum, G. K., and Jian, S. S., Design and optimization of photonic crystal fibers for broad-band dispersion compensation, *IEEE Photon. Technol. Lett.*, 15, 540–542, 2003.

Sugimoto, N., Shintaku, T., Tate, A., Tervi, H., Shimokozono, M., Ishii, E., and Inone, Y., Waveguide polarization-independent optical circulator, *IEEE Photon. Technol. Lett.*, 11, 355–357, 1999.

Yeh, C., *Handbook of fiber Optics: Theory and Applications*, Academic Press, San Diego, CA, 1990.

Yeh, C., *Applied Photonics*, Academic Press, San Diego, CA, 1994.

Zhang, L., and Yang, C. Polarization splitter based on photonic crystal fibers, *Opt. Express*, 11, 1015–1020, 2003.

15 Photonics Laboratory Safety and Safety Ethics

15.1 INTRODUCTION

Our lives are filled with hazards created by electrical power supply, lasers, chemicals and a diversity of equipment used in laboratories and classrooms. While studying science and engineering and performing experiments, students will learn to identify hazards and to protect themselves. Students will also learn to take care of their health and safety while working in laboratories. A safer and healthier learning and working environment should be created, so that students have the opportunity to live safely and more healthily.

The following list of safety reminders is a brief compilation of generally accepted practices and should be adopted or modified to suit the unique aspects of each working environment, school policy and local and/or set of Provincial and Federal codes. The intent of this chapter is to stimulate thinking about important safety considerations for students in laboratories.

15.2 ELECTRICAL SAFETY

The importance of electrical safety cannot be overstated. Electrical accidents can result in property damage, personal injury and sometimes death. Ensuring electrical safety in laboratories and classrooms is important for students and staff. Students can learn to have a healthy respect for electricity and to spot potential electrical hazards anywhere. Respecting electricity does not mean that one should fear it; rather, one should just use it properly and wear personal protective equipment.

15.2.1 Fuses/Circuit Breakers

The most common protection against property damage from circuit overloads (too much current) and overheating is the use of fuses and circuit breakers. All electrical circuits in laboratories are required to be protected by these means. When too much current flows in a circuit, the circuit becomes hot and could melt the wire insulation, emit caustic fumes and start a fire. An overload may also burn out and damage devices and instruments. Electronic equipment commonly has fuses to protect the components from overloads. A fuse is essentially a short strip of metal with a low melting point. When the current in a fused circuit exceeds the fuse rating, for example, 3, 5, 15 or 20 amps, the heat melts or vaporises the fuse strip. The fuse blows, and the circuit is opened. Figure 15.1 shows types of fuses used in most laboratory electric instruments.

Fuses and circuit breakers should be the correct current rating for the circuit. If the correct rating is unknown, a certified electrician can identify and label it. A fuse should always be replaced with another of the same rating. Determine the reason why a fuse blew or a circuit breaker tripped, before replacing the fuse or resetting the breaker. Figure 15.1 also shows a few types of fuses and a power supply. The fuse will need to be plugged in on the back of the power supply. Plug in the electrical cord; turn on the key switch on the front panel; and turn the power supply on, as shown in Figure 15.1. After finishing with the power supply, remember to turn off the key and unplug the fuse.

A common problem is that the insulation may become worn on, for example, an extension cord, device wire or instrument cord. If bare wires touch each other, or if a high-voltage or hot wire touches the ground, this is called a short circuit, because the path of the circuit is effectively shortened. A low-resistance path to ground is created, causing a large current, which blows the protecting fuse.

Circuit breakers are more commonly used today instead of fuses in large equipment and houses, as shown in Figure 15.2. If the current in a circuit exceeds a certain value, the breaker is activated, and a magnetic relay (switch) breaks or opens the circuit. The circuit breaker switch can be reset or closed manually.

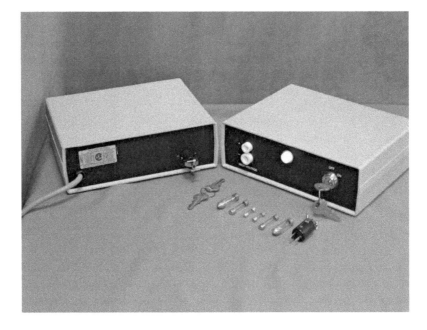

FIGURE 15.1   Types of fuses.

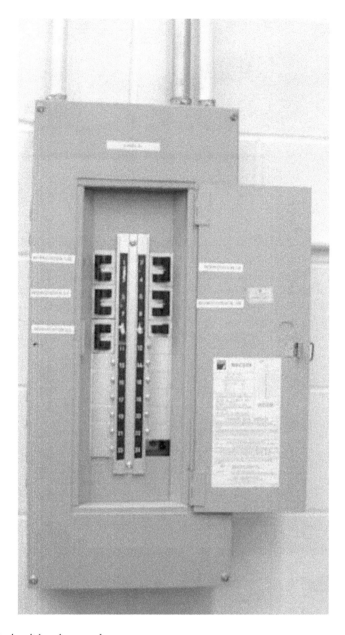

FIGURE 15.2 A circuit breaker panel.

In either case, whether a circuit is opened when a fuse blows or when a circuit breaker trips, steps should be taken to remedy the cause. Remember, fuses and circuit breakers are safety devices. When fuses blow and open a circuit, they are indicating that the circuit is overloaded or shorted. Or they may be indicating the presence of another problem. In any case, a certified technician must investigate the source of the problem.

15.2.2 SWITCHES ON/OFF

Figure 15.3 shows samples of ON/OFF switches, which are used in computers, lighting systems and instruments.

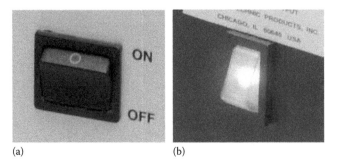

ON

OFF

(a)   (b)

FIGURE 15.3   Examples of switches. (a) Standard switch is ON and (b) Light switch is ON.

15.2.3 PLUGS

Switches, fuses and circuit breakers are always placed on the hot (high-voltage) side of the line, to interrupt power flow to the circuit element. Fuses and circuit breakers may not, however, always protect from electrical shock. To prevent shock, a grounding wire is used. The circuit is then completed (shorted) to ground, and the fuse in the circuit is blown. This is why many electrical tools and appliances have three-prong plugs, as shown in Figure 15.4. In the wall receptacle, this connection runs to ground.

When trying to plug-in a two-prong plug that will not fit, do not use force. Instead, turn the plug over and try again. Figure 15.5 shows a two-prong plug. One of the prongs is bigger than the

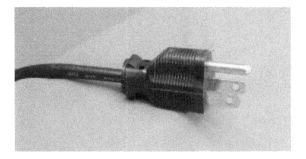

FIGURE 15.4   A three-prong plug.

FIGURE 15.5   Two-prong plugs.

other, making the plug polarised. Polarising in the electrical sense refers to a method of identification by which proper connections can be made. The original purpose of these types of plugs was to act as a safety feature. The small slit in the receptacle is the hot side, and the large slit is the neutral or ground side, if properly connected. The housing of an appliance could then be connected to the ground side all the time via a three-prong plug. A receptacle or appliance not wired (polarised) properly can be dangerous. The polarisation is ensured with a dedicated third grounding wire as in a three-prong plug system, which is the accepted safety system. The original two-prong polarised plug system remains as a general backup safety system, provided it is wired properly.

Ensure the plug type fits the receptacle. Never remove the ground pin (the third prong) to make a three-prong plug fit into a two-conductor outlet; doing so could lead to an electrical shock. Never force a plug into an outlet if it does not fit. Plugs should fit securely into outlets. Avoid overloading electrical outlets with too many devices.

15.2.4   WALL OUTLETS

Figure 15.6 shows a wall outlet, which is used to connect computer and extension cords. Avoid using wall outlets with loose fitting plugs. They can overheat and lead to fire. Ask a certified technician to replace any missing or broken wall plates.

FIGURE 15.6   A wall outlet.

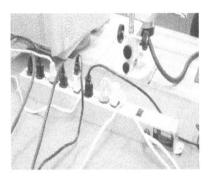 

FIGURE 15.7   An extension power bar.

15.2.5 CORDS

Ensure the cords are in good condition. Check cords for cut, broken or cracked insulation. Protect flexible cords and cables from physical damage. Ensure they are not placed in traffic areas. Cords should never be nailed or stapled to the wall, table, baseboard or to another object. Do not place cords under a device or computer; do not rest them under any object. Cords can create tripping hazards and may be damaged if walked upon. Allow slack in flexible cords to prevent tension on electrical terminals.

Check that extension power bars are not overloaded, as demonstrated in Figure 15.7. Figure 15.7a shows an overloaded extension power bar, whereas Figure 15.7b shows a bar not overloaded. Additionally, extension power bars should only be used on a temporary basis; they are not intended for use as permanent wiring. Ensure that the extension power bars have safety closures.

1. Extension power bar is overloaded.
2. Extension power bar is not overloaded.

15.2.6 GROUND FAULT CIRCUIT INTERRUPTERS

Ground fault circuit interrupters (GFCIs) can help prevent electrocution. They should be used in any area where water and electricity may come into contact, especially near a sink or basin. Water and electricity do not mix; they create an electrical shock. When a GFCI senses current leakage in an electrical circuit, it assumes that a ground fault has occurred. It then interrupts power quickly enough to help prevent serious injury due to electrical shock. GFCIs should be regularly tested according to the manufacturer's instructions to ensure they are working properly. Some benches are connected to true ground to be electrostatic discharge (ESD) compliant. This compliance is very important for devices and equipment that are very sensitive to ESD. Figure 15.8 shows ESD warning symbols and signs.

Figure 15.9 shows an ESD wrist strap and table mat used in handling an ESD-sensitive device. The straps and mats should be connected to the true ground before handling a sensitive device. The strength of the charge on a human body is enough to destroy an ESD-sensitive device. Each person should discharge his or her electrostatic charge before entering an environment sensitive to ESD. The discharge devices are usually located at the entrance of sensitive areas. An ESD heel strap is also available to wear when handling devices and walking in an environment sensitive to ESD.

Figure 15.10 shows ESD bags used to package devices sensitive to ESD. Available in various sizes, the bags have printed labels.

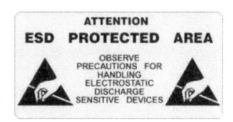

FIGURE 15.8 Electrostatic discharge warning symbols and signs.

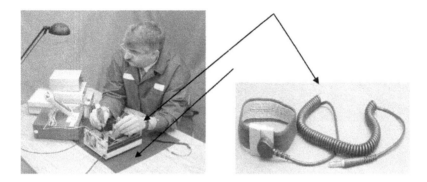

FIGURE 15.9 Working with ESD wrist strap and table mat.

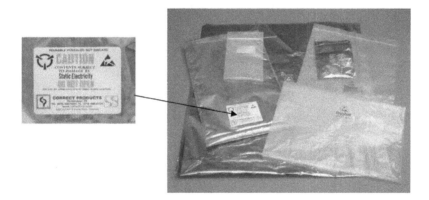

FIGURE 15.10 ESD bags.

15.3 LIGHT SOURCES

The wattage rating should be checked for all bulbs in light fixtures, table lamps and other light sources, to make sure they are the correct rating for the fixture. Bulbs must be replaced with another of the same wattage rating; bulbs' wattage rating must not be higher than recommended. If the correct wattage is unknown, check with the manufacturer of the fixture. Ensure that the bulbs are screwed in securely; loose bulbs may overheat. Different gas light sources (e.g. hydrogen, mercury, neon), as shown in Figure 15.11, are used in laboratories for light-loss measurements and for spectrometers and optical applications. These lamps operate at much higher temperatures than those of standard incandescent light bulbs. Never place a lamp where it could come in contact with any

FIGURE 15.11   A mercury light source.

combustible materials or the skin. Be sure to turn the lamp off before leaving the laboratory for an extended period of time. Note that laser light sources have special provisions and, therefore, special precautions must be taken to operate them.

15.4   DEVICES AND EQUIPMENT

If a device or piece of equipment repeatedly blows a fuse or trips a circuit breaker, or if it has given you a shock, report the incident immediately to your supervisor/instructor. Unplug the device and remove it to have it repaired or replaced.

15.5   AUDIO–VISUAL AND COMPUTER PERIPHERALS

Audio–visual and computer equipment must be checked and kept in good working condition. Ask the technician to load the printer with paper and replace the toner. Report the faulty equipment to the technician for repair.

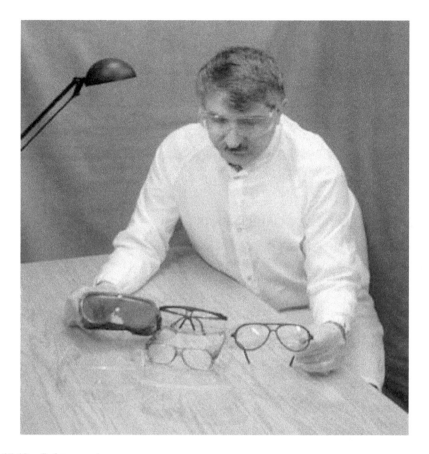

FIGURE 15.12   Safety goggles.

15.6   HANDLING OF FIBRE-OPTIC CABLES

Fibre-optic cables are made from a glass strand, covered with a polymer jacket. They are very thin and rigid, with sharp ends. Handle fibre-optic cable with care during inspection, cleaning and preparation of the fibre-optic cable ends. Fibre-optic cables should be cleaned using the cleanser recommended by the manufacturer. Follow the recommended procedure for each fibre-optic cable type during cleaning, handling, assembling, packaging and storage. When cleaving a fibre-optic cable, the loose scrap material is hard to see and can be very dangerous. Dispose of loose scrap immediately in a properly designated container. Do not touch the end of a stripped fibre-optic cable or a loose (scrap) piece of fibre. Fibre easily penetrates skin, and a fibre shard could break off. Do not rub your eyes when handling fibre-optic cables; this would be extremely painful and requires immediate medical attention. Follow all safety procedures and regulations, and always wear the required personal protective safety equipment. Use safety goggles with side shields and wear protective rubber gloves or finger cots, when handling fibre-optic cables. Figure 15.12 shows different types of safety goggles. Always treat fibre-optic cables as a potential hazard. Never look directly at the fibre-optic cable ends during fibre-optic assembly and testing.

15.7   EPOXY ADHESIVES AND SEALANTS

Epoxy adhesives and sealants are essential components in the manufacturing of optical devices. There are different types and colours, depending on the application. Epoxy adhesives come in several forms. One-part, two-part and ultraviolet (UV) systems are the most common. A graded index

(GRIN) lens can be glued to a beamsplitter with an epoxy. Sealant materials are used in the packaging of optical devices.

When using adhesives and sealant materials, be aware of their specifications. Specifications, applications and handling procedures of these materials are found on the material safety data sheets (MSDS), which are available from the manufacturer or distributor. They may also be downloaded from a number of websites. The adhesives and sealants are also very hazardous during storage, handling and application. Prolonged or repeated exposure may cause eye or skin irritation. If contact does occur, wash the contact area immediately and seek medical help. Use safety goggles with side shields and wear protective rubber gloves or finger cots when handling adhesives and sealants. Follow all safety procedures and regulations, read the MSDS carefully, and wear the required personal protective safety equipment.

15.8   CLEANING OPTICAL COMPONENTS

Optical surfaces have to be clean and free of dust and other particles, which can range in size from tenths to hundreds of microns in diameter. Their comparative size means that they can cover a part of the optical surfaces, and thus degrade the reflection or transmission quality of the data transmission in telecommunication systems. There are many standard procedures for cleaning optical surfaces. Before starting any cleaning procedure, locate the following standard equipment:

1. Cleaning material (Denatured Ethanol)
2. Cotton swabs
3. Tissue
4. Safety goggles
5. Finger cots or rubber gloves
6. Compressed air
7. Disposal container
8. Microscope with a magnification range of about 50X
9. Infrared sensor card
10. Additional cleaning equipment:
    Ultrasonic bath
    Warm water and liquid soap
    Pre-moistened cleaning wipes
    Polymer film

Some optical components (e.g. lenses, mirrors, prisms and beamsplitters) have special coatings, such as anti-reflection coatings, that are sensitive to solvents, grease, liquid and mechanical abrasion. Take extra care and choose appropriate cleaning liquid and swabs when cleaning optical components with these coatings. The following is the preferred cleaning procedure for optical components:

1. Wear rubber gloves and safety goggles.
2. Hold a lens or a mirror by the rim and a prism by the corners. Clean the optical component using a new dry or dampened swab with the recommended solvent. Rub the surfaces of the lens, using small circular movements, or one-directional movement on plane prism surfaces.
3. Blow away any remaining lint with compressed air. This step depends on the optical component size and surface conditions. Check the air quality from the compressor before using to clean optical components.

Some optical devices consisting of several optical components may not always be sealed completely. Therefore, use the recommended procedure to clean optical component surfaces without leaving any residue that could reduce the optical performance.

When cleaning any optical interface, disable all sources of power, such as the end of the ferrule on a fibre connector. Under no circumstances should you look into the end of an optical device in operation. Light from a laser device may not be visible, but it can seriously damage the human eye.

15.9 OPTIC/OPTICAL FIBRE DEVICES AND SYSTEMS

There has been a significant increase in the use of optic/optical-fibre devices and systems. As optic/optical fibre devices become more common, it is important to understand the associated hazards. Optical devices typically use a laser as a light source. Not all lasers are created equal. They are classified based on their output wavelength and power. As they operate over a wide range of wavelengths and power outputs, the hazards arising from their use vary substantially.

Lasers are classified into four classes. Laser sources conformant to Class 1 and Class 2 do not cause serious damage, but the use of eye protection should be taken into consideration. Class 3 and Class 4 lasers are powerful and can cause serious damage. Therefore, it is important to determine the class type of any optical equipment before working with it, assess the associated hazard and comply with the safety requirements.

It is always a good practice to handle optical devices and measuring instruments with care. Normally, these devices and instruments are very expensive and sensitive, and they may present a potential hazard if not used properly. Follow the recommended procedures for each device or instrument to ensure proper handling during assembly, testing, packaging and storage.

15.10 CLEANING CHEMICALS

Before the application of an epoxy or sealant, all surfaces should be treated using the recommended cleaning material. When using cleaning materials, be aware of appropriate precautions. Read all the information regarding cleaning materials in the MSDS. All types of cleaning materials are potentially hazardous; they may be flammable (even at low temperatures) and may pose other exposure risks. Use safety goggles with side shields and wear appropriate protective rubber gloves or finger cots. Follow all safety procedures and regulations. Use a ventilation hood when working with cleaning chemicals and epoxy adhesives, sealants or any material producing fumes.

15.11 WARNING LABELS

There are various types of warning labels used in buildings, transportation, services and industry to warn users about the level of danger ahead. Warning labels sometimes are called safety signs or safety messages. Safety signs clearly communicate by choosing the proper design and wording to suit safety needs. Standard signs, such as traffic warning signs and construction work labels, are available for general warnings.

Safety signs are divided into three general categories: danger, warning and caution. They are also available in different sizes and colours, and with different graphics. Sometimes, a standard header can be used to create a new sign to suit a specific need. It is very important to use warning labels in laboratories to alert students to any source of danger. These dangers may come from devices, instruments, chemicals, lasers, sounds, vibrations and biological hazards. Students should be introduced, in advance, to each source of danger in laboratories and be shown the required personal protective safety equipment. Everybody must remember to consider safety first.

15.12 LASER SAFETY

A laser beam is a parallel, narrow, coherent and powerful light source. It is increasingly powerful when concentrated by a lens. It is a hazard to human eyes and skin even at very low power.

All lasers are classified based on their potential power. These classifications are from the American National Standards Institute (ANSI Standard Z136.1-1993) entitled American National Standard for Safe Use of Lasers, and Z136.3 (1996), American National Standard for Safe Use of Lasers in Health Care Facilities, the Canada Labor Code and Occupational and Safety and Health Legislation (L-2-SOR/86-304).

Needing to be adhered to when using laser devices, these standards and codes are universally recognised as definitive documents for establishing an institution, such as a school, factory or hospital. Their basic classification system has been adopted by every major national and international standards board, including the Center for Devices and Radiological Health (CDRH) in the U.S. Federal Laser Product Performance Standard, which governs the manufacture of lasers in the United States.

Lasers are typed into four classes, with some subclasses: Class 1, Class 2, Class 2a, Class 3a, Class 3b and Class 4. Higher numbers reflect an increased potential to harm users. Figure 15.13 shows laser warning labels, which are required to identify hazard from laser light sources.

The following criteria are used to classify the hazard level of lasers:

1. *Wavelength*: If the laser is designed to emit multiple wavelengths, the classification is based on the most hazardous wavelength.
2. *Continuous wave*: For continuous wave (CW) or repetitively pulsed lasers, the average power output (Watts) and limiting exposure time inherent in the design are considered.
3. *Pulse*: For pulsed lasers, the total energy per pulse (Joule), pulse duration, pulse repetition frequency and emergent-beam radiant exposure are considered.

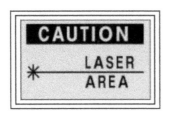

FIGURE 15.13 Laser warning labels.

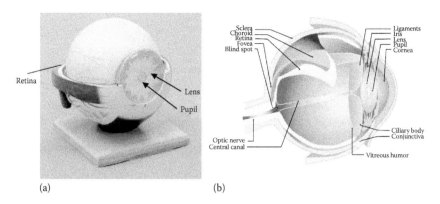

(a)   (b)

FIGURE 15.14   The human eye: (a) eye ball and (b) eye cross-section.

Details of the laser classifications are listed below:

Class 1 lasers are laser devices with very low output power (between 0.04 and 0.40 mW), and they operate in the lower part of the visible range (450 nm < λ < 500 nm). These lasers are generally considered to be safe when viewed indirectly. Some examples of Class 1 laser devices include CD players, scanners, laser pointers and small measurement equipment. Figure 15.14a shows the human eye, whereas Figure 15.14b shows an eye cross-section. Laser light in the visible range entering the human eye is focused on the retina and causes damage. The most likely effect of intercepting a laser beam with the eye is a thermal burn, which destroys the retinal tissue. Never view any Class 1 laser beam directly.

Class 2 lasers are devices with low output power (<1 mW of visible CW), and operate in the visible range (400 nm < λ < 700 nm). This class of laser could cause eye damage, if the beam is directly viewed for a very short period of time (more than 0.25 s). Some examples of Class 2 lasers include classroom demonstration laser sources and laser-source devices for testing and telecommunications. Never view any Class 2 laser beam directly.

Class 2a lasers are low-output power devices, which are considered to be visible-light lasers. This class of laser causes injury only when viewed directly for more than 0.25 second. This class must be designed so that intentional viewing of the laser beam is not anticipated. A supermarket bar-code scanner is a typical example of a Class 2a laser device. Never view any Class 2a laser beam directly.

Class 3 lasers are divided into two subgroups (Class 3a and Class 3b lasers).

Class 3a lasers are intermediate power devices; they are allowed to exceed the output power limit of Class 2 lasers by no more than a factor of five, or have visible light power less than 5 mW. They are considered CW lasers. Often they will have an expanded beam diameter so that no more than 1 mW can enter a fully dilated pupil, which is 7 mm in diameter. Some examples of Class 3a laser devices are laser scanners, laser printers and laser-source devices for testing and telecommunications. Direct viewing of a laser in this class could be hazardous to the eyes. Never view any Class 3a laser beam directly. Although the beam wavelength may not be visible to the human eye, it can cause damage to the eye and skin. Laser safety goggles for appropriate wavelength are required when working with this class.

Class 3b lasers are intermediate power devices; they output between 5 and 500 mW of CW, or else pulsed 10 J/cm² power. They are considered to be CW lasers. Scattered energy (diffuse reflection) is not considered hazardous in most situations, unless the laser source is operating near its upper power limit and the diffuse target is viewed at close range. Some examples of Class 3b lasers are laser-source devices for testing. Never view any Class 3b laser beam directly or indirectly (by viewing any reflection from the surrounding surfaces). The laser beam wavelength

TABLE 15.1
Institutional Programme Requirements

Class	Power	Class Control Measures	Medical Surveillance	Safety and Training Programme
1	No more than 0.04–0.40 mW	Not applicable	Not applicable	Not required
2	Less than 1 mW of visible, continuous wave light	Applicable	Not applicable	Recommended
2a	Less than 1 mW of visible, continuous wave light	Applicable	Not applicable	Recommended
3a	From 1 to 5 mW of continuous wave light	Applicable	Not applicable	Required
3b	From 5 to 500 mW of continuous wave light	Applicable	Applicable	Required
4	More than 500 mW of continuous wave light	Applicable	Applicable	Required

may not be visible to the human eye, but it causes damage to the eye and skin immediately, with no time to react. Laser safety goggles for the appropriate wavelength are required when working with this class of laser.

Class 4 lasers are high-power devices; they output more than 500 mW of CW, or else pulsed 10 J/cm^2 power. They are considered to be very high-power lasers. Some applications of Class 4 laser devices include the following: surgery, drilling, cutting, welding and micromachining. For the use of Class 4 lasers, all types of reflections (whether direct, specular or diffuse) are extremely hazardous to the eyes and skin. Class 4 laser devices can also be a fire hazard. Much greater control is required to ensure the safe operation of this type of laser device. Never view any Class 4 laser beam directly or indirectly (any reflection by surrounding surfaces). Be cautious of this type of laser. The laser beam wavelength may not be visible to the human eye, but it can immediately cause damage to the eye and skin, with no time to react. Laser safety goggles for the appropriate wavelength are required when working with this class of laser.

Always follow all safety procedures and regulations, and wear the required the appropriate personal protective safety equipment when using lasers. Never look directly or indirectly at a laser beam. Each institute should create appropriate safety procedure to guide students and staff towards the creation of a safe working environment. Each laser laboratory has to be controlled by a designated instructor/professor certified in laser safety. All laser safety requirements should be implemented in a laser laboratory. It is recommended to have an introduction course and workshop in laser safety for each laser classification.

Knowing the classification of a particular device and comparing the information in Table 15.1 will usually eliminate the need to measure laser radiation or perform complex analyses of hazard potential.

15.13   LASER SAFETY TIPS

1. Do not enter the Nominal Hazard Zone (NHZ). This zone is established according to the procedures described in ANSI Z136.1-1993. Enter this area accompanied by a designated instructor/professor certified in laser safety. Do not put any body part or clothing in the way of a laser beam.

2. Notice and comply with the signs and labels (shown in Figure 15.13) posted on laboratory door, devices and equipment.

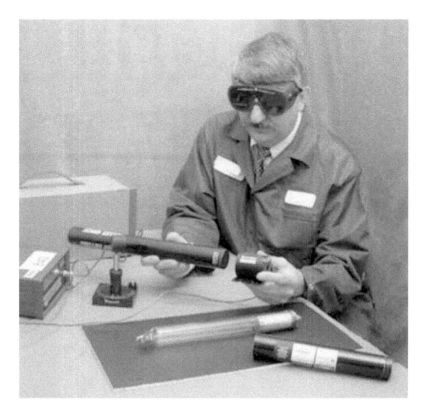

FIGURE 15.15 Wear laser safety goggles.

3. Wear the recommended eyewear and other protective equipment. Use laser safety goggles when you are in a laser laboratory, or in the vicinity of one, as shown in Figure 15.15.
4. Comply with the laser safety controls in the facility.
5. Attend laser safety training and workshops.
6. Update laser safety training and workshops, as needed.
7. While assembling and operating laser devices, it is important to remember that laser beams can cause severe eye damage. Keep your head well above the horizontal plane of the laser beams at all times. Use white index cards to locate beamspots along the various optical paths.
8. When moving optical components, mirrors or metal tools through the laser beams, the beam may reflect laser light momentarily at your lab partner or you. If there is a possibility of an accidental reflection during a particular step in an operation, then temporarily block or attenuate the laser beam until all optical components are in their proper place. It is a good policy to be aware of any stray laser beam reflections, and to warn anybody of any danger. If you are unsure of how to proceed safely with a given step in the operation of the laser device, ask the professor/instructor for assistance.

15.14 INDOOR AIR QUALITY

Concerns with indoor air quality (IAQ) have increased since energy conservation measures were instituted in office buildings during the 1970s. These measures minimised the infiltration of outside air and contributed to the build-up of indoor air contaminants. IAQ generally refers to the quality of the air in a work environment. Other terms related to IAQ include indoor environmental quality (IEQ) and sick-building syndrome. Complaints about IAQ range from simple complaints, such as the air smelling odd, to more complex situations, where the air quality causes illness and lost work time. It may not be easy to identify a single reason for IAQ complaints because of the number and variety of possible sources, causes and varying individual sensitivities.

IAQ problems can be caused by ventilation system deficiencies, overcrowding, the presence of tobacco smoke, microbiological contamination, outside air pollutants and off-gassing from materials in the building and mechanical equipment. Related problems may also include comfort problems caused by improper temperature and relative humidity conditions, poor lighting and unacceptable noise levels, as well as adverse ergonomic conditions, and study-related psycho-social stressors. Typical symptoms may include headaches, unusual fatigue, itching or burning eyes, skin irritation, nasal congestion, dry or irritated throats and nausea.

Ventilation is one of the most common engineering controls used to control emissions, exposures and chemical hazards in the workplace. Other workplace environmental factors, including temperature, humidity and odours, are also controlled with non-industrial ventilation systems commonly known as heating, ventilating and air-conditioning (HVAC) systems.

Management should have created guidelines for the following:

1. IAQ
2. Building air quality (BAQ)
3. Investigations, recommendations on sampling instrumentation and methods
4. Guidelines for management to prevent or alleviate
5. IAQ problems and take acute health effects of major indoor air contaminants.

Management should have an overview of

1. Sources of indoor air pollution, and health problems
2. Ventilation, control, ventilation standards and building codes and ventilation system problems
3. Solutions for air cleaners and resolving problems.

15.15   OTHER CONSIDERATIONS

These considerations apply to all students, staff and management.

1. Laboratory injuries and illnesses are usually preventable by simply following safety precautions in school throughout the year.
2. Never overload circuits, power bars or connectors.
3. Lead innovative and cooperative efforts to improve laboratory safety and health and the quality of student life.
4. Do not use or work with any device or equipment until it has been checked by qualified and authorised personnel in charge of the laboratory operation, as shown in Figure 15.16.
5. Everyone must wear personal protective equipment (e.g. safety goggles, protective gloves, ground connection, insulated tools) when working with electrical or laser equipment and chemicals.
6. Immediately report any damaged electrical or laser devices and equipment to the professor/instructor for immediate corrective action.
7. Staff should promote safety awareness amongst students.
8. Staff should teach safe work practices, at the beginning of each new laboratory session.
9. The NHZ should be established for each laser system.
10. Management should create and maintain a safe and healthy work, and study environment.

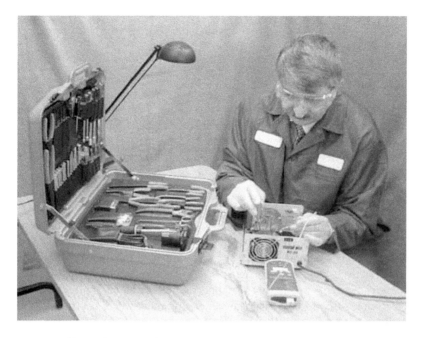

FIGURE 15.16 A certified technical staff member.

11. Management, staff and students should understand the human and economic impact of poor safety and health in laboratories and classrooms.
12. Management should create a safety checklist, and maintenance and auditing programmes for each laboratory.
13. Eye protection should be worn at all times.
14. Eating, drinking and smoking are not allowed in laboratories.
15. Unauthorised personnel should not be present in the laboratory or area, whether lasers are operating or not.
16. Laboratory coats must be worn when handling and cleaning, corrosive, toxic or flammable materials. Gloves should be worn when necessary, especially when handling corrosive and highly toxic materials.
17. Never work alone in a laboratory or workshop.
18. If a colleague is doing something dangerous, point out the action immediately and inform the supervisor.
19. Know where safety equipment (e.g. eyewash, shower, extinguishers, emergency exits, first aid kit) is located and how to use it.
20. Know where the MSDS and Workplace Hazardous Materials Information System (WHMIS) are located and how to use them.
21. Know where the emergency phones and alarms are located and how to use them.
22. Know how to clean up chemical spills using the appropriate agents.
23. Pre-planned experiments and a properly organised work area can eliminate a lot of potential safety problems. Clean-up and decontamination must be routine parts of the experimental procedure for all students.
24. Wash your hands after handling chemicals and before leaving the laboratory.
25. Ensure the laboratory safety programme complements science.

15.16 SAFETY ETHICS IN SCHOOLS

Safety ethics can be enforced by

15.16.1 LAWS AND REGULATIONS

Laws and regulations mainly inform individuals to follow them to achieve the job requirements with high standards and safety. Ethics instruct individuals and organisations to do the right activities lined up with the laws, regulations and standards. Students and management must follow each safety instruction that is created for a particular classroom, laboratory and workshop in a school.

15.16.2 MANAGEMENTS AND ETHICS

There are many aspects of traditional approaches to safety management. It could be viewed as less than ethical, but safety programs based on *doing the right thing* are a better and more sustaining in order to manage safety programs. Management is ultimately responsible for building an ethical system that can effectively examine and control school safety.

Perhaps the greatest economic reason to support an ethics-based approach to safety management within a school is that success generates an environment where continuing students present and eliminate risk can be achieved all year around. To enable management to use an ethical safety management systems approach, safety professionals need to promote a more ethical approach. Management should include all level of students and employees in any approach.

An established school safety plan will produce ethical outcomes such as integrity, open communication between management and students. Management credibility is to keep students in each study, practice and play places safely. The school safety plan will drive an organisation towards credibility, profitability and sustainability. Management must lead cultural or professional practices that value safe behaviour for the benefit of ethics and for the purpose of increasing students' morale and productivity in their study time.

15.16.3 OCCUPATIONAL SAFETY AND HEALTH ADMINISTRATION

Proventil and Federal safety procedures and legislations help to minimise risk and control hazards. Any violations may bring serious consequences, such as fines and legal consequences, and even loss of a business licence. Ethical principles demand that managements, employee, students and student associations do more to protect their students, employees and buildings and a create culture of constructive daily practices. Managers are ethically obliged to use their authority to create a healthily working and study environment. However, proper attention to school safety can result in improved students' and employees' morale.

Students and employees expect management to act responsibly and to put students' and employees' health and safety first by establishing and practicing the safety workplace plan. Students and employees health and safety should have priority over business profits. In order to create a safe working environment, students, employers and managements can practice the following principles:

1. Set up a safety motivation programme to inspire students and employees to comply with classroom, laboratory and playground safety guidelines, roles and regulations.
2. Adopt a zero-tolerance policy on school violence and safety violations.
3. Ongoing safety training for managers, employees and students.
4. Update safety guidelines, roles and regulations regularly.
5. Have easy documentation access to all safety guidelines, roles and regulations.
6. Ensure employees that work hand-in-hand with management to achieve safety goals.

These practices, in the long run, will save time, money and eliminate school injuries. For optimal effectiveness of school safety, managers should always take necessary actions to check on employees and students operating in the school. These actions should be followed up with an analysis for the improvement of future plan modification.

A school safety has healthier students and employees, reduces costs and increases performance. This leads to higher quality and students' satisfaction. Employers must go above and beyond of what is required and create a culture of care and respect for their valuable assets: students and employees in school.

15.17 SOCIETY, FAMILY AND PERSON SAFETY ETHICS

A human society is defined as group of people living in a particular country or region and having shared customs, laws, cultures, ethics and organisations. The society has responsibility towards their members. Many societies have a tradition of ethics, which transfer from the old to the new generation. These ethics are built on traditional culture and religious fundamentals. The new generation should respect these values and try to maintain it. Each person in the society should practice good values in her/his house, school and work. There are many societies around the world that represent a group of students and professionals. For example, student associations and professional student societies uphold and advance the integrity, honour and dignity of the members and their profession by using their knowledge and skills for the enhancement of human safety and environment.

FURTHER READING

Agrawal, G. P. and Dutta, N. K., *Semiconductor Lasers*, 2nd ed., Van Nostrand, New York, 1993.

Black, E., *An Introduction to Pound-Drever-Hall Laser Frequency Stabilization LIGO*, California Institute of Technology and Massachusetts Institute of Technology, 3rd ed., McGraw-Hill Pub, New York, 2000.

Canadian Health and Safety Legislation, Ecolog Canadian Health and Safety Legislation, Federal, Provincial, and Territorial Acts, Regulations, Guidelines, Codes, Objectives, Workers' Compensation, and WHMIS Legislation, 2000.

Charschan, S., *Lasers in Industry*, Van Nostrand, New York, 1972.

Cornsweet, T. N., *Visual Perception*, Academic Press, New York, 1970.

Davis, C. C., *Lasers and Electro-Optics, Fundamental and Engineering*, Cambridge University Press, New York, 1996.

Duarte, F. J., *Tunable Lasers Handbook*, Elsevier, Amsterdam, 1999.

Duarte, F. J. and Piper, J. A., Narrow-linewidth, high prf copper laser-pumped dye laser oscillators, *Appl. Opt.*, 23, 1391–1394, 1984.

Hood, D. C. and Finkelstein, M. A., Sensitivity to light handbook of perception and human performance, In *Sensory Processes and Perception*, Boff, K. R., Kaufman, L., and Thomas, J. P., Eds., Vol. 1, Wiley, Toronto, Canada, 1986.

Kuhn, K., *Laser Engineering*, Prentice Hall, Englewood Cliffs, NJ, 1998.

McComb, G., *The laser Cookbook—88 Practical Projects*, McGraw-Hill, New York, 1988.

Nanni, C. A. and Alster, T. S., Laser-assisted hair removal: Side effects of Q-switched Nd:YAG, long-pulsed ruby, and alexandrite lasers, *J. Am. Acad. Dermatol.*, 2 (1), 165–171, 1999.

Nichols, D. R., *Physics for Technology with Applications in Industrial Control Electronics*, Prentice Hall, Englewood Cliffs, NJ, 2002.

Salah, B. E. A. and Teich, M. C., *Fundamentals of Photonics*, Wiley, New York, 1991.

SETON, Signs, labels, tags, and workplace safety, Catalog, 2006.

Tao, W. K. and Janis, R. R., *Mechanical and Electrical Systems in Buildings*, Prentice Hall, Englewood Cliffs, NJ, 2001.

Thompson, G. H. B., *Physics of Semiconductor Laser Device*, Wiley, Chichester, 1980.

Topping, A., Linge, C., Gault, D., Grobbelaar, A., and Sanders, R., A review of the ruby laser with reference to hair depilation, *Ann. Plast. Surg.*, 44, 668–674, 2000.

Venkat, V., *Introduction to Laser Safety*, Photonics Research, Ontario, Canada, 2002.

Yeh, C., *Handbook of Fiber Optics: Theory and Applications*, Academic Press, San Diego, CA, 1990.

Appendix A: Details of the Devices, Components, Tools and Parts

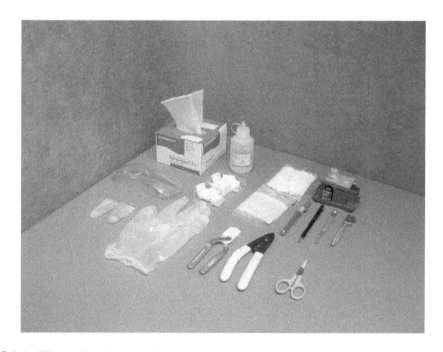

FIGURE A.1 Fibre-cable end preparation kit.

FIGURE A.2   250 μm diameter/500 metres long fibre-optic cable.

FIGURE A.3   2 × 2 ft breadboard.

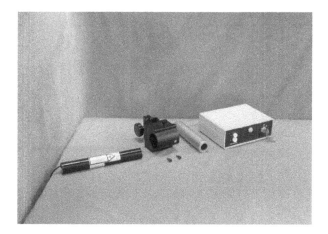

FIGURE A.4 He–Ne laser source, laser power supply and laser mount assembly.

FIGURE A.5 Fibre-optic cable holder/positioner assembly.

FIGURE A.6   Lens/fibre-cable holder/positioner assembly.

FIGURE A.7   Lens and lens holder/positioner assembly.

FIGURE A.8   Fibre-cable holders.

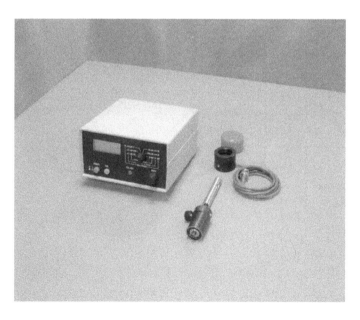

FIGURE A.9   Laser sensors.

FIGURE A.10 Laser power meter with matching laser power detector.

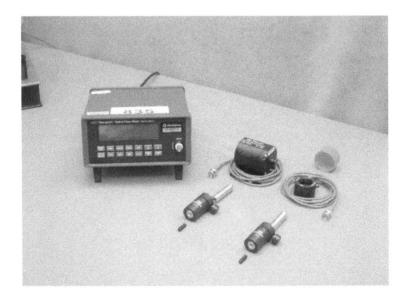

FIGURE A.11 Laser power meter and laser power detectors.

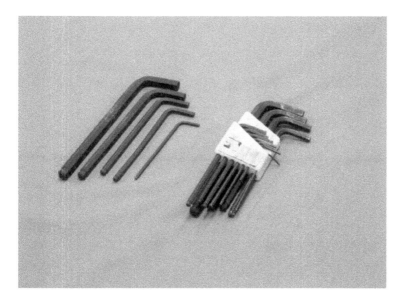

FIGURE A.12   Allen key set.

FIGURE A.13   Black/white card and cardholder.

FIGURE A.14   GRIN lens/fibre-cable holder/positioners assembly.

FIGURE A.15   Rotation stage.

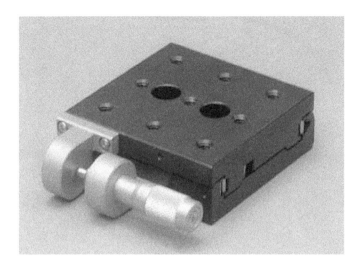

FIGURE A.16   Translation stage.

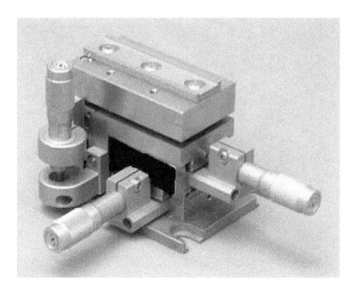

FIGURE A.17   XYZ translation stage.

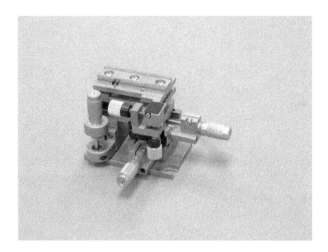

FIGURE A.18 Multi-translation stage.

FIGURE A.19 Lab jack.

FIGURE A.20   He–Ne laser clamp.

FIGURE A.21   Connector holder/positioner.

FIGURE A.22   Cube prism holder/positioner assembly.

FIGURE A.23   Convex lens and lens holder/positioner assembly.

FIGURE A.24 Prism and prism holder/positioner assembly.

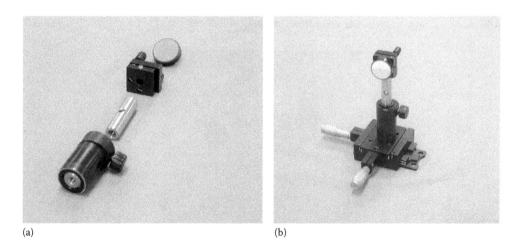

(a)   (b)

FIGURE A.25 (a) Mirror and mirror holder and (b) positioner assembly.

FIGURE A.26   Slide holder/positioner assembly.

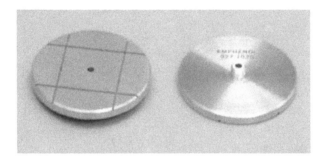

FIGURE A.27   Connector polishing disks (front and back sides).

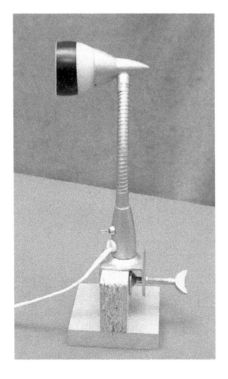

FIGURE A.28   Spot light source.

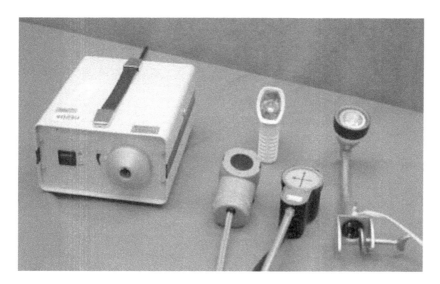

FIGURE A.29 Light sources.

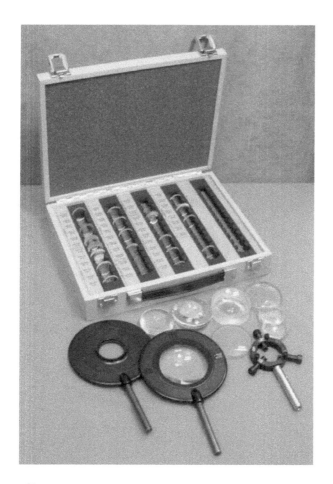

FIGURE A.30 Types of lenses.

FIGURE A.31   Types of prisms.

FIGURE A.32   Types of mirrors.

FIGURE A.33   Optical spectrum analyser.

Appendix B: International System of Units (SI)

International System of Units (SI) (It is also called metric system). The modern form of the metric system, which has been developed by international standards. The SI is constructed from seven base units for independent physical quantities. The following tables showing these values are included below and are currently used worldwide (Tables B.1 through B.6).

TABLE B.1
The Common Metric SI Prefixes

Multiplication Factor	Prefix	Prefix Name	Symbol
1 000 000 000 000 000 000 000 000	10^{24}	Yotta	Y
1 000 000 000 000 000 000 000	10^{21}	Zetta	Z
1 000 000 000 000 000 000	10^{18}	Exa	E
1 000 000 000 000 000	10^{15}	Peta	P
1 000 000 000 000	10^{12}	Tera	T
1 000 000 000	10^{9}	Giga	G
1 000 000	10^{6}	Mega	M
1 000	10^{3}	Kilo	k
100	10^{2}	Hecto	h
10	10^{1}	Deka	da
0.1	10^{-1}	Deci	d
0.01	10^{-2}	Centi	c
0.001	10^{-3}	Milli	m
0.000 001	10^{-6}	Micro	μ
0.000 000 001	10^{-9}	Nano	n
0.000 000 000 001	10^{-12}	Pico	p
0.000 000 000 000 001	10^{-15}	Femto	f
0.000 000 000 000 000 001	10^{-18}	Atto	a
0.000 000 000 000 000 000 001	10^{-21}	Zepto	z
0.000 000 000 000 000 000 000 001	10^{-24}	Yecto	y

TABLE B.2
Base Units

Quantity	Unit Name	Unit Symbol
Length	Metre	M
Mass	Kilogram	Kg
Time	Second	S
Electric current	Ampere	A
Thermodynamic temperature	Kelvin	K
Amount of substance	Mole	Mol
Luminous intensity	Candela	Cd

TABLE B.3
SI Derived Units

Quantity	Unit Name	Unit Symbol	Expression in Terms of Other SI Units
Absorbed dose, specific energy imparted, kerma, absorbed dose index	Gray	Gy	J/kg
Activity (of a radionuclide)	Becquerel	Bq	1/s
Celsius temperature	Degree Celsius	°C	K
Dose equivalent	Sievert	Sv	J/kg
Electric capacitance	Farad	F	C/V
Electric charge, quantity of electricity	Coulomb	C	A s
Electric conductance	Siemens	S	A/V
Electric inductance	Henry	H	Wb/A
Electric potential, potential difference, electromotive force	Volt	V	W/A
Electric resistance	Ohm	Ω	V/A
Energy, work, quantity of heat	Joule	J	N m
Force	Newton	N	Kg m/s^2
Frequency (of a periodic phenomenon)	Hertz	Hz	1/s
Illuminance	Lux	Lx	lm/m^2
Luminous flux	Lumen	Lm	cd sr
Magnetic flux	Weber	Wb	V s
Magnetic flux density	Tesla	T	Wb/m^2
Plane angle	Radian	Rad	m/m
Power, radiant flux	Watt	W	J/s
Pressure, stress	Pascal	Pa	N/m^2
Solid angle	Steradian	Sr	m^2/m^2

Derived units are formed by combining base units and other derived units according to the algebraic relations linking the corresponding quantities. The symbols for derived units are obtained by means of the mathematical signs for multiplication, division and use of exponents. Some derived SI units were given special names and symbols, as listed in this table.

TABLE B.4
Conversion Factors from U.S. Customary Units to Metric Units

To Convert from	Multiply by	To Find
Inches	25.4	Millimetres
	2.54	Centimetres
Feet	30.48	Centimetres
Yards	0.91	Metres
Miles	1.61	Kilometres
Teaspoons	4.93	Millilitres
Tablespoons	14.79	Millilitres
Fluid ounces	29.57	Millilitres
Cups	0.24	Litres
Pints	0.47	Litres
Quarts	0.95	Litres
Gallons	3.79	Litres
Cubic feet	0.028	Cubic metres
Cubic yards	0.76	Cubic metres
Ounces	28.35	Grams
Pounds	0.45	Kilograms
Short tons (2000 lbs)	0.91	Metric tons
Square inches	6.45	Square centimetres
Square feet	0.09	Square metres
Square yards	0.84	Square metres
Square miles	2.6	Square kilometres
Acres	0.4	Hectares

TABLE B.5
Conversion Factors from Metric Units to U.S. Customary Units

To Convert from	Multiply by	To Find
Millimetres	0.04	Inches
Centimetres	0.39	Inches
Metres	3.28	Feet
	1.09	Yards
Kilometres	0.62	Miles
Millilitres	0.2	Teaspoons
Litres	0.06	Tablespoons
	0.03	Fluid ounces
	1.06	Quarts
	0.26	Gallons
	4.23	Cups
	2.12	Pints
Cubic metres	35.32	Cubic feet
	1.35	Cubic yards
Grams	0.035	Ounces
Kilograms	2.21	Pounds
Metric ton (1000 kg)	1.1	Short ton
Square centimetres	0.16	Square inches
Square metres	1.2	Square yards
Square kilometres	0.39	Square miles
Hectares	2.47	Acres

Temperature conversion between Celsius and Fahrenheit $°C = (F - 32)/1.8$, $°F = (°C \times 1.8) + 32$

TABLE B.6
The Common Natural Temperatures

Condition	Fahrenheit (°)	Celsius (°)
Boiling point of water	212	100
A very hot day	104	40
Normal body temperature	98.6	37
A warm day	86	30
A mild day	68	20
A cool day	50	10
Freezing point of water	32	0
Lowest temperature Fahrenheit could be obtained by mixing salt and ice	0	−17.8

Glossary

aberration: distortion in an image produced by a lens or mirror caused by limitations inherent to some degree in all optical systems

absorption of radiation: the loss of light energy as it passes through a material. Loss is converted to other energy forms, which is usually heat (rise in temperature). The absorption process is dependent on the wavelength of the light and on the absorbing material

acceptance angle: the maximum angle over which the core of a fibre-optic cable accepts incoming light. The angle is measured from the centreline of the core

active medium: collection of atoms or molecules that can be stimulated to a population inversion, and emit electromagnetic radiation in a stimulated emission

adapter: it is a passive device used to connect two different connector types together

amplification: the process in which the electromagnetic radiation inside the active medium within the laser optical cavity increases by the process of stimulated emission

amplitude: the maximum value of a wave, measured from its equilibrium

angle of incident (θ_i): the angle formed by an incident ray and the normal line to the optical surface at the point of incident

angle of reflection (θ_{refl}): the angle formed by a reflected ray and the normal line to the optical surface at the point of reflection

angle of refraction (θ_{refr}): the angle formed by a refracted ray and the normal line to the optical surface at the point of penetration. The ray is refracted (bent) while passing from one transparent medium to another having different refractive indices

angstrom: a unit of measurement, equalling 10^{-10} m or 10^{-8} cm, usually used to express short wavelengths

aperture: an adjustable opening in an instrument (like a camera) that controls the amount of light that can enter

attenuation: the decrease in magnitude of the power of a signal in transmission media between points, normally measured in decibels (dB). Similarly, attenuation in a fibre-optic cable is a measure of how much of the signal injected into an optical fibre cable actually reaches the other end, usually expressed in decibels per kilometre (dB/km)

backreflection: reflection of light in the direction opposite to that in which the light was originally propagating

bandwidth: the measure of the information-carrying capacity of a fibre-optic cable, normalised to a unit of MHz-km. This term is used to specify the capacity of a fibre cable

beam: a bundle of light rays that are diverging, converging or parallel

beamsplitter: an optical device that divides incident light into two components (magnetic and electric)

beat-length: the length over which polarisation rotates through 360° within an optical fibre and, therefore, a fundamental measure of the polarisation-maintaining ability of a polarisation-maintaining fibre

bel (B): unit of intensity of sound, named after Alexander Graham Bell. The threshold of hearing is 0 B (10^{-12} W/m^2). The intensity is often measured in decibels (dB), which is one-tenth of a bel

bend radius: the amount that a fibre-optic cable can bend before the risk of breakage or increase in attenuation. The minimum bend radius is dependent on the fibre-optic cable diameter and type

bi-directional device: an optical device that operates in both directions

binary code: code based on the binary number system (that uses a base of 2). In binary code, any number can be expressed as a succession of ones and zeros. For example, the number 7 is 0111. These ones and zeros can then be interpreted and transmitted electronically as a series of *on* and *off* pulses, the basis for all computers and other digital equipment

birefringence: the fundamental principle by which polarisation-maintaining fibre works

birefringent: a birefringent material has distinct indices of refraction. The separation of light beam, as it passes through a calcite crystal object, into two diverging beams, commonly known as ordinary and extraordinary beams

bundle of fibre cables: a group of fibre cables packaged together in a unit

candle or candela (cd): a unit of luminous intensity

cascade: an arrangement of devices, each of which feeds into the next

chromatic dispersion: a pulse-broadening and, therefore, bandwidth-limiting phenomenon that occurs because different wavelengths of light travel at different velocities

cladding: the layer of glass or other transparent material surrounding the light-carrying core of a fibre-optic cable. The cladding has a lower refractive index than the core. The difference between the refractive indexes creates an interface that confines light propagation in the core

coating: one or more thin layers of optical material applied to an optical surface to reduce reflection, create a mirror surface, absorb light or protect the surface

coherence: a property of electromagnetic waves that are in phase in both time and space. Coherent light has monochromaticity and low beam divergence, and can be concentrated to high power densities. Coherence is needed for interference processes like holography

collimate: to cause light rays to become parallel

concentricity: core-cladding concentricity is the distance between the geometric centre of the core and the geometric centre of the cladding. Sometimes called eccentricity or concentricity error

connector: a device mounted on the end of a fibre-optic cable that mates to a similar device to couple light into fibre cables. A connector joins a fibre-cable end and a light source or detector in an optical device

continuous wave (CW): it is the output of a laser, which is operated in a continuous rather than pulsed mode

control area: it is an area in which the occupancy and activity of those present is subject to control and supervision for the purpose of protection from hazards like radiation, chemical and electrical

conversion efficiency: in an erbium-doped fibre amplifier (EDFA), it is the ratio between the amplified signal output power and the power input from the pump laser

convex lens: curved outward. A lens with a surface shaped like the exterior surface of a sphere

core: the central part of a fibre-optic cable that conducts light. The core is made up of glass or plastic. It has a higher refractive index than the cladding

cornea: it is the transparent outer coat of the human eye, covering the iris and the crystalline lens. The cornea is the main refraction element of the eye

coupler: a device that connects two or more output ends, dividing one input between two or more outputs, or combining two or more inputs into one output. Couplers are used in telecommunication systems

coupling: transfer of light into or out of a fibre-optic cable. A coupling is used to launch light source into a fibre-optic cable

cube beamsplitter: cube beamsplitters consist of matched pairs of right angle prisms cemented together along their hypotenuses

cut-off wavelength: the wavelength at which an optical fibre becomes single-moded. Below cut-off, the fibre will transmit more than one mode. Above cut-off, the strength of the guidance is gradually reduced

dB: abbreviation for decibel. One decibel is one tenth of one bel (a unit of power ratio, equal to 10 decibels)

decibel (dB): the standard unit used to express loss. Decibel is defined as 10 times the base-10 logarithm of the ratio of the output signal to the input signal power

detector: a light-sensitive device that produces electrical signals when illuminated

diffraction grating: a grooved optical element that has been deformed to reflect or transmit light of many colours. It acts like a prism to produce a spectrum

diffuse reflection: it is the change in the spatial distribution of a beam of radiation when the beam is reflected in many directions by a rough surface or by a medium

diffusion: it is the flow of particles of a given species from high to low concentration regions by virtue of their random motions

dispersion: the separation of a light beam into its various wavelength components. All transparent materials have different indices of refraction for light of different wavelengths

distortion: an aberration in an optical element that causes straight lines in the object, which are off the axis, to appear as curved lines in the image

EDFA: erbium-doped fibre amplifier. A device incorporating erbium-doped fibre to provide direct amplification of optical signals when pumped at either 980 or 1480 nm

electron volt (eV) unit of energy: the amount of energy that the electron acquires while accelerating through a potential difference of 1 V. 1 eV $= 1.6 \times 10^{-19}$ J

excess loss: loss within a four-directional coupler. It is defined as 10 times the base-10 logarithm of the ratio of the signal power between Ports 2 and 3, and Port 1

extinction ration: in a polarisation-maintaining fibre, it is the ratio between the wanted and unwanted polarisation states, expressed in decibels (dB). It is highly dependent on operating environment

ferrule: a part of a connector with a central hole that contains and aligns a stripped fibre cable in a connector assembly

fibre cable bundle: a group of fibre cables packaged together in a unit

fibre distributed data interface (FDDI): a standard developed by ANSI for data transmissions through fibre optics systems. FDDI is capable of data transmissions of 100 Mb/s and above through a LAN incorporating 500 stations in a ring topology with a length of up to 100 km

fibre grating: a selective reflector formed by inducing a periodic variation of refractive index within the core of an optical fibre

fibre laser: a laser in which the gain element is a length of rare-earth-doped optical fibre

fibre-optic cable: an optical cable used for light transmission in telecommunications. Fibre-optic cables come in a great variety of configurations

focal length (f): the distance between the second principal plane or equivalent refracting plane of a lens and the lens focal point when the lens is imaging an object at infinity. In a positive lens, the focal length is measured on the side of the lens opposite to the object. In a negative lens, the focal length is measured on the same side as the object

focal point: the point on the optical axis of a lens where light rays from a distant object point will converge after being refracted by the lens

focus: the plane at which light rays from object points form a sharp image after being refracted by a lens

Fresnel reflection: the reflection that occurs between parallel optical surfaces or at the interface where two materials have different refractive indices

Fresnel reflection loss: loss of signal power due to Fresnel reflection

fused fibre: two fibre cables are heated, placed under tension and caused to create a taper coupler join

fusion splice: a splice made by melting the ends of two fibre cables together by a spark shot so as to form a permanent connection

gain: in an erbium-doped fibre amplifier (EDFA), it is the ratio between the amplified signal output and the (un-amplified) signal input, expressed in dB

graded index fibre cable: a fibre-optic cable whose core has a non-uniform index of refraction. The core is composed of concentric rings of glass whose refractive indices decrease radially from the centre of the core

GRIN lenses: an acronym for gradient index lenses. They have a cylindrical shape with one end polished at an angle of 2°, 6°, 8° or 12° and the other end polished at an angle of 2° or 90°

hertz (Hz): the unit used to measure frequency. 1 Hz equals one wave or cycle per second

homogeneous: a term used to describe any medium that is uniform in composition throughout

image: a likeness of an object formed by an optical element or system

image distance: the distance between the equivalent refracting plane or second principal plane and the focal point measured on the optical axis

index of refraction (n): the ratio of the speed of light in a vacuum to the speed of light in a material

index-matching gel: a gel or fluid with a refractive index, which is matched to the refractive index of two fibre-optic cores. It fills in the air gap between the fibre cable ends and reduces the Fresnel reflection, which occurs in the gap

intensity: the light energy per unit area

interferometer: the interferometer invented by the American physicist A. A. Michelson (1852–1931) is an ingenious device, which splits a light beam into two parts and then recombines them to form an interference pattern. The device can be used for obtaining accurate measurements of wavelength, precise length measurement and to measure accuracy of an optical surface

jacket: a polymer (plastic, PVC, etc.) layer that covers the cladding/core layers of a fibre-optic cable. The jacket has different colours and used as a mechanical protection layer

kevlar: a strong synthetic material used as a strength member in a fibre-optic cable. A pull wire can be fastened to the kevlar during fibre-optic cable installation

laser: an acronym for *Light Amplification by the Stimulated Emission of Radiation*. Lasers produce the coherent source of light for fibre-optic telecommunication systems

laser source: an instrument that produces monochromatic, coherent and collimated light

lens: one or more optical elements having flat or curved surfaces. If used to converge light rays, it is a positive lens; if used to diverge light rays, it is a negative lens. Usually made of optical glass, but may be moulded from transparent plastic. Lenses are sometimes made from a natural or synthetic crystalline substance to transmit very short wavelengths (UV) or very long wavelengths (IR)

light: the form of electromagnetic radiation with a wavelength ranging from ~400 to ~700 nm. It generally travels in straight-line and exhibits the characteristics of both a wave and a particle

lightguide: a fibre-optic cable or fibre-optic cable bundle

light ray: the path of a single beam of light. In graphical ray tracing, a straight line represents the path along which the light travels

local area network: high-speed and high-capacity computer links used over relatively short distances (a few kilometres) to connect several buildings

loss (dB): attenuation of the power of a signal when it travels through an optical component. Normally measured in decibels (dB)

lumen (lm): a SI unit of luminous flux. One lumen is the luminous flux emitted per unit solid angle by a light source having an intensity of one candela (cd)

luminous intensity (I): luminous intensity measures the brightness of a light source. The unit of measure is the candle or candela (cd)

lux (lx): a unit of luminance equal to one lumen per square metre

majority carriers: they are electrons in an n-type and holes in a p-type semiconductor

mechanical splice: a process whereby two optical fibres are joined together using mechanical means

metropolitan area networks (MAN): an optical fibre or cable backbone interconnecting a number of LANs in a specific area

microbending: tiny bends in a fibre-optic cable, which allow light to leak out of the core and introduce loss

micrometre (μm): one-millionth of a metre

minority carriers: they are electrons in a p-type and holes in an n-type semiconductor

mirror: an optical element with a smooth, highly polished surface (plane or curve) for reflecting light. The reflecting surface is produced by a thin coating of gold, silver or aluminium

mode: a term used to describe a light path(s) passing through a fibre-optic cable as in single mode or multi-mode

mode scramble: mode scrambling is accomplished by bending a fibre-optic cable in a corrugated pattern. This causes light to leak out and attenuation is increased

mode scrambler: a device that bends a fibre-optic cable to increase loss

monochromatic light: light is at one specific wavelength. The light out of a laser device is the monochromatic light

multi-mode fibre: a fibre-optic cable in which light travels in multiple modes

nanometre (nm): one-billionth of a metre. The unit usually used in specifying the wavelength of light

noise: in an erbium-doped fibre amplifier (EDFA), it is typically based on signal-to-noise ratio in a regime in which signal-spontaneous beat-noise and amplified signal shot-noise dominate

normal line: a reference line constructed perpendicular to an optical surface

numerical aperture (NA): a measure of the divergence of the light emitted from the fibre, determined by the refractive index difference between the core and the cladding. The sine of the half the angle in which the core of a fibre-optic cable can accept or transmit light

optical coatings: coatings specifically made for optical components (lenses, prisms, etc.) in light sensitive devices. There are many types of coating the materials. One coating helps to protect the optical components from scratches and wear. Some optical components are coated with anti-reflective (AR) layer(s) to reduce backreflection

optical path: the sum of the optical distances along a specified light ray

optical pumping: the excitation of the active medium in a laser by the application of light, rather than electrical discharge. Light can be from a conventional source like Xenon or Krypton lamp, or from another laser

optical radiation: ultraviolet, visible and infrared spectrum (0.35–1.4 μm) that falls in the region of transmittance of the human eye

optical resonator: the mirrors (or reflectors) making up the laser cavity including the laser rod or tube. The mirrors reflect light back and forth to build up amplification

optical surface: the reflecting or refracting surface of an optical element

phase: the position of a wave in its oscillation cycle

photon: a particle or packet of radiant electromagnetic energy representing a quanta of light

photonics: the field of science and engineering encompassing the physical phenomena and associated with the generation, transmission, manipulation, detection and utilisation of light

plastic fibre cable: a fibre-optic cable having a plastic core and plastic cladding. Plastic fibre cables are typically used in applications where sensitivity and loss are not important

plenum: the air space between walls, under structural floors and above drop ceilings, which can be used to route interconnection cabling in a building

polarisation: alignment of the electric and magnetic fields, which comprise an electromagnetic wave. If all light waves from a source have the same alignment, then the light is said to be polarised

population inversion: an excited state of matter, in which more atoms (or molecules) are in an upper state than in a lower one. This is a required situation for a laser action

prism: an optical element, which is used to change the direction and orientation of a light beam. A prism has polished faces that are used to transmit and reflect light

proximity sensor: a device that senses distance from a reflecting surface

pulsed laser: laser that delivers energy in the form of a single or sequence of laser pulses

PVC: polyvinyl chloride, a material used in the manufacture of fibre-optic cable jackets

quantum efficiency: in the erbium-doped fibre amplifier (EDFA), it is the actual conversion efficiency, expressed as a percentage of the maximum possible conversion efficiency (equal to the ratio of the pump and signal wavelengths)

ray: straight lines that represent the path of a light ray

Rayleigh scattering: the scattering of light, which results from small impurities in a material or composition

recombination: recombination of an electron hole pair involves an electron in the conduction band (CB) falling in energy down into an empty state (hole) in the valence band (VB) to occupy it. The result is the annihilation of the EHP

recombination current: recombination current flows under forward bias to replenish the carriers recombining in the space charge (depletion) layer

rectangular beamsplitters: three prisms carefully cemented together along their hypotenuses. Polarisation beamsplitters are used in optical devices where the output components are required to exit from the opposite side to the input signal. It also produces a lateral displacement between the two output components

reflection: the change in the direction of a light ray when it bounces off of a reflecting surface

refraction: the bending of a light ray as it passes from one transparent medium to another of different refractive index

refractive index (n): the ratio of the speed of light in a vacuum to the speed of light in a specific material

ribbon fibre cables: cables in which many optical fibres are embedded in a plastic flat ribbonlike structure

right angle prism: a prism whose cross-section is a right angle triangle with two 45° interior angles. The prism faces, which are at right angles, are transmitting surfaces, whereas the hypotenuse face is a reflecting surface

saturation: in an erbium-doped fibre amplifier (EDFA), it is the performance under conditions of total population inversion, which occur at high input powers

scattering: loss of light due to the presence of atoms in a transparent material

sensor: a device that responds to the presence of energy

sheath: an outer protective layer of a fibre-optic cable. Fibre-optic cables having an outer protective layer are suitable for indoor and outdoor cable installations

simplex fibre-optic cable: a term sometimes used to describe a single fibre-optic cable

single-mode fibre: A fibre-optic cable in which the signal travels in one mode

Snell's law: describes the path that a light ray takes as it goes from one optical medium to another. It is also called Law of Refraction

spectrometer: a spectroscope equipped with the ability to measure wavelengths

spectra, spectrum: spectra is the plural of spectrum, which is a series of energies (like light) arranged according to wavelength or frequency. The electromagnetic spectrum is an array of radiation that is divided into a number of sub-portions, where the boundaries are only vaguely defined. They extend from the shortest cosmic rays, through gamma rays, X-rays, ultraviolet light, visible light, infrared radiation, microwave and all other wavelengths of radio energy

speed of light: in vacuum, approximately 3×10^8 m/s

splice: a method for joining the ends of two fibre-optic cables. There are two primary methods for splicing fibre-optic cables: fusion and mechanical

spontaneous emission: random emission of a photon by decay of an excited state to a lower level. Determined by the lifetime of the excited state

spot size: a measure of the diameter of the beam of laser radiation

stimulated emission: coherent emission of radiation, stimulated by a photon absorbed by an atom (or molecule) in its excited state

strength member: the part of a fibre-optic cable composed of Kevlar armed yarn, steel strands or fibreglass filaments. Strength members increase the tensile strength of a cable

switch: a device that regulates or directs a signal in telecommunication systems

total internal reflection: total internal reflection of light occurs when light rays in a high-index medium exceed the critical angle (to the normal to a surface). This is the principal theory for explaining how light travels in the core of a fibre-optic cable

transparent: the adjective used to describe a medium through which light can pass in a percentage

transverse electro-magnetic (TEM) mode: used to designate the shape of a cross-section of a laser beam

transverse mode: the geometry of the power distribution in a cross-section of a laser beam

ultraviolet light (UV) (extreme ultraviolet and far ultraviolet): a portion of the complete electromagnetic spectrum, which has a shorter wavelength than visible light; roughly, with a Å wavelength interval from 100 to 4000 Å. Ultraviolet radiation from the Sun is responsible for many complex photochemical reactions like the formation of the ozone layer. Extreme and far ultraviolet wavelengths are different portions of the ultraviolet portion of the spectrum, with extreme being between 55.8 and 118 nm and far being between 110 and 190 nm

V-value: also called normalised frequency. The fundamental relationship between numerical aperture, cut-off wavelength and core diameter

visible light: electromagnetic radiation that is visible to the human eye. It has a wavelength range between 400 and 700 nm

wave: one complete cycle of a signal with a fixed period

waveguide: a structure, which guides an electromagnetic wave along its length. A fibre-optic cable is an example of optical waveguide

wavelength (λ): the period of a wave. Distance between successive crests, troughs or identical parts of a wave

wavelength division multiplexing (WDM): the process whereby multiple optical carriers of different wavelengths utilise the same optical fibre cable

Index

Note: Page numbers followed by f and t refer to figures and tables, respectively.

Milton Keynes UK
Ingram Content Group UK Ltd.
UKHW032233151223
434481UK00022B/549

9 780367 878856